PROJECT MANAGEMENT METRICS, KPIs, AND DASHBOARDS

A Guide to Measuring and Monitoring Project Performance

Second Edition

Harold Kerzner, Ph.D.

Sr. Executive Director for Project Management
The International Institute for Learning

Cover image: © Wiley
Cover design: © Ventyx, an ABB company

This book is printed on acid-free paper.

Published by John Wiley & Sons, Inc., Hoboken, New Jersey
Published simultaneously in Canada

Library of Congress Cataloging-in-Publication Data
Kerzner, Harold.
Project management metrics, KPIs, and dashboards : a guide to measuring and monitoring project performance / Harold Kerzner, Ph.D., Sr. Executive Director for Project Management, The International Institute for Learning.—Second edition.
pages cm.
Includes index.
ISBN 978-1-118-52466-4 (cloth); 978-111-8-65895-6 (ebk.); 978-111-8-65899-4 (ebk.)
1. Project management. 2. Project management—Quality control. 3. Performance standards. 4. Work measurement. I. Title.
HD69.P75K492 2013
658.4'04—dc23

2013004686

Printed in the United States of America
10 9 8 7 6 5 4 3 2 1

CONTENTS

PREFACE

The ultimate purpose of metrics and dashboards is not to provide more information but to provide the right information to the right person at the right time, using the correct media and in a cost-effective manner. This is certainly a challenge. As computer technology has grown, so has the ease with which information can be generated and presented to management and stakeholders. Today, everyone seems concerned about information overload. Unfortunately, the real issue is non-information overload. In other words, there are too many useless reports that cannot easily be read and that provide readers with too much information, much of which may have no relevance. It simply distracts us from the real issues.

Insufficient or ineffective metrics prevent us from understanding what decisions really need to be made. In traditional project review meetings, emphasis is placed upon a detailed schedule analysis and a lengthy review of the cost baseline versus actual expenditures. The resulting discussion and explanation of the variances are most frequently pure guesswork. Managers who are upset about the questioning by senior management then make adjustments that do not fix the problems but limit the time they will be grilled by senior management at the next review meeting. They then end up taking actions that may be counterproductive to the timely completion of the project and real issues are hidden.

You cannot correct or improve something that cannot be effectively identified and measured. Without effective metrics, managers will not respond to situations correctly and will end up reinforcing undesirable actions by the project team. Keeping the project team headed in the right direction cannot be done easily without effective identification and measurement of metrics.

When all is said and done, we wonder why we have studies like the Chaos Report, which has shown us over the past 15 years that only about 30 percent of IT projects are completed successfully. We then identify hundreds of causes as to why projects fail, but neglect what is now being recognized as perhaps the single most important cause: a failure in metrics management.

Metrics management should be addressed in all of the areas of knowledge in the PMBOK® Guide, especially communications management. We are now struggling to find better ways of communicating on projects. Our focus today is on the unique needs of the receiver of the information. The need to make faster and better decisions mandates better information. Human beings have a variety of ways in which they can absorb information. We must address all of these ways in the selection of the metrics and the design of the dashboards that convey this information.

The three most important words in a stakeholder's vocabulary are, "making informed decisions." This is usually the intent of effective stakeholder relations management. Unfortunately, this cannot be accomplished without an effective information system based upon meaningful and informative metrics and key performance indicators (KPIs).

All too often, we purchase project management software and reluctantly rely upon the report generators, charts, and graphs to provide the necessary information, even when we realize that this information is either not sufficient or has limited value. Even those companies that create their own project management methodologies neglect to consider the metrics and KPIs that are needed for effective stakeholder relations management. Informed decisions require effective information. We all seem to understand this, yet it has only been in recent years that we have tried to do something about it.

For decades we believed that the only information that needed to be passed on to the client and the stakeholders was information related to time and cost. Today, we realize that the true project status cannot be determined from time and cost alone. Each project may require its own unique metrics and key performance indicators. The future of project management may very well be metric-driven project management.

Information design has finally come of age. Effective communications is the essence of information design. Today, we have many small companies that are specialists in business information design. Larger companies may maintain their own specialist team and call these people graphic designers, information architects, or interaction designers. These people maintain expertise in the visual display of both quantitative and qualitative information necessary for informed decision making.

Traditional communications and information flow has always been based upon tables, charts, and indexes that were hopefully organized properly by the designer. Today, information or data graphics combines points, lines, charts, symbols, images, words, numbers, shades, and a symphony of colors necessary to convey the right message easily. What we know with certainty is that dashboards and metrics are never an end in themselves. They go through continuous improvement and are constantly updated. In a project management environment, each receiver of information can have different requirements and may request different information during the life cycle of the project.

With this in mind, the book is structured as follows:

- Chapters 1 and 2 identify how project management has changed over the last few years and more pressure is being placed upon the organization for effective metrics management.
- Chapter 3 provides an understanding of what metrics are and how they can be used.
- Chapter 4 discusses key performance indications and explains the difference between metrics and KPIs.
- Chapter 5 focuses on the value-driven metrics and value-driven key performance indicators. Stakeholders are asking for more metrics related to the project's ultimate value. The identification and measurement of value-driven metrics can be difficult.
- Chapter 6 describes how dashboards can be used to present the metrics and KPIs to the stakeholders. Examples of dashboards are included together with some rules for dashboard design.
- Chapter 7 identifies dashboards that are being used by companies.
- Chapter 8 provides various techniques for the actual measurement of the metric and the KPI. Metrics and KPIs serve no viable purpose if they cannot be effectively measured.

HAROLD KERZNER, PH.D.
Sr. Executive Director for Project Management
The International Institute for Learning

1 THE CHANGING LANDSCAPE OF PROJECT MANAGEMENT

CHAPTER OVERVIEW

The way we managed projects in the past will not suffice for many of the projects we are managing now, as well as for the projects of the future. The complexity of these projects will place pressure on organizations to better understand how to identify, select, measure, and report project metrics. The future of project management may very well be metric-driven project management.

CHAPTER OBJECTIVES

- To understand how project management has changed
- To understand the need for project management metrics
- To understand the need for better, more complex project management metrics

KEY WORDS

- Certification Boards
- Complex Projects
- Engagement Project Management
- Frameworks
- Governance
- Project Management Methodologies
- Project Success

1.0 INTRODUCTION

For more than 50 years project management has been in use but perhaps not on a worldwide basis. What differentiated companies that were using project management in the early years was whether or not they used project management, not how well they used it. Today, almost every company uses project management, and the differentiation is whether they are simply good at project management or whether they truly excel at project management. The difference between using project management and being good at project management is relatively small, and most companies can become good at project management in a relatively short time period, especially if they have executive-level support. A well-organized project management office (PMO) can also accelerate the maturation process. The difference, however, between being good and excelling at project management is quite large. One of the critical differences is that excellence in project management on a continuous basis requires more metrics than just time and cost. The success of a project cannot be determined just from the time and cost metrics, yet we persist in the belief that this is possible.

Companies such as IBM, Microsoft, Siemens, Hewlett-Packard, Computer Associates, and Deloitte, just to name a few, have come to the realization that they must excel at project management. This requires additional tools and metrics to support project management. IBM has more than 300,000 employees with more that 70 percent outside of the United States. This includes some 30,000 project managers. Hewlett-Packard (HP) has more than 8000 project managers and 3500 Project Management Professionals (PMP®s). HP desires 8000 project managers and 8000 PMP®s. These numbers are now much larger with HP's acquisition of Electronic Data Systems (EDS).

1.1 EXECUTIVE VIEW OF PROJECT MANAGEMENT

The companies mentioned previously perform strategic planning for project management and are focusing heavily on the future. Several of the things that these companies are doing will be discussed in this chapter, beginning with senior management's vision of the future. Years ago, senior management provided lip service to project management, reluctantly supporting it to placate the customers. Today, senior management appears to have recognized the value in using project management effectively and maintains a different view of project management as shown in Table 1–1.

TABLE 1-1 The Executive View of Project Management

OLD VIEW	NEW VIEW
Project management is a career path.	Project management is a strategic or core competency necessary for the growth and survivability of the company.
We need our people certified as Project Management Professionals (PMP®s).	We need our people to undergo multiple certifications, at a minimum, to be certified in project management and corporate business processes.
Project managers will be used for project execution only.	Project managers will participate in the portfolio selection of projects and capacity-planning activities.
Business strategy and project execution are separate activities.	Part of the project manager's job is to bridge strategy and execution.
Project managers make solely project-based decisions.	Project managers make both project and business decisions.

Project management is no longer regarded as a part-time occupation or even a career path position. It is now viewed as a strategic competency needed for the survival of the firm. Superior project management capability can make the difference between winning and losing a contract.

For more than 20 years, becoming a PMP® was seen as the light at the end of the tunnel. Today, that has changed. Becoming a PMP® is the light at the entryway to the tunnel. The light at the end of the tunnel may require multiple certifications. As an example, after becoming a PMP®, a project manager may desire to become certified in:

- Business Analyst Skills or Business Management
- Program Management
- Business Processes
- Managing Complex Projects
- Six Sigma
- Risk Management

Some companies have certification boards, which meet frequently and discuss what certification programs would be of value for their project managers. Certification programs that require specific knowledge of company processes or company intellectual property may be internally developed and taught by the company's own employees.

Executives have come to the realization that there is a return on investment in project management education. Therefore, executives are now investing heavily in customized project management training, especially in the behavioral courses. As an example, one executive commented that he felt that presentation skills training was the highest priority for his project managers. If a project manager makes a highly polished presentation before the client, the client believes that the project is being managed the same way. If the project manager makes a poor presentation, then the client might believe the project is managed the same way. Other training programs that executives feel would be beneficial for the future include:

- Establishing metrics and key performance indicators (KPIs)
- Dashboard design
- Managing complex projects
- How to perform feasibility studies and cost–benefit analyses
- Business analysis
- Business case development
- How to validate and revalidate project assumptions
- How to establish project governance
- How to manage multiple stakeholders
- How to design and implement "fluid" or adaptive enterprise project management methodologies
- How to develop coping skills and stress management skills

Project managers are now being brought on board projects at the beginning of the initiation phase rather than at the end of the initiation phase. To understand the reason for this, consider the following situation:

SITUATION: A project team is assembled at the end of the initiation phase of a project to develop a new product for the company. The project manager is given the business case for the project together with a listing of the assumptions and constraints. Eventually, the project is completed, somewhat late and significantly over budget. When asked by marketing and sales why the project costs were so large, the project manager responds, "According to my team's interpretation of the requirements and the business case, we had to add in more features than we originally thought."

Marketing then replies, "The added functionality is more than what our customers actually need. The manufacturing costs for what you developed will be significantly higher than anticipated and that will force us to raise the selling price. We may no longer be competitive in the market segment we were targeting."

"That's not our problem," responds the project manager. "Our definition of project success is the eventual commercialization of the product. Finding customers is your problem, not our problem."

Needless to say, we could argue about what the real issues were in this project that created the problems. For the purpose of this book, there are two issues that stand out. First and foremost, project managers today are paid to make business decisions as well as project decisions. Making merely project-type decisions could result in the development of a product that is either too costly to build or overpriced for the market at hand. Second, the traditional metrics used by project managers over the past several decades were designed for project rather than business decision making. Project managers must recognize that, with the added responsibilities of making business decisions, a new set of metrics may need to be included as part of the project manager's responsibility. Likewise, we could argue that marketing was remiss in not establishing and tracking business-related metrics throughout the project and simply waited until the project was completed to see the results.

1.2 COMPLEX PROJECTS[1]

For more three decades, project management has been used to support traditional projects. Traditional projects are heavily based upon linear thinking; we have well-structured life cycle phases and templates, forms,

1. Adapted from Harold Kerzner and Carl Belack, *Managing Complex Projects*, John Wiley & Sons and the International Institute for Learning (IIL) Co-publishers, 2010, Chapter 1.

TIP Today's project manager sees himself/herself as managing part of a business rather than simply managing a project. Therefore, additional metrics may be required for informed decision making to happen.

guidelines, and checklists for each phase. As long as the scope is reasonably well defined, traditional project management works well.

Unfortunately, only a small percentage of all of the projects within a company fall into this category. Most nontraditional or complex projects use seat-of-the-pants management because they are largely based upon business scenarios where the outcome or expectations can change from day to day. Therefore, project management techniques were neither required nor used on these complex projects that were more business oriented and aligned to five-year or ten-year strategic plans that were constantly updated.

Now, we are finally realizing that project management can be used on these complex projects, but the traditional project management processes may be inappropriate or must be modified. This includes looking at project management metrics and KPIs in a different light. The leadership style for complex projects may not be the same as that for traditional projects. Risk management is significantly more difficult on complex projects, and the involvement of more participants and stakeholders is necessary.

Now that we have become good at traditional projects, we are focusing our attention on the nontraditional or complex projects. Unfortunately, there is no clear-cut definition of a complex project. Some of the major differences between traditional and nontraditional or complex projects, in the author's opinion, are shown in Table 1–2.

TABLE 1-2 Traditional versus Nontraditional Projects

TRADITIONAL PROJECTS	NONTRADITIONAL PROJECTS
The time duration is 6–18 months.	The time duration can be several years.
The assumptions are not expected to change over the duration of the project.	The assumptions can and will change over the project's duration.
Technology is known and will not change over the project's duration.	Technology will most certainly change.
People that started on the project will remain through to completion (the team and the project sponsor).	People who approved the project and are part of the governance may not be there at the project's conclusion.
The statement of work is reasonably well defined.	The statement of work is ill defined and subject to numerous scope changes.
The target is stationary.	The target may be moving.
There are few stakeholders.	There are multiple stakeholders.
There are few metrics and key performance indicators.	There can be numerous metrics and key performance indicators.

Comparing Traditional and Nontraditional Projects

The traditional project that most people manage is usually less than 18 months. In some companies, the traditional project might be six months or less. The length of the project is usually dependent on the industry. In the auto industry, for example, a traditional project is three years.

With projects that are 18 months or less, we assume that technology is known with some degree of assuredness and technology may undergo little change over the life of the project. The same holds true for the assumptions. We tend to believe that the assumptions made at the beginning of the project will remain intact for the duration of the project unless a crisis occurs.

People that are assigned to the project will most likely stay on board the project from beginning to end. The people may be full-time or part-time. This includes the project sponsor as well as the team members.

Because the project lasts 18 months or less, the statement of work is usually reasonably well defined and the project plan is based upon reasonably well-understood and proven estimates. Cost overruns and schedule slippages can occur, but not to the degree that they will happen on complex projects. The objectives of the project, as well as critical milestone or deliverable dates, are reasonably stationary and not expected to change unless a crisis occurs.

The complexities of nontraditional projects seem to have been driven in the past by time and cost. Some people believe that these are the only two metrics that need to be tracked on a continuous basis. Complex projects may run as long as 10 years, or even longer. Because of the long time duration, the assumptions made at the initiation of the project will most likely not be valid at the end of the project. The assumptions will have to be revalidated throughout the project. There can be numerous metrics, and the metrics can change over the duration of the project. Likewise, technology can be expected to change throughout the project. Changes in technology can create significant and costly scope changes to the point where the final deliverable does not resemble the initially planned deliverable.

People on the governance committee and in decision-making roles most likely are senior people and may be close to retirement. Based upon the actual length of the project, the governance structure can be expected to change throughout the project if the project's duration is 10 years or longer.

Because of scope changes, the statement of work may undergo several revisions over the life cycle of the project. New governance groups and new stakeholders can have their own hidden agendas and demand that the scope be changed or they might even cancel their financial support for the project. Finally, whenever you have a long-term complex project where continuous scope changes are expected, the final target may move. In other words, the project plan must be constructed to hit a moving target.

SITUATION: A project manager was brought on board a project and provided with a project charter than included all of the assumptions made in the selection and authorization of the project. Part way through the project, some of the business assumptions changed. The project manager assumed that the project sponsor would be monitoring the enterprise environmental factors for changes in the business assumptions. That did not happen. The project was eventually completed, but there was no real market for the product.

Given the premise that project managers are now more actively involved in the business, we must track the assumptions the same way that we track budgets and schedules. If the assumptions are wrong or no longer valid, then we may need to either change the statement of work or even consider canceling the project. We should also track the expected value at the end of the project because unacceptable changes in the final value may be another reason for project cancellation.

Examples of assumptions that are likely to change over the duration of a project, especially on a long-term project, include:

- The cost of borrowing money and financing the project will remain fixed.
- Procurement costs will not increase.
- The breakthrough in technology will take place as scheduled.
- The resources with the necessary skills will be available when needed.
- The marketplace will readily accept the product.
- Our competitors will not catch up to us.
- The risks are low and can be easily mitigated.
- The political environment in the host country will not change.

The problem with having faulty assumptions is that they can lead to bad results and unhappy customers. The best defense against poor assumptions is good preparation at project initiation, including the development of risk mitigation strategies and tracking metrics for critical assumptions. However, it may not be possible to establish metrics for the tracking of all assumptions.

Most companies either have or are in the process of developing an enterprise project management methodology (EPM). EPM systems are usually rigid processes designed around policies and procedures, and work efficiently when the statement of work is well defined. With the new type of projects expected over the next decade, however, these rigid and inflexible processes may be more of a hindrance.

EPM systems must become more flexible in order to satisfy business needs. The criteria for good systems will lean toward forms, guidelines, templates, and checklists rather than policies and procedures. Project managers will be given more flexibility in order to make decisions necessary to satisfy the business needs of the project. The situation is further complicated in that all active stakeholders may wish to use their own methodology, and having multiple methodologies on the same project is never a good idea. Some host countries may be quite knowledgeable in project management, whereas other may have just cursory knowledge.

In the future, having a fervent belief that the original plan is correct may be a poor assumption. As the project's business needs change, the need to change the plan will be evident. Also, decision making based entirely upon the triple constraints, with little regard for the final value of the project, may result in a poor decision. Simply stated, today's view of project management is quite different from the views in the past, and this is partially the result of recognizing the benefits of project management over the past two decades.

TIP Metrics and key performance indicators must be established for those critical activities that can have a direct impact on the success or failure of the project. This includes the tracking of assumptions and value.

TIP The more flexibility the methodology contains, the greater the need for additional metrics and key performance indicators.

We can now summarize some of the differences between managing traditional and complex projects. These are shown in Table 1–3. Perhaps the primary difference is whom the project manager must interface with on a daily basis. With traditional projects, the project manager interfaces with the sponsor and the client, both of whom may provide the only governance on the project. With complex projects, governance is by committee and there can be multiple stakeholders whose concerns need to be addressed.

TABLE 1-3 Summarized Differences between Traditional and Nontraditional Projects

MANAGING TRADITIONAL PROJECTS	MANAGING NONTRADITIONAL PROJECTS
Single-person sponsorship	Governance by committee
Possibly a single stakeholder	Multiple stakeholders
Project decision making	Both project and business decision making
An inflexible project management methodology	Flexible or "fluid" project management methodology
Periodic status reporting	Real-time reporting
Success is defined by the triple constraints.	Success is defined by competing constraints, value, and other factors.
Metrics and KPIs are derived from the earned value measurement system.	Metrics and KPIs may be unique to the particular project and even to a particular stakeholder.

Defining Complexity

Complex projects can differ from traditional projects for a multitude of reasons, including:

- Size
- Dollar value
- Uncertain requirements
- Uncertain scope
- Uncertain deliverables
- Complex interactions
- Uncertain credentials of the labor pool
- Geographical separation across multiple time zones
- Use of large virtual teams
- Other differences

There are numerous definitions of a complex project, based upon the interactions of two or more of the preceding elements. Even a small, two-month infrastructure project can be considered complex according to the definition. This can create havoc when selecting and using metrics. The projects that you manage within your own company can be regarded as complex projects if the scope is large and the statement of work is only partially complete. Some people believe that R&D projects are always complex because, if you can lay out a plan for R&D, then you probably do not have R&D. R&D is when you are not 100 percent sure where you are heading, you do not know what it will cost, and you do not know if and when you will get there.

Complexity can be defined according to the number of interactions that must take place for the work to be executed. The greater the number of functional units that must interact, the harder it is to perform the integration. The situation becomes more difficult if the functional units are dispersed across the globe and if cultural differences makes integration difficult. Complexity can also be defined according to size and length. The larger the project is in scope and cost, and the greater the time frame, the more likely it is that scope changes will occur significantly, affecting the budget and schedule. Large, complex projects tend to have large cost overruns and schedule slippages. Good examples of this are Denver International Airport, the Channel between England and France, and the "Big Dig" in Boston.

Tradeoffs

Project management is an attempt to improve efficiency and effectiveness in the use of resources by getting work to flow multidirectionally through an organization. This holds true for both traditional projects and complex projects. Initially, this might seem easy to accomplish, but there are typically a number of constraints imposed upon a project. The most common constraints are time, cost, and performance (also referred to as scope or quality) and are known as "the triple constraints."

From an executive-level perspective, the goal of project management may be meeting the triple constraints of time, cost, and performance, while maintaining good customer relations. Unfortunately, because most projects have some unique characteristics, highly accurate estimates may not be possible and tradeoffs between the triple constraints may be necessary. As will be discussed later, there may be significantly more than three constraints on a project, and metrics may have to be established to track each of the constraints. The metrics provide the basis for informed tradeoff decision making. Executive management, functional management, and key stakeholders must be involved in almost all tradeoff discussions to ensure that the final decision is made in the best interests of the project, the company, and the stakeholders. If multiple stakeholders are involved, as there are on complex projects, then agreement from all of the stakeholders may be necessary. Project managers may possess sufficient knowledge for some technical decision making but may not have sufficient business or technical knowledge to adequately determine the best course of action to address the interests of the parent company as well as the individual stakeholders on the project.

TIP Because of the complex interactions of the elements of work, a few simple metrics may not provide a clear picture of project status. The combination of several metrics may be necessary in order to make informed decisions.

Skill Set

All project managers have skills, but not all project managers will have the right skills for the given job. For projects internal to a company, it may be possible to develop a company-specific skill set or company-specific body of knowledge. Specific training courses can be established to support company-based knowledge requirements.

For complex projects with a multitude of stakeholders, all from different countries with different cultures, finding the perfect project manager may be an impossible task. Today, we are in the infancy stage of understanding complex projects and the accompanying metrics, and we may not be able to determine the ideal skill set for managing complex projects. We must remember that project management existed for more than three decades before we created the first Project Management Body of Knowledge (*PMBOK® Guide*), and even now with the fifth edition, it is still referred to as a "guide."

We can, however, conclude that there are certain skills required to manage complex projects. Some additional skills needed might be: how to manage virtual teams; understanding cultural differences; managing multiple stakeholders, each of whom may have a different agenda;

understanding the impact of politics on project management; and selecting and measuring project metrics.

Governance

Cradle-to-grave user involvement in complex projects is essential. What is unfortunate is that user involvement can change because of politics and the length of the project. It is not always possible to have the same user community attached to the project from beginning to end. Promotions, changes in power and authority positions because of elections, and retirements can cause a shift in user involvement.

Governance is the process of decision making. On large complex projects, governance will be in the hands of the many rather than the few. Each stakeholder may either expect or demand to be part of all critical decisions on the project. This must be supported by proper metrics that provide meaningful information. The channels for governance must be clearly defined at the beginning of the project, possibly before the project manager is assigned. Changes in governance, which are increasingly expected, the longer the project takes, can have a serious impact on the way the project is managed, as well as on the metrics used.

Decision Making

Complex projects have complex problems. All problems generally have solutions, but not all solutions may be good or even practical. Good metrics can make decision making easier. Also, some solutions to problems can be more costly than other solutions. Identifying a problem is usually easy. Identifying alternatives may require the involvement of many stakeholders, and each stakeholder may have a different view of the actual problem and the possible alternatives. To complicate matters, some host countries have very long decision-making cycles, for the identification of the problem as well as for the selection of the best alternative. Each stakeholder may select an alternative that is in the best interests of that particular stakeholder rather than in the best interests of the project.

Obtaining approval can take just as long, especially if the solution requires that additional capital be raised and if politics play an active role. In some emerging countries, every complex project may require the signature of a majority of the ministers and senior government leaders. Decisions may be based upon politics and religion as well.

Fluid Methodologies

With complex projects, the project manager needs a fluid or flexible project management methodology capable of interfacing with multiple stakeholders. The methodology may need to be aligned more with business processes

TIP Completing a project within the triple constraints is not necessarily success if perceived stakeholder value is not there at the conclusion of the project.

TIP The more complex the project, the more time is needed to select metrics, perform measurements, and report on the proper mix of metrics.

TIP The longer the project, the greater the flexibility needed for metrics to change.

than with project management processes, since the project manager may need to make business decisions as well as project decisions. Complex projects seem to be dictated more by business decisions than by pure project decisions.

Complex projects are driven more by the project's end value than by the triple or competing constraints. Complex projects tend to take longer than anticipated and cost more than originally budgeted because of the need to guarantee that the final result will have the value desired by the customers and stakeholders. Simply stated, complex projects tend to be value-driven rather than driven by the triple or competing constraints.

1.3 GLOBAL PROJECT MANAGEMENT

Every company in the world has complex projects that they would have liked to undertake but were unable to because of limitations such as:

- No project portfolio management function to evaluate projects
- A poor understanding of capacity planning
- A poor understanding of project prioritization
- A lack of tools for determining project value
- A lack of project management tools and software
- A lack of sufficient resources
- A lack of qualified resources
- A lack of support for project management education
- A lack of a project management methodology
- A lack of knowledge in dealing with complexity
- A fear of failure
- A lack of understanding of metrics needed to track the project

Because not every company has the capability to manage these complex projects, they must look outside for suppliers of project management services. Companies that provide these services on a global basis consider themselves to be business solution providers and differentiate themselves from localized companies according to the elements in Table 1–4.

Those companies that have taken the time and effort to develop flexible project management methodologies and become solution providers are companies that are competing in the global marketplace. Although these companies may have as part of their core business the providing of products and services, they may view their future as being a global solution provider for the management of complex projects.

TABLE 1-4 Global versus Nonglobal Companies

FACTOR	NONGLOBAL	GLOBAL
Core business	Sell products and services	Sell business solutions
PM satisfaction level	Must be good at project management	Must excel at project management
PM methodology	Rigid	Flexible and fluid
Metrics/KPIs	Minimal	Extensive
Supporting tools	Minimal	Extensive
Continuous Improvement	Follow the leader	Capture best practices and lessons learned
Business knowledge	Know your company's business	Understand the client's business as well as your company's business
Type of team	Co-located	Virtual

TIP Competing globally cannot be accomplished effectively with the same mindset as competing locally. An effective project management information system based upon possibly project-specific metrics may be essential.

For these companies, being good at project management is not enough; they must excel at project management. They must be innovative in their processes to the point that all processes and methodologies are highly fluid and easily adaptable to a particular client. They have an extensive library of tools to support the project management processes. Most of the tools were created internally with ideas discovered through captured lessons learned and best practices.

1.4 PROJECT MANAGEMENT METHODOLOGIES AND FRAMEWORKS

Most companies today seem to recognize the need for one or more project management methodologies but either create the wrong methodologies or misuse the methodologies that have been created. Many times, companies rush into the development or purchasing of a methodology without any understanding of the need for one other than the fact that their competitors have a methodology. Jason Charvat states:[2]

> Using project management methodologies is a business strategy allowing companies to maximize the project's value to the organization. The methodologies must evolve and be "tweaked" to accommodate a company's changing focus or direction. It is almost a mind-set, a way that reshapes

2. Jason Charvat, *Project Management Methodologies*, John Wiley & Sons Publishers, Hoboken, 2003; p.2.

entire organizational processes: sales and marketing, product design, planning, deployment, recruitment, finance, and operations support. It presents a radical cultural shift for many organizations. As industries and companies change, so must their methodologies. If not, they're losing the point.

There are significant advantages to the design and implementation of a good, flexible methodology:

- Shorter project schedules
- Reduces and/or provides better control of costs
- Prevents unwanted scope changes
- Can plan for better execution
- Can predict results more accurately
- Improves customer relations during project execution
- Can adjust the project during execution to fit changing customer requirements
- Provides senior management with better visibility of status
- Provides standardization in execution
- Captures best practices

Rather than using policies and procedures, some methodologies are constructed as a set of forms, guidelines, templates, and checklists that can and must be applied to a specific project or situation. It may not be possible to create a single enterprise-wide methodology that can be applied to each and every project. Some companies have been successful doing this, but there are still many companies that successfully maintain more than one methodology. Unless the project manager is capable of tailoring the enterprise project management methodology to his/her needs, more than one methodology may be necessary.

There are several reasons why good intentions often go astray. At the executive levels, methodologies can fail if the executives have a poor understanding of what a methodology is and believe that a methodology is:[3]

- A quick fix
- A silver bullet
- A temporary solution
- A cookbook approach for project success

At the working levels, methodologies can also fail if they:[4]

- Are abstract and high level
- Contain insufficient narratives to support these methodologies

3. Ibid., p.4.

4. Ibid., p.5.

- Are not functional or do not address crucial areas
- Ignore the industry standards and best practices
- Look impressive but lack real integration into the business
- Use nonstandard project conventions and terminology
- Compete for similar resources without addressing this problem
- Don't have any performance metrics
- Take too long to complete because of bureaucracy and administration

Other reasons why methodologies can fail include:

- The methodology must be followed exactly even if the assumptions and environmental input factors have changed.
- The methodology focuses on linear thinking.
- The methodology does not allow for out-of-the-box thinking.
- The methodology does not allow for value-added changes that are not part of the original requirements.
- The methodology does not fit the type of project.
- The methodology is too abstract (rushing to design it).
- The methodology development team neglects to consider bottlenecks and the concerns of the user community.
- The methodology is too detailed.
- The methodology takes too long to use.
- The methodology is too complex for the market, clients, and stakeholders to understand.
- The methodology does not have sufficient or correct metrics.

Deciding on what type of methodology is not an easy task. There are many factors to consider such as:[5]

- The overall company strategy—how competitive are we as a company?
- The size of the project team and/or scope to be managed
- The priority of the project
- How critical the project is to the company
- How flexible the methodology and its components are

There are numerous other factors that can influence the design of a methodology. Some of these factors include:

- Corporate strategy
- Complexity and size of the projects in the portfolio
- Management's faith in project management
- Development budget
- Number of life cycle phases

5. Ibid., p.66.

- Technology requirements
- Customer requirements
- Training requirements and costs
- Supporting tools and software costs

Project management methodologies are created around the project management maturity level of the company and the corporate culture. If the company is reasonably mature in project management and has a culture that fosters cooperation, effective communication, teamwork, and trust, then a highly flexible methodology can be created based upon guidelines, forms, checklists, and templates. As stated previously, the more flexibility that is added into the methodology, the greater the need for a family of metrics and KPIs. Project managers can pick and choose the parts of the methodology and metrics that are appropriate for a particular client. Organizations that do not possess either of these two characteristics rely heavily upon methodologies constructed with rigid policies and procedures, thus creating significant paperwork requirements with accompanying cost increases, and removing the flexibility that the project manager needs to adapt the methodology to the needs of a specific client. These rigid methodologies usually rely upon time and cost as the only metrics and can make it nearly impossible to determine the real status of the project.

Jason Charvat describes these two types as light methodologies and heavy methodologies:[6]

Light Methodologies

Ever-increasing technological complexities, project delays, and changing client requirements brought about a small revolution in the world of development methodologies. A totally new breed of methodology—which is agile, adaptive, and involves the client every part of the way—is starting to emerge. Many of the heavyweight methodologists were resistant to the introduction of these "lightweight" or "agile" methodologies (Fowler, 2001[7]). These methodologies use an informal communication style. Unlike heavyweight methodologies, lightweight projects have only a few rules, practices, and documents. Projects are designed and built on face-to-face discussions, meetings, and the flow of information to the clients. The immediate difference of using light methodologies is that they are much less documentation-oriented, usually emphasizing a smaller amount of documentation for the project.

6. Ibid, pp.102–104.

7. Martin Fowler, *The New Methodology, Thought Works*, 2001. Available at www.martinfowler.com/articles.

Heavy Methodologies

The traditional project management methodologies (i.e., SDLC approach) are considered bureaucratic or "predictive" in nature and have resulted in many unsuccessful projects. These heavy methodologies are becoming less popular. These methodologies are so laborious that the whole pace of design, development, and deployment slows down—and nothing gets done. Project managers tend to predict every milestone because they want to foresee every technical detail (i.e., software code or engineering detail). This leads managers to start demanding many types of specifications, plans, reports, checkpoints, and schedules. Heavy methodologies attempt to plan a large part of a project in great detail over a long span of time. This works well until things start changing, and the project managers inherently try to resist change.

Frameworks

More and more companies today, especially those that wish to compete in the global marketplace as a business solution provider, are using frameworks rather than methodologies.

- **Framework:** The individual segments, principles, pieces, or components of the processes needed to complete a project. This can include forms, guidelines, checklists, and templates.
- **Methodology:** The orderly structuring or grouping of the segments or framework elements. This can appear as policies, procedures, or guidelines.

Frameworks focus on a series of processes that must be done on all projects. Each process is supported by a series of forms, guidelines, templates, checklists, and metrics that can be applied to a particular client's business needs. The metrics will be determined jointly by the project manager, the client, and the various stakeholders.

As stated previously, a methodology is a series of processes, activities, and tools that are part of a specific discipline, such as project management, and designed to accomplish a specific objective. When the products, services, or customers have similar requirements and do not require significant customization, companies develop methodologies to provide some degree of consistency in the way that projects are managed. With these methodologies, the metrics, once established, usually remain the same for every project.

As companies become reasonably mature in project management, the policies and procedures are replaced by forms, guidelines, templates, and checklists. This provides more flexibility for the project manager in how to apply the methodology to satisfy a specific customer's requirements. This leads to a more informal application of the project management methodology, and significantly more metrics are now required.

Today, this informal project management approach has been somewhat modified and called a framework. A framework is a basic conceptual structure that is used to address an issue, such as a project. It includes a set of assumptions, project-specific metrics, concepts, values, and processes that provide the project manager with a means for viewing what is needed to satisfy a customer's requirements. A framework is a skeletal support structure for building the project's deliverables.

Frameworks work well as long as the project's requirements do not impose severe pressure upon the project manager. Unfortunately, in today's chaotic environment, this pressure appears to be increasing because:

- Customers are demanding low-volume, high-quality products with some degree of customization.
- Project life cycles and new product development times are being compressed.
- Enterprise environmental factors are having a greater impact on project execution.
- Customers and stakeholders want to be more actively involved in the execution of projects.
- Companies are developing strategic partnerships with suppliers, and each supplier can be at a different level of project management maturity.
- Global competition has forced companies to accept projects from customers that are all at a different level of project management maturity.

These pressures tend to slow down the decision-making processes at a time when stakeholders want the processes to be accelerated. This slowdown is the result of:

- The project manager being expected to make decisions in areas where he/she has limited knowledge.
- The project manager hesitating to accept full accountability and ownership for the projects.
- Excessive layers of management being superimposed on the project management organization.
- Risk management is being pushed up to higher levels in the organizational hierarchy.
- The project manager demonstrates questionable leadership ability.

Both methodologies and frameworks are mechanisms by which we can obtain best practices and lessons learned in the use of metrics and KPIs. Figure 1–1 illustrates the generic use of a methodology or framework. Once we identify the clients and stakeholders, we then input the requirements, business case, and accompanying assumptions. The methodology then guides us through the *PMBOK® Guide* process groups of initiation (I), planning (P), execution (E), monitoring and controlling (M), and closure (C).

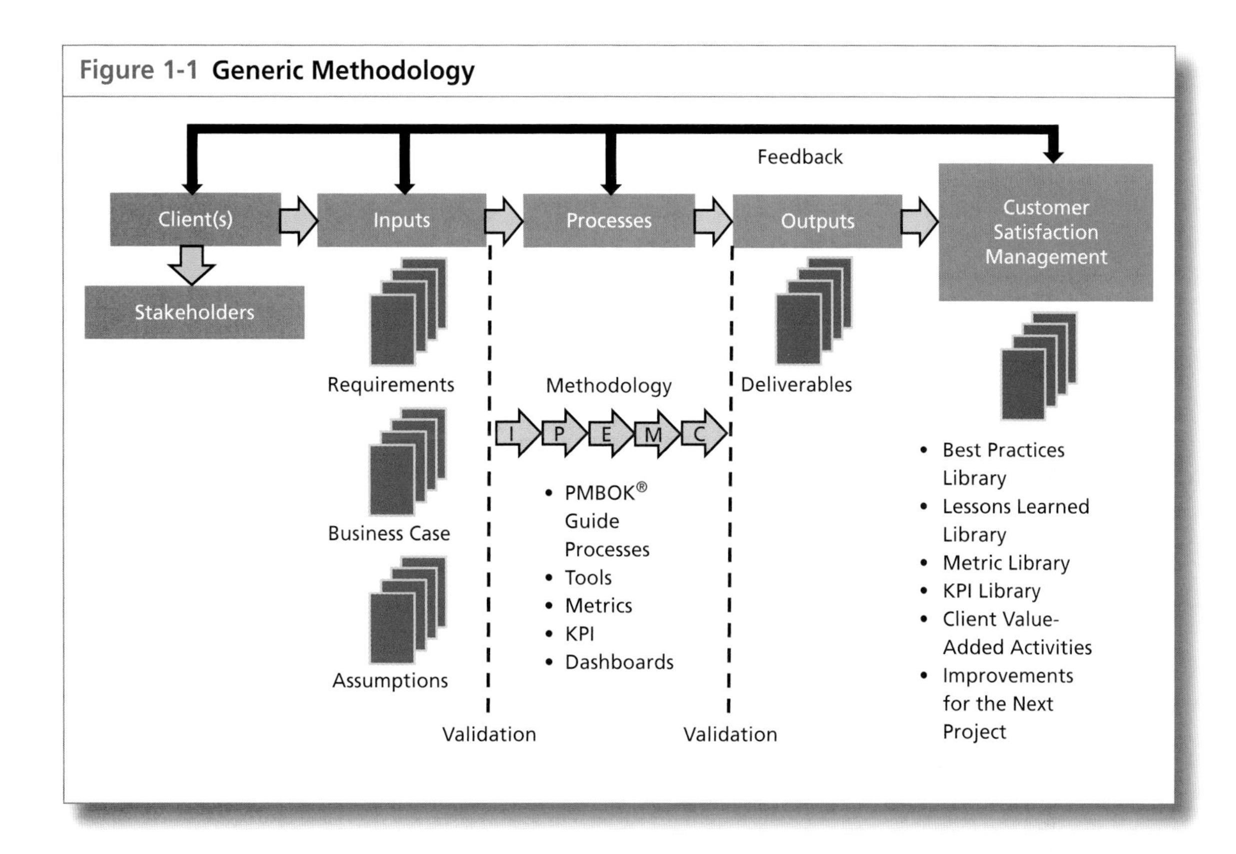

Figure 1-1 Generic Methodology

The methodology also provides us with guidance in the identification of metrics, KPIs, and dashboard reporting techniques for a particular client.

Some people believe that, once the deliverables are provided to the client and project closure takes place, the project is completed. This is not the case. More companies today are adding, at the end of the life cycle phases of the methodology, another life cycle phase, entitled "Customer Satisfaction Management." The purpose of this phase is to meet with the client and the stakeholders and discuss what was learned on the project regarding best practices, lessons learned, metrics, and KPIs. The intent is to see what can be done better for that client on future projects. Today, companies maintain metric and KPI libraries the same way that they maintain libraries for best practices and lessons learned.

1.5 THE NEED FOR EFFECTIVE GOVERNANCE

The problems described previously can be resolved by using effective project governance. Project governance is actually a framework by which decisions are made. Governance relates to decisions that define expectations, accountability, responsibility, the granting of power, or the verifying of performance. Governance relates to consistent management, cohesive policies, processes, and decision-making rights for a given area of responsibility. Governance enables efficient and effective decision making to take place.

Every project can have different governance, even if each project uses the same enterprise project management methodology. The governance function can operate as a separate process or as part of project management leadership. Governance is not designed to replace project decision making but to prevent undesirable decisions from being made. Effective governance must be supported by a good project management information system (PMIS). The PMIS must have agreed upon metrics and key performance indicators such that informed decision making is possible rather than seat-of-the-pants decision making.

SITUATION: At the onset of a project, the governance committee agreed to make certain decisions to assist the project manager. Unfortunately, metrics were not established to support the governance committee. The result was a schedule slippage and a cost overrun due to delayed decision making.

Historically, governance was provided by the project sponsor. Today, governance is provided by a committee. The membership of the committee can change from project to project and industry to industry. The membership may also vary according to the number of stakeholders and whether the project is for an internal or external client.

1.6 ENGAGEMENT PROJECT MANAGEMENT

With project management viewed as a strategic competency today, it is natural for companies that wish to compete in a global marketplace to be strong believers in "engagement project management" or "engagement selling." Years ago, the sales force would sell a product or services to a client and then move on to find another client. Today, the emphasis is on staying with the clients and looking for additional work from the same clients.

In a marital context, an engagement can be viewed as the beginning of a lifelong partnership. The same holds true with engagement project management. Companies like IBM and Hewlett-Packard no longer view themselves as selling products or services. Instead, they see themselves as business solution providers for their clients, and you cannot remain in business as a business solution provider without having superior project management capability.

As part of engagement project management, you must convince the client that you have the project management capability to provide solutions to their business needs on a repetitive basis. In exchange for this, you want the client to treat you as a strategic partner rather than as just another contractor. This is shown in Figure 1–2.

Figure 1-2 "Engagement" Project Management *International Institute for Learning, Inc.*

Previously, we stated that those companies that wish to compete in a global environment must have superior project management capability. This capability must appear in the contractor's response to a request for proposal issued by the client. Clients today are demanding the following in their proposal:

- Show us the number of PMP®s in your company and identify which PMP® will manage this contract if you are the winner through competitive bidding.
- Show us that you have an enterprise project management methodology or framework, and that it has a history of providing repeated successes.
- Show us that you are willing to customize the framework or methodology to fit the client's environment.
- Show us the maturity level of project management in your company and identify which project management maturity model you used to perform the assessment.
- Show us that you have a best practices library for project management and your willingness to share this knowledge with us, as well as the best practices you discover on our project.

Decades ago, the sales force (and marketing) had very little knowledge about project management. The role of the sales force was to win contracts, regardless of the concessions that had to be made. The project manager then "inherited" a project with an underfunded budget and an impossible schedule. Today, sales and marketing must understand project management and be able to sell it to the client as part of engagement selling. The sales force must sell the company's project management methodology or framework and the accompanying best practices. Sales and marketing are now involved in project management.

Engagement project management benefits both the buyer and the seller, as shown in Table 1–5.

TABLE 1-5 Before and after Engagement Project Management

BEFORE ENGAGEMENT PROJECT MANAGEMENT	AFTER ENGAGEMENT PROJECT MANAGEMENT
Continuous competitive bidding	Sole-source or single-source contracting (fewer suppliers to deal with)
Focus on the near-term value of the deliverable	Focus on the lifetime value of the deliverable
Contractor provides minimal lifetime support for clients with their customers	Contractor provides lifetime support for customer value analyses (CVA) and customer value measurement (CVM)
Utilize one inflexible system	Access to contractor's many systems
Limited metrics	Use of the contractor's metrics library

The benefits of engagement project management are clear:

- Both the buyer and the seller save on significant procurement costs by dealing with single-source or sole-source contracts without having to go through a formalized bidding process for each project.
- Because of the potential long-term strategic partnership, the seller is interested in the lifetime value of the business solution rather than just the value at the end of the project.
- You can provide lifelong support to your client as they try to develop value-driven relationships with their clients.
- The buyer will get access to many of the project management tools used by the seller. The corollary is also true.

There is a risk in hiring consultants to manage your projects if they bring their own methodology and accompanying metrics that are not compatible to your business or your needs. You must make sure that the business solution providers demonstrate that:

- Their approach is designed to your business model and strategy.
- The metrics they bring with them fit your business model and strategy.
- You understand the metrics they are proposing.
- If necessary, they are willing to create additional metrics that fit your needs.

1.7 CUSTOMER RELATIONS MANAGEMENT

Engagement project management is forcing project managers to become active participants in customer relations management (CRM) activities. CRM activities focus on:

- Identifying the right customers
- Developing the right relationship with the customers
- Maintaining customer retention

TABLE 1-6 Engagement Manager versus Project Manager

CUSTOMER VALUE MANAGEMENT	ENGAGEMENT MANAGER	PROJECT MANAGER
Phase 1: Identifying the right customers	• Strategic marketing • Proposal preparation • Engagement selling	• Assist in proposal preparation • May report to engagement manager
Phase 2: Developing the right relationship	• Defining acceptance criteria (metrics/KPIs) • Risk mitigation planning • Client briefings • Client invoicing • Soliciting satisfaction feedback and CRM	• Supporting CRM • Establishing performance metrics • Measuring customer value and satisfaction • Improving customer satisfaction management
Phase 3: Maintaining Retention	• Conducting customer satisfaction management meeting • Updating client metrics and KPIs	• Attending customer satisfaction management meetings • Looking for future areas of improvement

This cannot be done entirely by the project manager. Some companies have both engagement managers and project managers. These two individuals must work together to maintain customer satisfaction. Table 1–6 below shows the partial responsibilities of each.

1.8 OTHER DEVELOPMENTS IN PROJECT MANAGEMENT

For companies to be successful at managing complex projects on a repetitive basis and function as a solution provider, the project management methodology and accompanying tools must be fluid or adaptive. This means that you may need to develop a different project management approach when interfacing with each stakeholder, given the fact that each stakeholder may have different requirements and expectations, and the fact that most complex projects have long time spans. Figure 1–3 illustrates some of the new developments in project management. This applies to both traditional and nontraditional projects.

The five items in the figure fit together when done properly.

- **New success criteria:** At the initiation of the project, the project manager will meet with the client and the stakeholders to come to stakeholder agreements on what constitutes success on the project. Initially, many of the stakeholders may have their own definition of success, but the project manager must forge an agreement, if possible.
- **Key performance indicators:** Once the success criteria are agreed upon, the project manager and the project team will work with the stakeholders

Figure 1-3 **New Developments in Project Management**

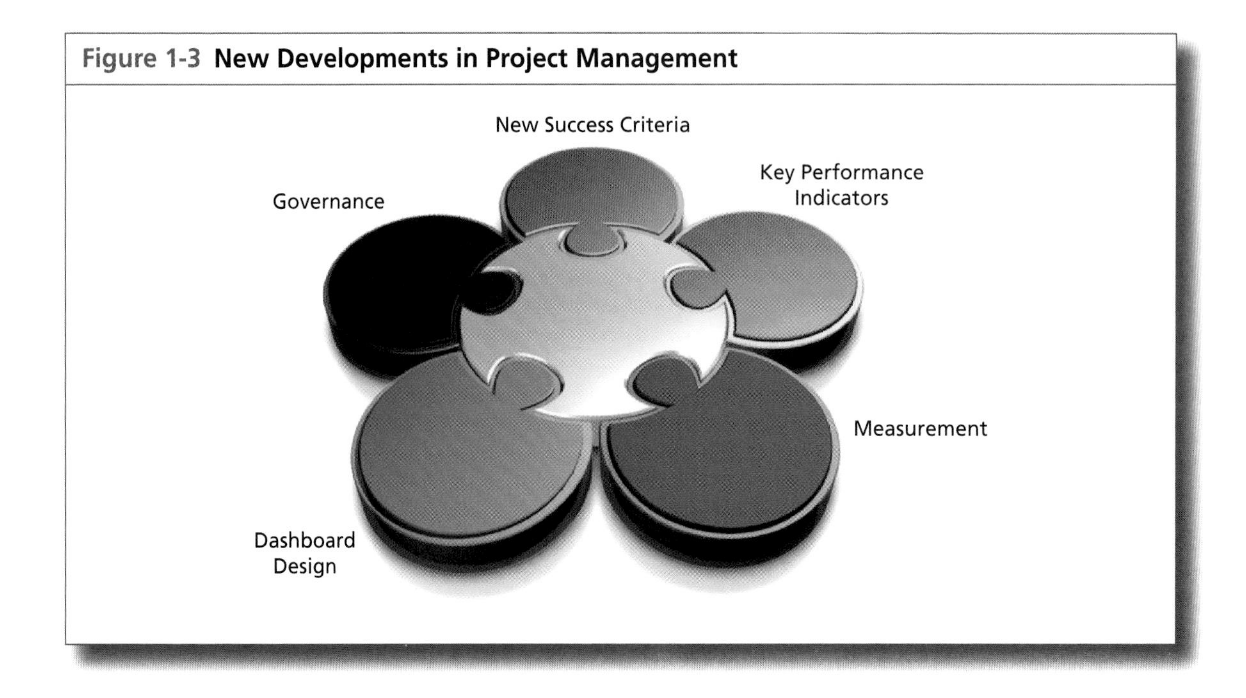

to define the metrics and key performance indicators that each stakeholder wishes to track. It is possible that each stakeholder will have different KPI requirements.

- **Measurement:** Before the metrics and KPIs are agreed to and placed on the dashboards, we must be sure we know how to perform the measurements. This is the hardest part because not all team members or strategic partners may have the capability or skills to measure all of the KPIs.
- **Dashboard design:** Once the KPIs are identified and measurement techniques are identified, the project manager, along with the appropriate project team members, will design a dashboard for each stakeholder. Some of the KPIs in the dashboards will be updated periodically, whereas others may be updated on a real-time basis.
- **Governance:** Once the measurements are made, critical decisions may have to be supervised by the governance board. The governance board can include key stakeholders, as well as stakeholders who are functioning just as observers.

1.9 A NEW LOOK AT DEFINING PROJECT SUCCESS

The ultimate purpose of project management is to create a continuous stream of project successes. This can happen provided that you have a good definition of "success" on each project.

SITUATION: Many years ago, as a young project manager, I asked a vice president in my company, "What is the definition of success on my project?" He responded, "The only definition in this company is meeting the target profit margin in the contract." I then asked him, "Does our customer have the same definition of success?" That ended our conversation.

For years, customers and contractors were each working toward different definitions for success. The contractor focused on profits as the only success factor, whereas the customer was more concerned with the quality of the deliverables. As project management evolved, all of that was about to change.

Success Is Measured by the Triple Constraints

The triple constraints can be defined as a triangle with the three sides representing time, cost, and performance (which may include quality, scope, and technical performance). This was the basis for defining success during the birth of project management. This definition was provided by the customer, where cost was intended to mean within the contracted cost. The contractor's interpretation of cost was profit.

Customer Satisfaction Must Be Considered As Well

Managing a project within the triple constraints is always a good idea, but the customer must be satisfied with the end result. A contractor can complete a project within the triple constraints and still find that the customer is unhappy with the end result. So, we have now placed a circle around the triple constraints, entitled "customer satisfaction." The president of an aerospace company stated, "The only definition of success in our business is customer satisfaction." That brought the customer and the contractor a little closer together. Aerospace and defense contractors were incurring large cost overruns, and it was almost impossible to define success according to the triple constraints. Numerous scope changes were initiated by both the customer and the contractor. Because the scope changes were numerous, the only two metrics used on projects were related to time and cost. Success, however, was measured by follow-on business, which was an output of customer satisfaction.

Other (or Secondary) Factors Must Be Considered As Well

SITUATION: Several years ago, I met a contractor that had underbid a job for a client by almost 40 percent. When I asked them why they were willing to lose money on the contract, they responded, "Our definition of success on this project is being able to use the client's name as a reference in our sales brochures."

There can be secondary success factors that, based upon the project, are more important than the primary factors. These secondary factors include using the customer's name as a reference, corporate reputation and image, compliance with government regulations, strategic alignment, technical superiority, ethical conduct, and other such factors. The secondary factors may now end up being more important than the primary factors of the triple constraints.

Success Must Include a Business Component

By the turn of the century, companies were establishing project management offices (PMOs). One of the primary activities for the PMO was to make sure that each project was aligned to strategic business objectives. The definition of success, thus, included a business component as well as a technical component. As an example, consider the following components included in the definition of success provided by a spokesperson from Orange Switzerland:[8]

> The delivery of the product within the scope of time, cost, and quality characteristics
>
> The successful management of changes during the project life cycle
>
> The management of the project team
>
> The success of the product against criteria and target during the project initiation phase (e.g., adoption rates, ROI, . . .)

As another example, consider the following provided by Colin Spence, project manager/partner at Convergent Computing (CCO):[9]

General guidelines for a successful project are as follows:

> Meeting the technology and business goals of the client on time, on budget and on scope
>
> Setting the resource or team up for success, so that all participants have the best chance to succeed and have positive experiences in the process
>
> Exceeding the client's expectations in terms of abilities, teamwork, and professionalism and generating the highest level of customer satisfaction.
>
> Winning additional business from the client, and being able to use them as a reference account and/or agree to a case study.

---

8. Kerzner, H., *Project Management Best Practices; Achieving Global Excellence*, Hoboken, NJ: John Wiley & Sons Publishers, 2006, pp.22–23.

9. Ibid. p.23.

Creating or fine-tuning processes, documentation, and deliverables that can be shared with the organization and leveraged in other engagements.

Our definition of the role of the project manager also changed. Project managers were managing part of a business rather than merely a project, and they were expected to make sound business decisions as well as project decisions. There must be a business purpose for each project. Each project is expected to make a contribution of business value to the company when the project is completed.

Prioritization of Success Constraints May Be Necessary

Not all project constraints are equal. The prioritization of constraints is performed on a project-by-project basis. Sponsors' involvement in this decision is essential. Secondary factors are also considered to be constraints and may be more important than the primary constraints. For example, years ago, at Disneyland and Disney World, the project managers designing and building the attractions at the theme parks had six constraints:

- Time
- Cost
- Scope
- Safety
- Aesthetic value
- Quality

At Disney, the last three constraints, those of safety, aesthetic value, and quality, were considered locked in constraints that could not be altered during tradeoffs. All tradeoffs were made on time, cost, and scope.

The importance of the components of success can change over the life of the project. For example, in the initiation phase of a project, scope may be the critical factor for success, and all tradeoffs are made on the basis of time and cost. During the execution phase of the project, time and cost may become more important, and then tradeoffs will be made on the basis of scope.

SITUATION: The importance of the components of success at a point in time can also determine how decisions are made. As an example, a project sponsor asked a project manager when the project's baseline schedules will be prepared. The project manager responded, "As soon as you tell me what is most important to you, time, cost, or risk, I'll prepare the schedules. I can create a schedule based upon least time, least cost, or least risk. I can give you only one of those three in the preparation of the schedule." The project sponsor was somewhat irate because he wanted all three. The project manager knew better, however, and held his ground. He told the sponsor that he would prepare one and only one schedule, not three schedules. The project sponsor finally said, rather reluctantly, "Lay out the schedule based upon least time."

Previously we stated that the definition of project success has a business component. That's true for both the customer and contractor's definition of success. Also, each project can have a different definition of success. There must be upfront agreement between the customer and the contractor at project initiation or even at the first meeting between them on what constitutes success at the end of or during the project. In other words, there must be a common agreement on the definition of success, especially the business reason for working on the project.

The Definition of Success Must Include a "Value" Component

We stated previously that there must be a business purpose for working on a project. Now, however, we understand that, for real success to occur, there must be value achieved at the completion of the project. Completing a project within the constraints of time and cost does not guarantee that business value will be there at the end of the project. In the words of Warren Buffett, one of the world's most successful investors and chairman and CEO of Berkshire Hathaway, "Price is what you pay. Value is what you get."

One of the reasons why it has taken us so long to include a value component in the definition of success is that it is only in the last several years we have been able to develop models for measuring the metrics to determine the value on a project. These same models are now being used by PMOs in selecting a project portfolio that maximizes the value the company will receive. Also, as part of performance reporting, we are now reporting metrics on time at completion, cost at completion, value at completion, and time to achieve value.

TIP The definition of success must be agreed upon between the customer and the contractor.

Determining the value component of success at the completion of the project can be difficult, especially if the true value of the project cannot be determined until well after the project is completed. We may have to establish some criteria on how long we are willing to wait to assess the true value.

Multiple Components for Success

Today, we have come to the realization that there are multiple constraints on a project. We are now working on more complex projects, where the traditional triple constraints success factors are constantly changing. For example, in Figure 1–4, for traditional projects, time, cost, and scope may be a higher priority than the constraints within the triangle. However, for more complex projects, this is reversed.

The fourth edition of the *PMBOK® Guide* no longer uses the term "triple constraints." Because there can be more than three constraints, we are now using the term "competing constraints," where the exact number of success constraints and their relative importance can change from project to project. What is important is that metrics must be established for each constraint on a project. However, not all of the metrics on the constraints will be treated as key performance indicators.

The Future

So, what does the future look like? The following list is representative of some of the changes that are now taking place:

- The project manager will meet with the client at the very beginning of the project and they will come to an agreement on what constitutes project success.
- The project manager will meet with other project stakeholders and get their definition of success. There can and will be multiple definitions of success for each project.

Figure 1-4 From Triple to Competing Constraints

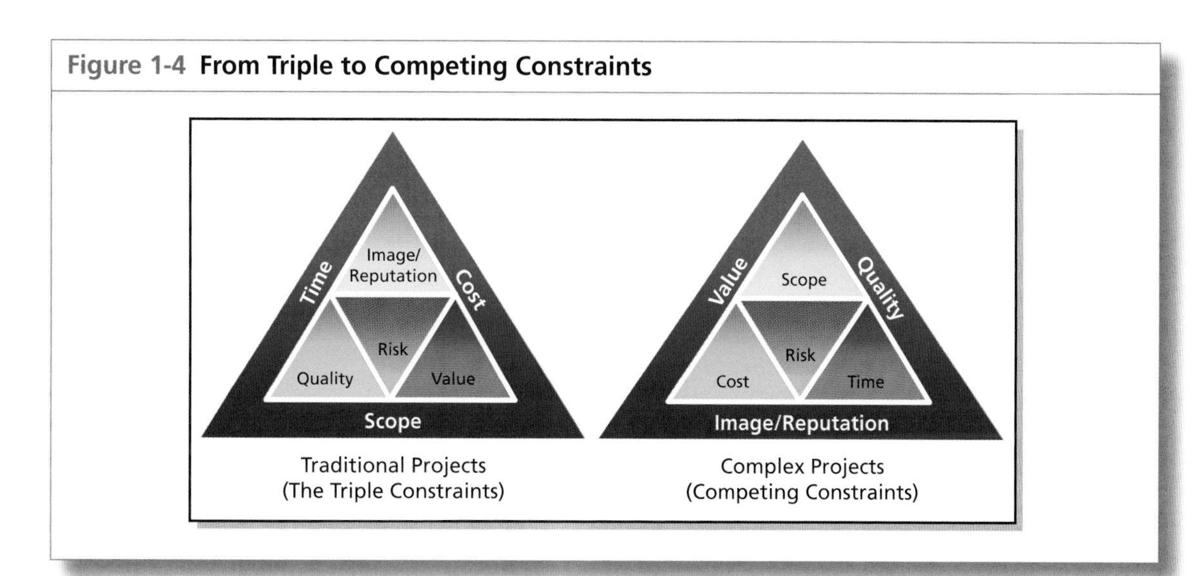

- The project manager, the client, and the stakeholders will come to an agreement on what metrics they wish to track to verify that success will be achieved. Some metrics will be treated as key performance indicators.
- The project manager, assisted by the PMO, will prepare dashboards for each stakeholder. The dashboards will track each of the requested success metrics in real time, rather than relying on periodic reporting.
- At project completion, the PMO will maintain a library of project success metrics that can be used on future projects.

In the future, we can expect the PMO to become the guardian of all project management intellectual property. The PMO will create templates to assist project manages in defining success and establishing success metrics.

1.10 THE GROWTH OF PAPERLESS PROJECT MANAGEMENT

Making informed decisions requires information. In the early years of project management, we relied heavily upon legacy systems for the information we needed. Over the past several decades, other information systems have emerged as seen in Figure 1-5. Project management information systems (PMIS) evolved to provide information solely for the project at hand. Later, enterprise resource planning (ERP) systems and customer relations management (CRM) systems appeared that provided project management with sufficient information such that they could now make business as well as

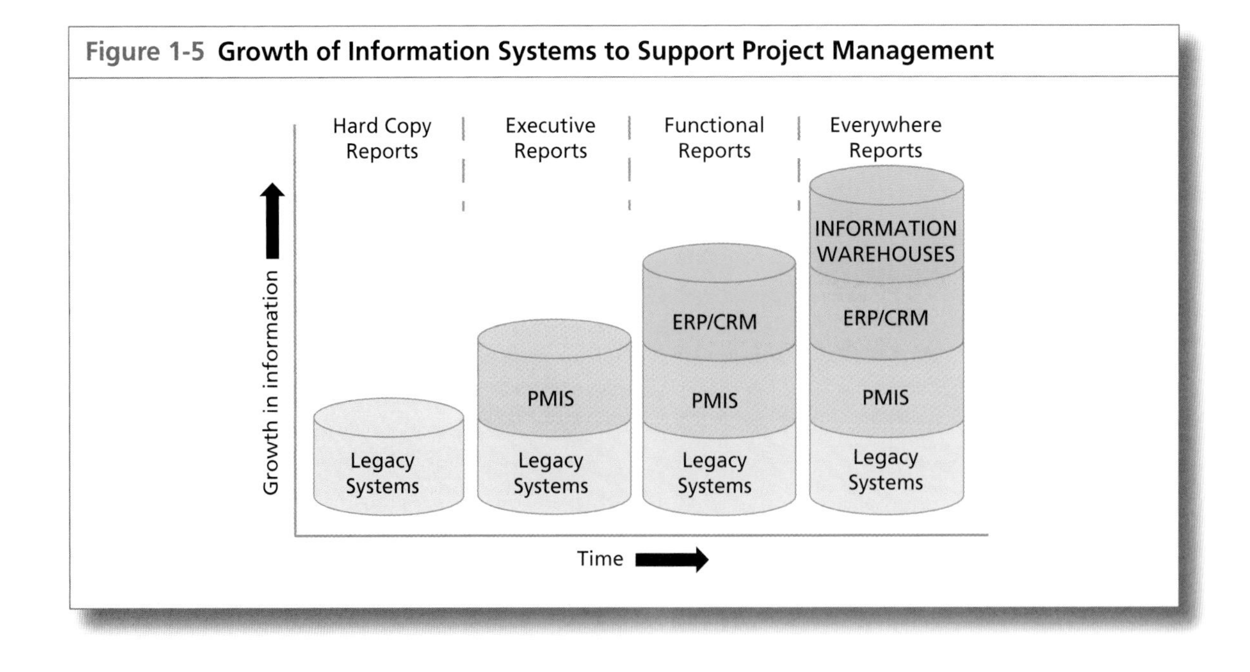

Figure 1-5 Growth of Information Systems to Support Project Management

project based decisions. Today, the amount of information that a company can generate is overwhelming, and all of this information will be stored in data or information warehouses. With pure legacy systems that tracked business metrics the information was reported mainly vertically up the organizational hierarchy. Today, project-based information can be reported everywhere including organizations external to your company.

Having more information comes with a price: more costly reporting and larger and more frequent reports. This is shown in Figure 1–6. As the cost of paperwork grew, companies began looking at the possibility of paperless project management. This would necessitate identification of only the critical information and presenting the information using dashboards.

Initially, reporting was done at the end of each life cycle phase. Unfortunately, some customers would not see project status until the end-of-phase gate review meetings. To solve this problem, we created policy and procedure manuals that dictated how and when reporting should take place. Unfortunately, this placed restriction on the project managers, and eventually the policies and procedures were replaced with guidelines. Today, the focus is on dashboards.

Figure 1-6 Growth of Information Systems to Support Project Management

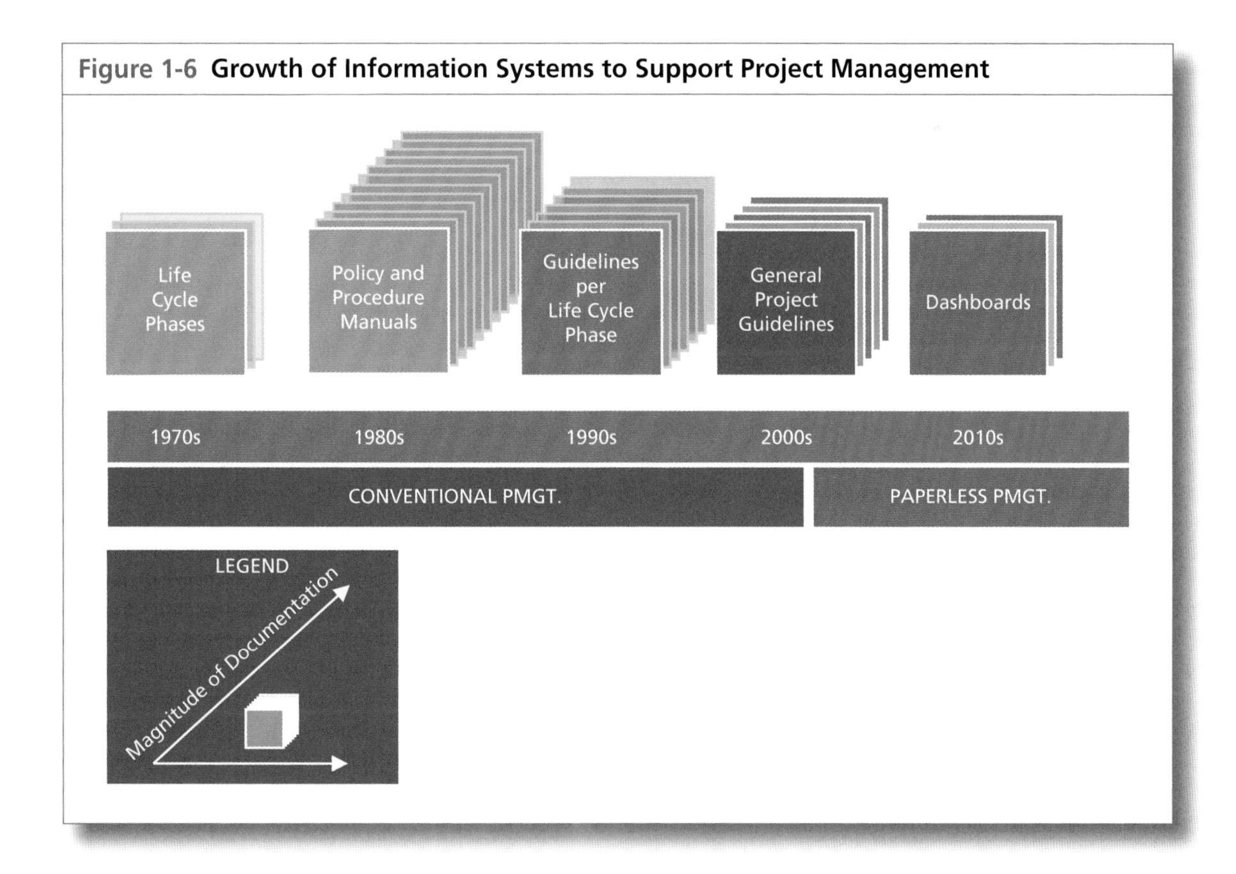

1.11 PROJECT MANAGEMENT MATURITY AND METRICS

All companies desire maturity and excellence in project management. Unfortunately, not all companies recognize that the time frame can be shortened by performing strategic planning for project management maturity and excellence. The simple use of project management, even for an extended period of time, does not lead to excellence. Instead, it can result in repeated mistakes and, what's worse, learning from your own mistakes rather than the mistakes of others.

Strategic planning for project management is unlike other forms of strategic planning in that it is most often performed at the middle and lower levels of management. Executive management is still involved, mostly in a supporting role, and provides funding together with employee release time for the effort.

There are models that can be used to assist in achieving excellence. One such model is the Project Management Maturity Model (PMMM), shown in Figure 1–7. Each of the five levels represents a different degree of maturity in project management.

Figure 1-7 Project Management Maturity and Metrics

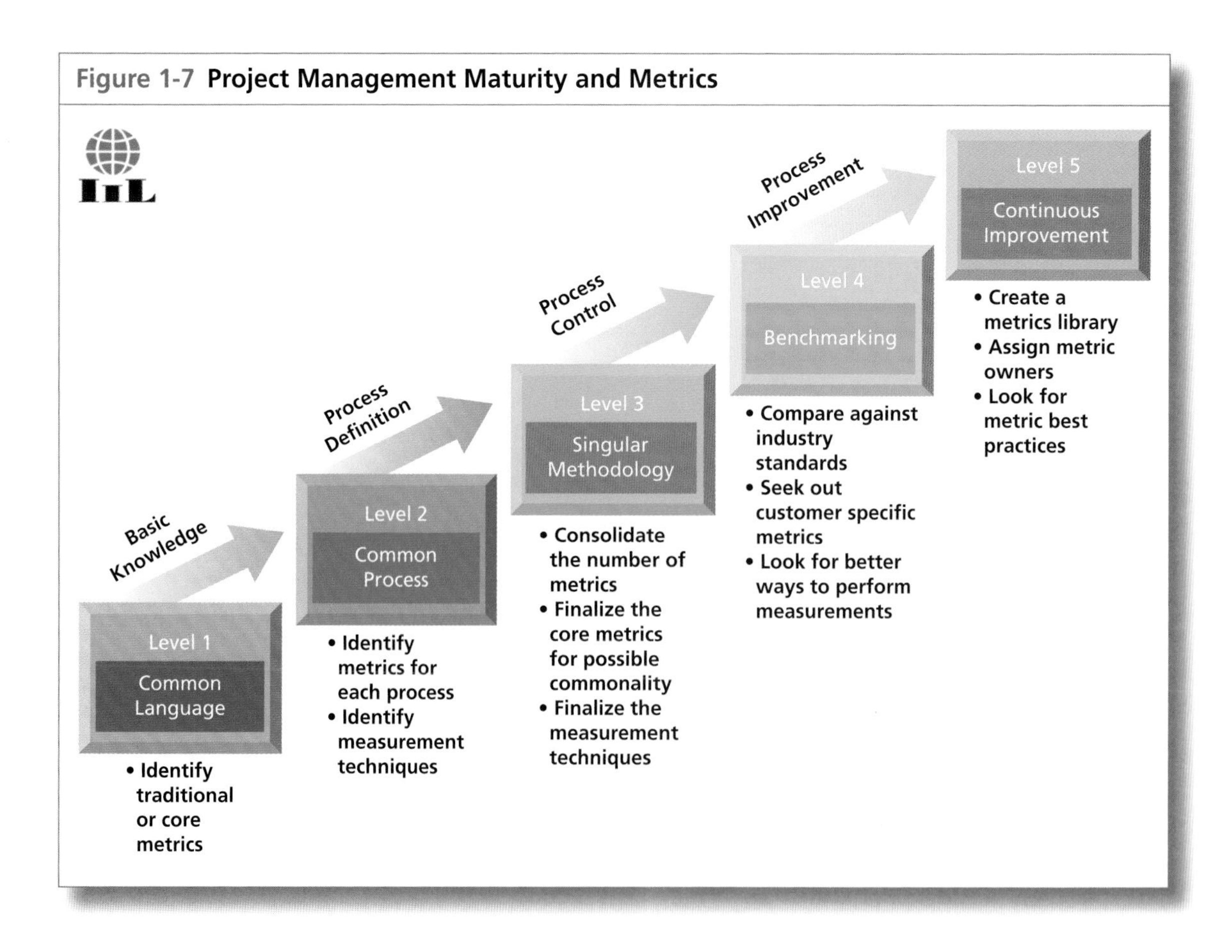

Level 1—Common Processes: In this level, the organization recognizes the importance of project management and the need for a good understanding of the basic knowledge on project management, along with the accompanying language and terminology.

Level 2—Common Processes: In this level, the organization recognizes that common processes need to be defined and developed such that the successes on one project can be repeated on other projects. Also included in this level is the recognition that project management can be applied to and support other methodologies employed by the company.

Level 3—Singular Methodology: In this level the organization recognizes the synergistic effect of combining all corporate methodologies and processes into a singular methodology, the center of which is project management. The synergistic effects also make process control easier with a single methodology than with multiple methodologies.

Level 4—Benchmarking: This level contains the recognition that process improvement is necessary to maintain a competitive advantage. Benchmarking should be performed on a continuous basis. The company must decide who to benchmark against and what to benchmark.

Level 5—Continuous Improvement: In this level, the organization evaluates the information obtained through benchmarking and must then decide whether or not this information will enhance the singular methodology.

Although these five levels are normally accomplished with forms, guidelines, templates, and checklists, the growth in metrics management has allowed us to further enhance PMMM by including in each level the necessity for metrics. This is shown in Figure 1–7. Metrics can serve as a sign of organizational maturity. The need for paperless project management will require that more emphasis be placed upon metrics management as part of the project management maturity process.

Maturity in project management allows companies to recognize that project management is a strategic competency as shown in Figure 1–8. For companies that promote their project management capabilities to external clients, competency in project management is viewed as a sustained competitive advantage (SCA). However, ineffective metrics management can increase the risks in maintaining a sustained competitive advantage as shown in Figure 1–9. These risks will be covered in detail in later chapters.

In Figure 1–8 we showed that excellence in project management is achieved when project management is seen as a strategic competency and the company recognizes that its project management capability has become a competitive advantage. Unfortunately, competitive advantages are not always sustainable as can be seen from Figure 1–10. As you exploit your competitive advantage, the competitors counterattack to reduce or eliminate your competitive advantage. Therefore, as illustrated in Figure 1–11, you must have continuous improvement for the competitive advantage to grow into a sustained competitive advantage.

Figure 1-8 **Project Management Competitiveness**

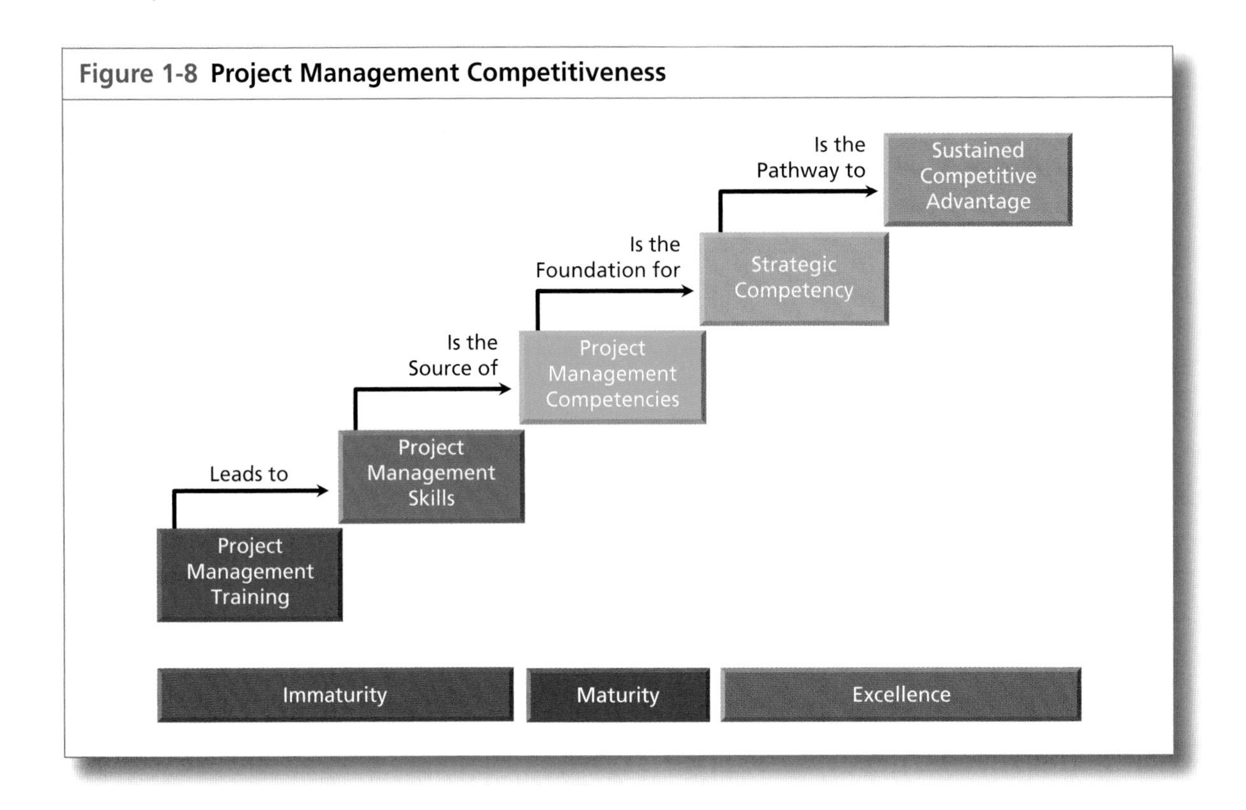

Figure 1-9 **Metric Risks to Maintain a SCA**

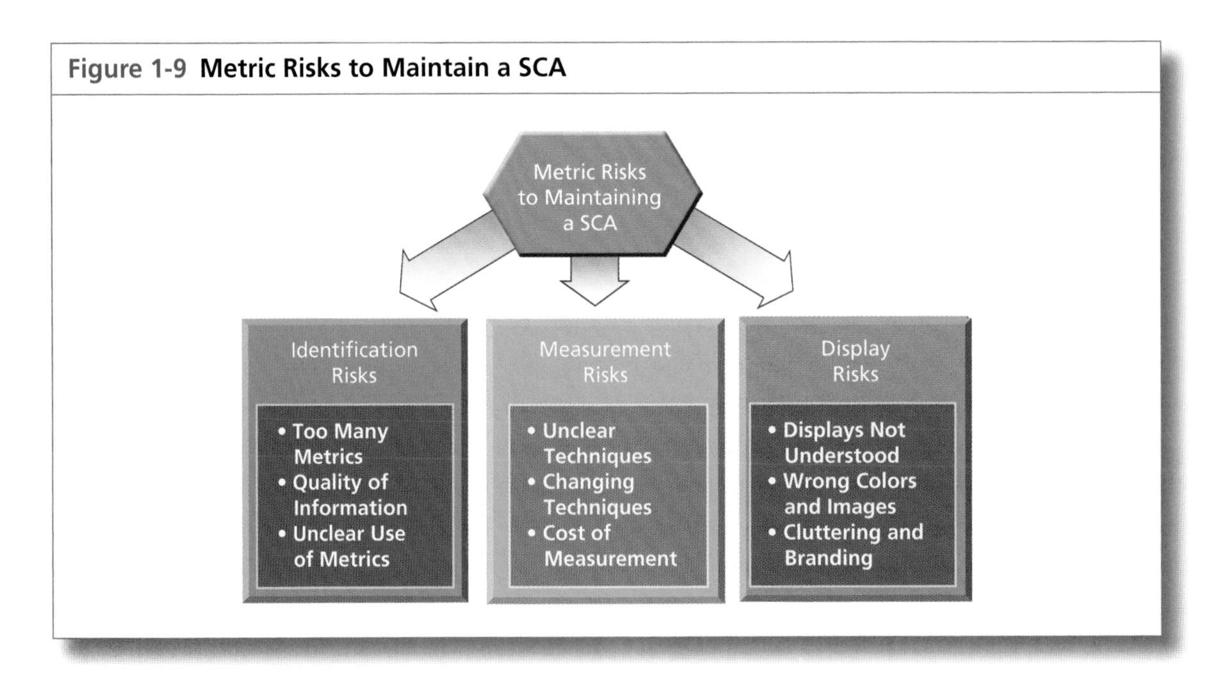

Having a sustained competitive advantage in project management does not come just from being on time and on budget at the end of each project. Rather, offering your clients something that your competitors cannot do may help. But in project management, a true competitive advantage occurs when your efforts are directly linked to the customers' perception of value, and whatever means you use to show this, such as through the use of value-reflective metrics, gives you a sustainable competitive advantage. Value-reflective metrics, which will be discussed in Chapter 5, show us how to create value. If these metrics undergo continuous improvement, then we may be adding value for the customers.

Figure 1-10 Nonsustainable Competitive Advantages

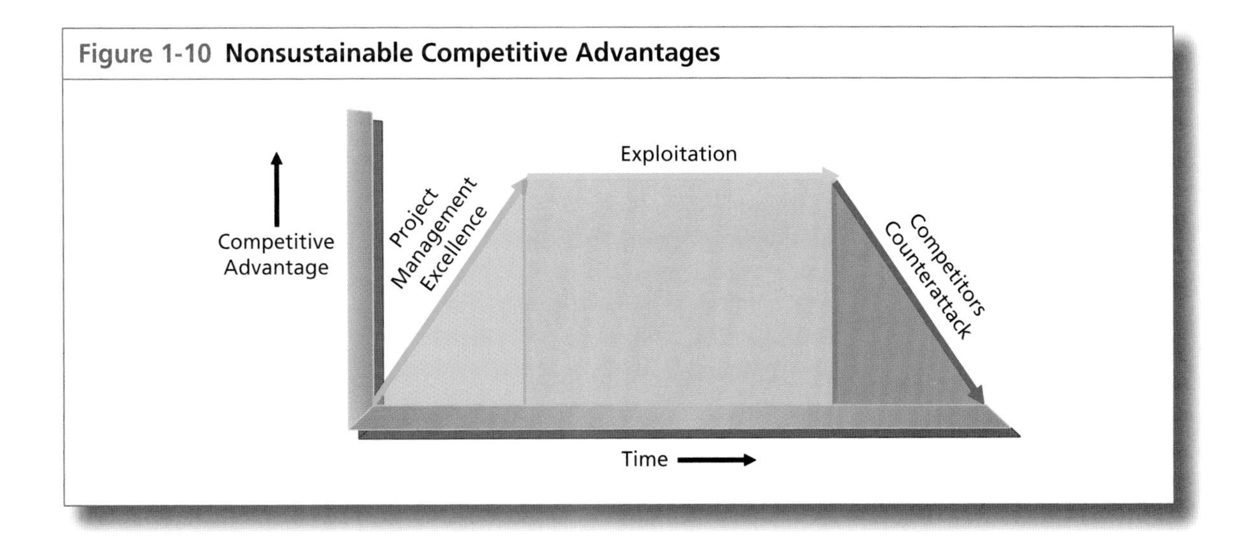

Figure 1-11 Sustainable Competitive Advantages

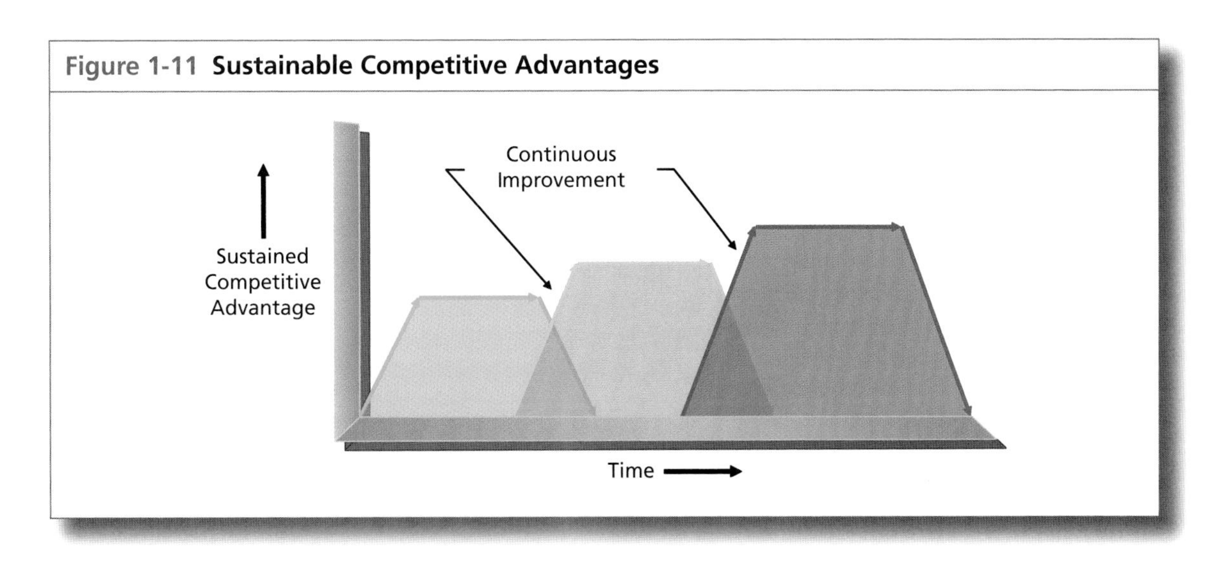

TABLE 1-7 Competitive Advantages from Value-Reflective Metrics

METRICS WITH VALUE ATTRIBUTES	POSSIBLE COMPETITIVE ADVANTAGE
Deliverables produced	Efficiency
Product functionality	Innovation
Product functionality	Product differentiation
Support response time	Service differentiation
Staffing and employee pay grades	People differentiation
Quality	Quality differentiation
Action items in the system and how long	Speed of problem resolution and decisions
Cycle time	Speed to market
Failure rates	Quality differentiation and innovation

There is no point in wasting resources on value metrics unless the client understands the metric and perceives the value that is being created. Therefore, client input into the selection of the attributes for the value metrics is essential. Table 1–7 shows some typical value-reflective metrics and the accompanying strategic competitive advantage.

1.12 PROJECT MANAGEMENT BENCHMARKING AND METRICS[10]

One of the fastest ways to reach maturity and excellence in project management is through the use of benchmarking. A benchmark is a measurement or standard against which comparisons can be made. Benchmarking is the process of comparing one's business processes and performance metrics to industry bests or best practices from other industries. Dimensions typically measured are quality, time, and cost. In the process of benchmarking, management identifies the best firms in their industry, or in another industry where similar processes exist, and compares the results and processes of those studied (the "targets") to their own company's results and processes. In this way, they learn how well the targets perform and, more importantly, the business processes that explain why these firms are successful.

Best Practice versus Proven Practice

In project management, we tend to use the terms "best practice benchmarking" or "process benchmarking," in which organizations evaluate various aspects of their processes in relation to best practice companies' processes, usually within a peer group defined for the purposes of comparison. This then allows organizations to develop plans on how to make improvements or adapt specific best practices, usually with the aim of increasing some aspect of project management performance. Benchmarking is often treated as a continuous process in which organizations continually seek to improve their practices.

For more than a decade, companies have been fascinated with the expression "best practices." Best practices are generally those practices that have been proven to produce superior results. But now, after a decade or more of use, we are beginning to scrutinize the term and realize that perhaps better expressions exist. When a company says that it has a best practice, it really means that there is a technique, process, metric, method, or activity that can be more effective at delivering an outcome than any other approach and provides the company with the desired outcome with fewer problems and unforeseen complications. As a result, the company ends up with the most efficient and effective way of accomplishing a task based upon a repeatable process that has been proven over time for a large number of people and/or projects.

There are several arguments why the words "best practice" should not be used. First, there is the argument that the identification of a best practice may lead some to believe that they were performing some activities incorrectly in the past, and that may not have been the case. This may simply be a more efficient and effective way of achieving a deliverable. Another argument is that some people believe that best practices imply that there is one and only one way of accomplishing a task. This also may be a faulty interpretation. Third, and perhaps most important, is the argument that a best practice is the "best" way of performing an activity and, since it is the best, no further opportunities for improvement are possible.

Once a best practice has been identified and been proven to be effective, we normally integrate the best practice into our project management processes so that it becomes a standard way of doing business. Therefore, after acceptance and proven use of the idea, the better expression possibly should be a "proven practice" rather than a best practice. This leaves the door open for further improvements.

These are just some arguments why best practices may be just buzzwords and should be replaced by proven practices. Perhaps in the future the expression best practices will be replaced by proven practices. However, for the remainder of this text, we will refer to the expression as best practices, but the reader must understand that other terms may be more appropriate.

Benchmarking Methodologies

There is no single benchmarking process that has been universally adopted. The wide appeal and acceptance of benchmarking has led to the emergence of benchmarking methodologies. However, benchmarking activities with regard to project management are usually easier to implement and accept because of the existence of the *PMBOK® Guide* and a project management office (PMO). The *PMBOK® Guide* helps us identify areas where benchmarking would be beneficial and people understand that the PMO is responsible for continuous improvements in project management.

The following is an example of a typical benchmarking methodology:

- **Identify problem areas:** Because benchmarking can be applied to any business process or function, a range of research techniques may be required. They include informal conversations with customers, employees, or suppliers; exploratory research techniques such as focus groups; and in-depth marketing research, quantitative research, surveys, questionnaires, reengineering analysis, process mapping, quality control variance reports, financial ratio analysis, or simply reviewing cycle times or other performance indicators.
- **Identify others that have similar processes:** Because project management exists in virtually every industry, benchmarking personnel should not make the mistake of looking only at their own industry.
- **Identify organizations that are leaders in these areas:** Look for the very best in any industry and in any country. Consult customers, suppliers, financial analysts, trade associations, and magazines to determine which companies are worthy of study. Symposiums and conferences sponsored by the Project Management Institute provide excellent opportunities to hear presentations from companies that are doing things exceptionally well. Even companies that are under financial distress may be outstanding is some areas of project management.
- **Visit the "best practice" companies to identify leading edge practices:** Companies typically agree to mutually exchange information beneficial to all parties in a benchmarking group and share the results within the group.
- **Implement new and improved business practices:** Take the leading edge practices and develop implementation plans that include identification of specific opportunities, funding the project, and selling the ideas to the organization for the purpose of gaining demonstrated value from the improvements.

Benchmarking Costs

The three main types of costs in benchmarking are:

- **Visitation costs:** This includes hotel rooms, travel costs, meals, a token gift, and lost labor time.
- **Time costs:** Members of the benchmarking team will be investing time in researching problems, finding exceptional companies to study, visits,

and implementation. This will take them away from their regular tasks for part of each day so additional staff might be required.

- **Benchmarking database costs:** Organizations that institutionalize benchmarking into their daily procedures find it is useful to create and maintain a database or library of best practices.

The cost of benchmarking can substantially be reduced through utilizing the many internet resources that have sprung up over the last few years. These aim to capture benchmarks and best practices from organizations, business sectors, and countries to make the benchmarking process much quicker and cheaper.

Types of Benchmarking

There are several types of benchmarking studies:

- **Process benchmarking:** The initiating firm focuses its observation and investigation of project management and business processes with a goal of identifying and observing the best practices from one or more benchmark firms. Activity analysis will be required where the objective is to benchmark cost and efficiency in executing the processes that are part of a project management methodology. This is the most common form of benchmarking in project management. Process benchmarking cannot be successful if you do not fully understand your own processes.
- **Metric benchmarking:** The process of comparing the different metrics that organizations are using for continuous improvements. Time, cost, and quality are just three of the metrics that are being used. We are now creating additional metrics to measure what is needed, not what is the easiest to measure. The intent is to identify the core metrics needed for project management. One of the biggest challenges for metric benchmarking is the variety of metric definitions used among companies or divisions. Definitions may change over time within the same organization due to changes in leadership and priorities. The most useful comparisons can be made when metrics definitions are common between compared units and do not change so improvements can be verified.
- **Financial benchmarking:** performing a financial analysis and comparing the results in an effort to assess your overall competitiveness and productivity.
- **Benchmarking from an investor perspective:** Extending the benchmarking universe to also compare to peer companies that can be considered alternative investment opportunities from the perspective of an investor.
- **Performance benchmarking:** Allows the initiator firm to assess their competitive position by comparing products and services with those of target firms.

- **Product benchmarking:** The process of designing new products or upgrades to current ones. This process can sometimes involve reverse engineering, which is taking apart competitors products to find strengths and weaknesses.
- **Strategic benchmarking:** This involves observing how others compete. This type is usually not industry specific, meaning it is best to look at other industries.
- **Functional benchmarking:** A company will focus its benchmarking on a single function to improve the operation of that particular function. Complex functions such as human resources, finance and accounting, and information and communication technology are unlikely to be directly comparable in cost and efficiency terms and may need to be disaggregated into processes to make valid comparison.
- **Best-in-class benchmarking:** This involves studying the leading competitor or the company that best carries out a specific function.
- **Internal benchmarking:** A comparison of a business process to a similar process inside the organization. This is a quest for internal best practices.
- **Competitive benchmarking:** This is a direct competitor-to-competitor comparison of a product, service, process or method.
- **Generic benchmarking:** This approach broadly conceptualizes unrelated business processes or functions that can be practiced in the same or similar ways regardless of industry.

Benchmarking Code-of-Conduct

There are numerous problems that can occur during benchmarking. Some problems result from misunderstandings, whereas other problems could involve legal issues. The Code-of-Conduct scripted by the International Benchmarking Clearinghouse is an excellent starting point.

- **Legality:** Avoid any discussions that could be interpreted as illegal for you or your benchmarking partners.
- **Exchange:** Be prepared to answer the same questions you are asking. Letting partners review the questions in advance is helpful.
- **Confidentiality:** All information should be treated as proprietary information. You may wish to consider having everyone sign a nondisclosure agreement.
- **Use of Information:** There must be an agreement, preferably in writing, on how the information will be used.
- **Contact:** Follow your partners' protocols and customs on who you are allowed to interface with.
- **Preparation:** Be fully prepared for partner interfacing and exchanges of information.
- **Completion:** Avoid making promises or commitments that cannot be kept.

Benchmarking Failures

There are benchmarking mistakes that can lead to benchmarking failures. Some of these mistakes include:

- Limiting benchmarking activities to just your own industry
- Benchmarking industry followers can provide just as much information as benchmarking industry leaders.
- Not all results may be applicable to your company, especially if organizational cultural differences exist.
- Failing to have a benchmarking plan and not knowing what you are looking for

Points to Remember

There are some critical points that must be remembered when performing benchmarking:

- It is necessary to understand the culture and circumstances behind the numbers to fully understand their meaning and use. The "how" is just as important as the "how much?"
- In project management, changes can occur quickly. It is important to set frequencies for the benchmarking studies, and each process studied may require different frequencies.
- The more rigorous the benchmarking process, the better the results.
- Regardless of how good you think your project management systems are, there is always room for improvement.
- Those who do not believe in continuous improvement soon become industry followers rather than leaders.
- Recognize that executives who are not familiar or supportive of benchmarking will always adopt the "not invented here" argument or "this is the way we have always done it."
- Successful benchmarking is "doing," not "knowing."
- Benchmarking allows you to learn from the mistakes of others rather than from your own mistakes.
- Because of the rate of change that takes place in project management, it is highly unlikely that the targets you benchmark with will be leaders in all areas of project management.
- Benchmarking can prevent surprises.
- You must get these people to recognize the need for change. This must be accomplished with benchmarking evidence rather than just claims or opinions.
- Change occurs quickly when the people who are needed to change or make the change are involved in the benchmarking studies.
- Implementing change requires a champion. Having a PMO is almost always the right idea.

1.13 CONCLUSIONS

The future of project management may very well rest in the hands of the solution providers. These providers will custom-design project management frameworks and methodologies for each client and possibly for each stakeholder. They must be able to develop metrics that go well beyond the current *PMBOK® Guide* and demonstrate a willingness to make business decisions as well as project decisions. The future of project management looks quite good, but it will be a challenge.

2 THE DRIVING FORCES FOR BETTER METRICS

CHAPTER OVERVIEW

Today, more than ever before, we are struggling with a great percentage of projects that are becoming distressed and possibly failing. Techniques such as project audits and health checks are encouraging the use of a more formalized metrics management system. Effective project decisions cannot be made without meaningful metrics. Stakeholder relations management thrives on meaningful metrics.

CHAPTER OBJECTIVES

- To understand the importance of metrics in dealing with stakeholders
- To understand the importance of metrics when conducting project audits
- To understand the importance of metrics when performing health checks
- To understand the metrics can and will change when trying to recover a distressed project

KEY WORDS

- Boundaries
- Distressed projects
- Project audits
- Project health checks
- Scope creep
- Stakeholder relations management

2.0 INTRODUCTION

Companies do not simply add more metrics or key performance indicators by choice. Usually, there are driving forces that make it evident that such changes are needed. Complacency works when things are going well or as planned. When we start accepting more complex projects, however, as was discussed in Chapter 1, things have a tendency to go poorly.

By performing audits and health checks, we can certainly prevent a project from becoming distressed, provided that the cause of the problem was detected early enough. Unfortunately, the existing metrics that we use might not act as an early warning system. By the time we establish new metrics for analysis of a potentially failing project, the damage may have been done and recovery may no longer be possible. The final result can be devastating if stakeholder relations management fails and future business is not forthcoming. All of this may be attributed to improper identification, selection, implementation, and measurement of the right metrics and key performance indicators.

2.1 STAKEHOLDER RELATIONS MANAGEMENT[1]

Stakeholders are, in one way or another, individuals, companies, or organizations that may be affected by the outcome of the project or the way in which the project is managed. Stakeholders may be either directly or indirectly involved throughout the project, or may function simply as observers. A stakeholder can shift from a passive role to being an active member of the team and participate in making critical decisions.

SITUATION: In order to impress the stakeholders, you agree to establish a multitude of metrics. Once the project begins and the stakeholders begin examining the measurements of the metrics, you realize that some of the stakeholders are now actively involved in the project to the point where they are trying to micromanage you.

SITUATION: As the project progresses, several of the stakeholders begin asking for additional metrics that were not part of the original plan. Your project management methodology does not provide data for these metrics, and the cost of changing the methodology at this point is prohibitive.

TIP Because of the potentially large number of stakeholders, do not attempt to establish metrics that can satisfy all of the stakeholders all of the time.

TIP Passive stakeholders can become active stakeholders when the situation merits it. The project manager must consider metrics for passive stakeholders as well, but perhaps not the same number of metrics that would be provided for the active stakeholders.

On small or traditional projects, project managers generally interface with just the project sponsor as the primary stakeholder, and the sponsor usually is assigned from the organization that funds the project. This is true for both internal and external projects. However, the larger the project, the greater the number of stakeholders you must interface with. The situation becomes even more potentially problematic if you have a large number of stakeholders, geographically dispersed, all at different levels of management in their respective hierarchy, each with a different level of authority, and language and cultural differences. Trying to interface with all of these people on a regular basis

1. Adapted from Harold Kerzner and Carl Belack, *Managing Complex Projects*, John Wiley & Sons and IIL Co-publishers, 2010, Chapter 10.

TIP Not all of the stakeholders will be in agreement on the interpretation of the metrics and have the same conclusions on what action, if any, is necessary.

and make decisions, especially on a large, complex project is very time-consuming.

One of the complexities of stakeholder relations management is figuring out how to do all of this without sacrificing your company's long-term mission or vision. Also, your company may have long-term objectives in mind for this project, and those objectives may not necessarily be aligned to the project's objectives or each stakeholder's objectives. Lining up all of the stakeholders in a row and getting them to uniformly agree to all decisions is more wishful thinking than reality. You may discover that it is impossible to get all of the stakeholders to agree, and you must simply hope to placate as many as possible at a given point of time.

Stakeholder relations management cannot work effectively without commitments from all of the stakeholders. Obtaining these commitments can be difficult if the stakeholders cannot see what's in it for them at the completion of the project, namely the value that they expect or other personal interest. The problem is that what one stakeholder perceives as value, another stakeholder may have a completely different perception or a desire for a different form of value. For example, one stakeholder could view the project as a symbol of prestige. Another stakeholder could perceive the value as simply keeping his/her people employed. A third stakeholder could see value in the final deliverables of the project and the inherent quality in it. A fourth stakeholder could see the project as an opportunity for future work with particular partners.

Another form of agreement involves developing a consensus on how stakeholders will interact with each other. It may be necessary for certain stakeholders to interact with each other and support one another with regard to sharing resources, providing financial support in a timely manner, and sharing intellectual property. While all stakeholders recognize the necessity for these agreements, they can be affected by politics, economic conditions, and other enterprise environmental factors that may be beyond the control of the project manager. Certain countries may not be willing to work with other countries because of culture, religion, views on human rights, and other such factors.

TIP Metrics systems, no matter how good, may not generate interaction between stakeholders. Metrics are not a replacement for effective project management communications.

TIP Changes in stakeholders may cause the creation of new metrics regardless how far the project has progressed.

For the project manager, obtaining these agreements right at the beginning of the project is essential. Some project managers are fortunate to be able to do this while others are not. Leadership changes in certain governments may make it difficult to enforce these agreements on complex projects.

It is important that everyone who has a stake in the project be willing to communicate what they believe are the factors critical for success. Success is easier to define and reach when there is broad agreement. However, there is no guarantee that there will be an agreement between the project manager, client(s) and stakeholders as to a core set of metrics and KPIs. Project managers must be prepared to construct a different set of dashboards possibly for each viewer. But there is a risk. If too many metrics are established, then too much data may be necessary and the project team may find it difficult to capture all of the information that is needed. The metrics selected must be scaled to fit the needs of the project. Also, the metrics selected must be used to ensure that the time and cost involved with selecting and measuring the metrics has been worthwhile.

There are some simple steps that can be followed:

- The project team must brainstorm what metrics are appropriate for this project. This is easy to do if a metrics library exists. This may be accomplished prior to stakeholder involvement.
- The team must then assess the metrics to make sure they are needed and look for ways of combining metrics if appropriate.
- The team must make sure that the metrics are presented in terms that every viewer can understand. (i.e., What is the meaning of a partial or incomplete deliverable? What is the meaning of an unacceptable deliverable?)
- The team must then decide what metrics they will recommend to the stakeholders.

Sometimes, we select the number of metrics based upon the personal whims of the stakeholders and the project team and then discover how costly it is to measure and report all of the metrics. The number of metrics selected should be based upon the size of the project, the complexity, the decisions that the governance committee must make, and the accompanying risks.

On small projects, there isn't sufficient time or funding to work with a large number of metrics. Success on a small project may be as simple as customer acceptance of the deliverables or no complaints by the customer. But for larger and more complex projects, success may not be able to be determined from a single metric. Core metrics may be necessary.

It is important for the project manager to fully understand the issues and challenges facing each of the stakeholders, especially their information needs. Although it may seem unrealistic, some stakeholders can have different views on the time requirements of the project. In some developing nations, the construction of a new hospital in a highly populated area may drive the commitment for the project even though the project could be late by a year or longer. People just want to know that the hospital will eventually be built.

In some cultures, workers cannot be fired. Because they believe they have job security, it may be impossible to get them to work faster or better.

In some countries, there may be as many as 50 paid holidays for the workers, and this can have an impact on the project manager's schedule.

Not all workers in each country have the same skill level even though they have the same title. For example, a senior engineer in an emerging nation may have the same skills as a lower-grade engineer in another country. In some locations where there may exist a shortage of labor, workers are assigned to tasks based upon availability rather than capability. Having sufficient headcount is not a guarantee that the work will get done in a timely manner and that the level of quality will be there.

In some countries, power and authority, as well as belonging to the right political party, are symbols of prestige. People in these positions may not view the project manager as their equal and may direct all of their communications to the project sponsor. In this case, it is possible that salary is less important than relative power and authority.

> It is important to realize that not all of the stakeholders may want the project to be successful. This will happen if stakeholders believe that, at the completion of the project, they may lose power, authority, hierarchical positions in their company, or in a worse case, even lose their job. Sometimes these stakeholders will either remain silent or even be supporters of the project until the end date approaches. If the project is regarded as unsuccessful, these stakeholders may respond by saying "I told you so." If it appears that the project may be a success, these stakeholders may suddenly be transformed from supporters or the silent majority to adversaries, and encourage failure.

It is very difficult to identify stakeholders with hidden agendas. These people can hide their true feelings and be reluctant to share information. There are often no tell-tale or early warnings signs that indicate their true belief in the project. However, if the stakeholders are reluctant to approve scope changes, provide additional investment, or assign highly qualified resources, this could be an indication that they may have lost confidence in the project.

TIP The project manager may find it necessary to establish country-specific metrics for the project manager's personal use.

Not all stakeholders understand project management. Not all stakeholders understand the role of a project sponsor. Not all stakeholders understand how to interface with a project or the project manager, even though they may readily accept and support the project and its mission. Simply stated, the majority of the stakeholders are never trained in how to properly function as a stakeholder. Unfortunately, this cannot be detected early on but may become apparent as the project progresses.

Some stakeholders may be under the impression that they are merely observers and need not participate in decision making or authorization of scope changes. For some stakeholders, who desire to be just observers, this could be a rude awakening. Some will accept the new role, whereas others will not. Those who do not accept the new role usually are fearful that participating in a decision that turns out to be wrong can be the end of their political career.

Some stakeholders view their role as that of micromanagers, often usurping the authority of the project manager by making decisions that they may not necessarily be authorized to make, at least not alone. Stakeholders who attempt to micromanage can do significantly more harm to the project than stakeholders who remain observers.

It may be a good idea for the project manager to prepare a list of expectations that he/she has of the stakeholders. This is essential even though the stakeholders visibly support the existence of the project. Role clarification for stakeholders should be accomplished early on the same way that the project manager provides role clarification for the team members at the initial kickoff meeting for the project.

The present view of stakeholder management in Table 2–1 results from the implementation of "engagement project management" practices. In the past, whenever a sale was made to the client, the salesperson would then move on to find a new client. Salespeople viewed themselves as providers of products and/or services.

TIP Providing too many metrics and key performance indicators may be an invitation for stakeholders to micromanage the project.

Today, salespeople view themselves as the providers of business solutions. In other words, salespeople now tell the client, "we can provide you with a solution to all of your business needs and what we want in exchange is to be treated as a strategic business partner." This benefits both the buyer and seller, as discussed previously.

Therefore, as a solution provider, the project manager focuses heavily on the future and establishing a long-term partnership agreement with the client and the stakeholders. This focus is heavily oriented toward value rather than near-term profitability.

TABLE 2-1 Changing Views in Stakeholder Relations Management

PAST VIEW	PRESENT VIEW
Manage existing relationships	Build relationships for the future; that is, engagement project management
Align the project to short term business goals	Align the project to long-term strategic business goals
Provide ethical leadership when suites	Provide ethical leadership throughout the project
The project is aligned to the profits	The project is aligned the stakeholders' expectation of value
Identify profitable scope changes	Identify value-added scope changes
Provide the stakeholders with the least number of metrics and KPIs	Provide the stakeholders with sufficient metrics and KPIs such that they can make informed decisions

On the micro-level, we can define stakeholder relations management using the six processes shown in Figure 2–1.

- **Identify the stakeholders:** This step may require support from the project sponsor, sales, and the executive management team. Even then, there is no guarantee that all of the stakeholders will be identified.
- **Stakeholder analysis:** This requires an understanding of which stakeholders are key stakeholders, those who have influence, the ability and authority to make decisions, and can make or break the project. This also includes developing stakeholder relations management strategies, based upon the results of the analysis.
- **Perform stakeholder engagements:** During this step, the project manager and the project team get to know the stakeholders.
- **Stakeholder information flow:** This step is the identification of the information flow network and the preparation of the necessary reports for each stakeholder.
- **Abide by agreements:** This step enforces stakeholder agreements made during the initiation and planning stages of the project.
- **Stakeholder debriefings:** This step occurs after contract or life cycle phase closure and is used to capture lessons learned and best practices for improvements on the next project involving these stakeholders or the next life cycle phase.

Figure 2-1 Stakeholder Relations Management

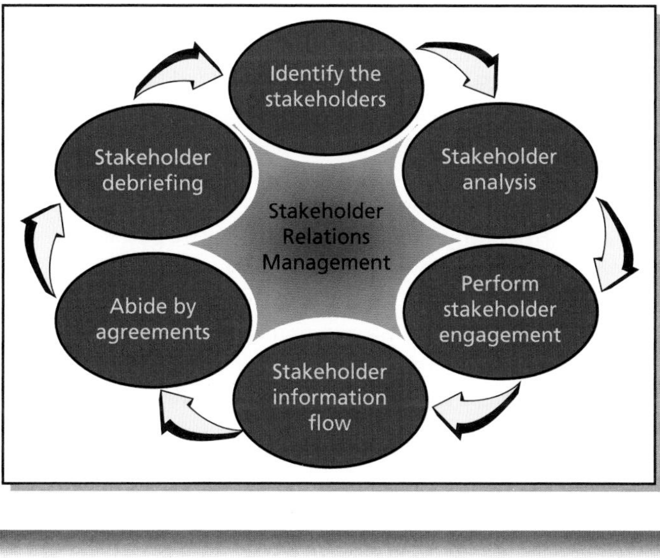

Stakeholder management begins with stakeholder identification. This is easier said than done, especially if the project is multinational. Stakeholders can exist at any level of management. Corporate stakeholders are often easier to identify than political or government stakeholders.

Each stakeholder is an essential piece of the project puzzle. Stakeholders must work together and usually interact with the project through the governance process. Therefore, it is essential to know which stakeholders will participate in governance and which will not.

As part of stakeholder identification, the project manager must know whether he/she has the authority or perceived status to interface with the stakeholders. Some stakeholders perceive themselves as higher stature than the project manager and, in this case, the project sponsor may be the person to maintain interactions.

There are several ways in which stakeholders can be identified. More than one way can be used on projects.

- **Groups:** This could include financial institutions, creditors, regulatory agencies, and the like.
- **Individuals:** These could be identified by name or title, such as the CIO, COO, CEO or just the name of the contact person in the stakeholder's organization.
- **Contribution:** This could include financial contributor, resource contributor, or technology contributor.
- **Other factors:** This could include the authority to make decisions or other such factors.

It is important to understand that not all stakeholders have the same expectations of a project. Some stakeholders may want the project to succeed at any cost, whereas other stakeholders may prefer to see the project fail even though they openly seem to support it. Some stakeholders view success as the completion of the project regardless of the cost overruns, whereas others may define success in financial terms only. Some stakeholders are heavily oriented toward the value they expect to see in the project, and this is the only definition of success for them. The true value may not be seen until months after the project has been completed. Some stakeholders may view the project as their opportunity for public notice and increased stature and, therefore, want to be actively involved. Others may prefer a more passive involvement.

On large, complex projects with a multitude of stakeholders, it may be impossible for the project manager to properly cater to all of the stakeholders. Therefore, the project manager must know who the most influential stakeholders are and who can provide the greatest support on the project. Typical questions to ask include:

- Who are powerful and who are not?
- Who will have or require direct, or indirect, involvement?

- Who has the power to kill the project?
- What is the urgency of the deliverables?
- Who may require more or less information than others?

Not all stakeholders are equal in influence, power, or the authority to make decisions in a timely manner. It is imperative for the project manager to know who sits on the top of the list as having these capabilities.

Finally, it is important to remember that stakeholders can change over the life of a project, especially if it is a long-term project. Also, the importance of certain stakeholders can change over the life of a project and in each life cycle phase. The stakeholder list is, therefore, an organic document subject to change.

Stakeholder mapping is most frequently displayed on a grid, comparing stakeholders' power and their level of interest. This is shown in Figure 2–2. The four cells can be defined as:

- **Manage closely:** These are high-powered, interested people who can make or break your project. You must put forth the greatest effort to satisfy them. Be aware that there are factors that can cause them to change quadrants rapidly.
- **Keep satisfied:** These are high-powered, less interested people who can also make or break your project. You must put forth some effort to

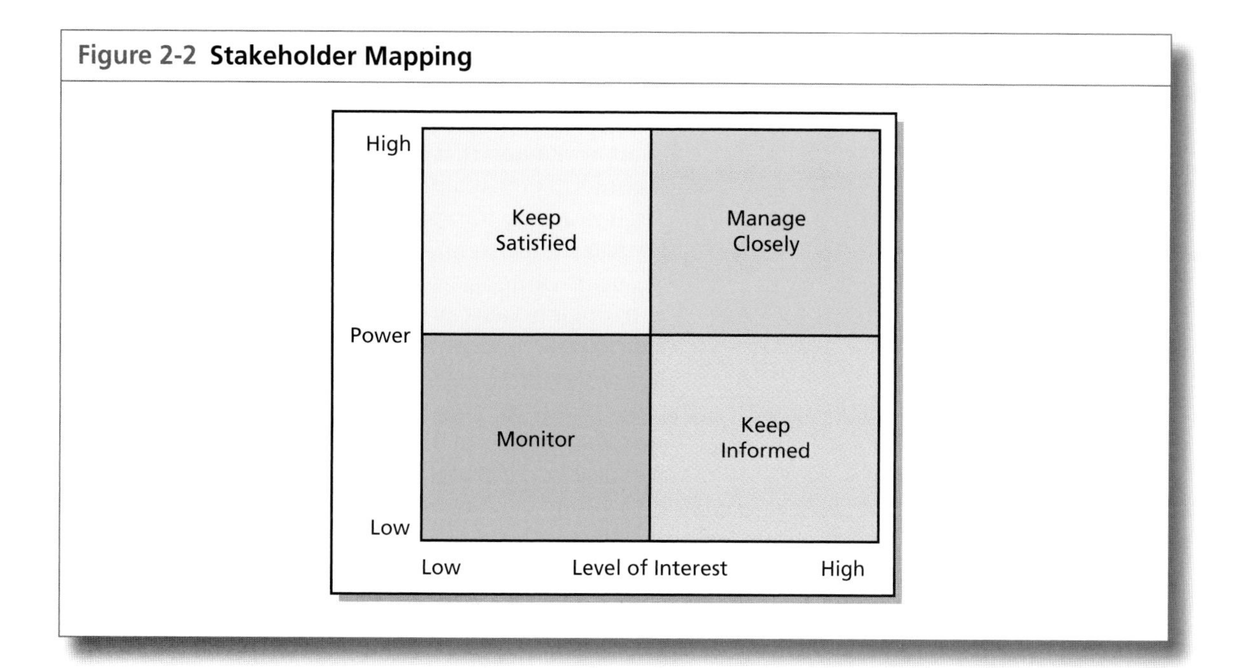

satisfy them but not with excessive detail that can lead to boredom and total disinterest. They may not get involved until the end of the project approaches.

- **Keep informed:** These are people with limited power but keen interest in the project. They can function as an early warning system of approaching problems and may be technically astute and able to assist with some technical issues. These are the stakeholders who often provide hidden opportunities.
- **Monitor only:** These are people with limited power and may not be interested in the project unless a disaster occurs. Provide them with some information but not with too much detail so that they will become disinterested or bored.

TIP Where the stakeholders are positioned on the grid could determine the number of metrics that they want or will be provided. As stakeholders change positions in the grid, so might the number of metrics.

The larger the project, the more important it becomes to know who is and is not an influential or key stakeholder. Although you must win the support of all stakeholders, or at least try to do so, the key stakeholders come first. Key stakeholders may be able to provide the project manager with assistance with the identification of enterprise environmental factors that can have an impact on the project. This could include forecasting on the host country's political and economic conditions, the identification of potential sources for additional funding, and other such issues. In some cases, the stakeholders may have software tools that can supplement the project manager's available organizational process assets.

Thus far, we have discussed the importance of winning over the key or influential stakeholders. There is also a valid argument for winning over the stakeholders who are considered to be unimportant. While some stakeholders may appear to be unimportant, this can change rapidly. For example, an unimportant stakeholder may suddenly discover that a scope change is about to be approved and that scope change can seriously affect the unimportant stakeholder, perhaps politically. Now, the unimportant stakeholder (originally deemed so for apparent lack of concern about the project) becomes a key stakeholder.

Another example occurs on longer-term projects, where stakeholders may change over time, perhaps because of politics, promotions, retirements, or reassignments. The new stakeholder may suddenly want to be an important stakeholder, whereas his/her predecessor was more of an observer. Finally, stakeholders may be relatively quiet in one life cycle phase because of limited involvement but become more active in other life cycle phase, where they must participate. The same may hold true for people who are key stakeholders in early life cycle phases and just observers in later phases. The project team must know who the stakeholders are. The team must also be able to determine which stakeholders are critical stakeholders at specific points in time. The criticality of the stakeholder determines what metrics will appear on the stakeholder's dashboard.

Stakeholder engagement is the point when you physically meet with the stakeholders and determine their needs and expectations. As part of this, you must:

- Understand them and their expectations
- Understand their needs
- Value their opinions
- Find ways to win their support on a continuous basis
- Identify any stakeholder problems early-on that can influence the project

Even though stakeholder engagement follows stakeholder identification, it is often through stakeholder engagement that we determine which stakeholders are supporters, advocates, neutral, or opponents. This may also be viewed as the first step in building a trusting relationship between the project manager and the stakeholders.

As part of stakeholder engagement, it is necessary for the project manager to understand each stakeholder's interests. One of the ways to accomplish this is to ask the stakeholders (usually the key stakeholders) what information they would like to see in performance reports. This information will help identify the key performance indicators (KPI) needed to service that stakeholder.

Each stakeholder may have a different set of KPI interests. It then becomes a costly endeavor for the project manager to maintain multiple KPI tracking and reporting flows, but it is a necessity for successful stakeholder relations management. Getting all of the stakeholders to agree on a uniform set of KPI reports and dashboards may be almost impossible.

There must be an agreement on what information is needed for each stakeholder, when the information is needed, and in what format the information will be presented. Some stakeholders may want a daily or weekly information flow, whereas others may be happy with monthly data. For the most part, the information will be provided via the Internet.

Project managers should use a Communications Matrix to carefully lay out planned stakeholder communications. Information in this matrix might include the following: the definition or title of the communication (e.g., status report, risk register), the originator, the intended recipients, the medium to be used, the rules for the access of information, and the frequency of publication or updates.

Previously, we discussed the complexities of determining the metrics for each stakeholder. Some issues that need to be addressed include:

- The potential difficulty in getting customer and stakeholder agreements on the metrics

TIP There is a high likelihood that stakeholders will change during the execution of the project, especially on long-term projects. This does not mean that changes in the metrics should also take place. Communication with these stakeholders is essential to see if their information needs have changed.

- Determining if the metric data is in the system or needs to be collected
- Determining the cost, complexity, and timing for obtaining the data
- Considering the risks of information system changes and/or obsolescence that can affect metric data collection over the life of the project

Metrics have to be measureable, but some metric information may be difficult to quantify. For example, customer satisfaction, goodwill, and reputation may be important to some stakeholders, but they may be difficult to quantify. Some metric data may need to be measured in qualitative terms rather than quantitative terms.

The need for effective stakeholder communications is clear. This includes:

- Communicating with stakeholders on a regular basis is a necessity.
- Knowing the stakeholders may allow you to anticipate their actions.
- Effective stakeholder communications builds trust.
- Virtual teams thrive on effective stakeholder communications.
- Although we classify stakeholders by groups or organizations, we still communicate with people.
- Ineffective stakeholder communications can cause a supporter to become a blocker.

Part of the process of stakeholder engagement involves the establishment of agreements between the individual stakeholders and the project manager, and among other stakeholders as well. These agreements must be enforced throughout the project. The project manager must:

- Identify any and all agreements among stakeholders (i.e., funding limitations, sharing of information, approval cycle for changes, etc.)
- Identify how politics may change stakeholder agreements
- Identify which stakeholders may be replaced during the project (i.e., retirement, promotion, change of assignment, politics, etc.)

The project manager must be prepared for the fact that not all agreements will be honored.

There are three additional critical factors that must be considered for successful stakeholder relations management:

- Effective stakeholder relations management takes time. It may be necessary to share this responsibility with sponsor, executives, and members of the project team.
- Based upon the number of stakeholders, it may not be possible to address their concerns face to face. You must maximize your ability to communicate via the Internet. This is also important when managing virtual teams.

- Regardless of the number of stakeholders, documentation on the working relationships with the stakeholders must be archived. This is critical for success on future projects.

Effective stakeholder management can be the difference between an outstanding success and a terrible failure. Successful stakeholder management can result in binding agreements. The resulting benefits may be:

- Better decision making and in a more timely manner
- Better control of scope changes; prevention of unnecessary changes
- Follow-on work from stakeholders
- End user satisfaction and loyalty
- Minimize the impact that politics can have on your project

Sometimes, regardless of how hard we try, we will fail at stakeholder relations management. Typical reasons include:

- Inviting stakeholders to participate too early, thus encouraging scope changes and costly delays
- Inviting stakeholders to participate too late so that their views cannot be considered without costly delays
- Inviting the wrong stakeholders to participate in critical decisions, thus leading to unnecessary changes and criticism by key stakeholders
- Key stakeholders becoming disinterested in the project
- Key stakeholders who are impatient with the lack of progress
- Allowing the key stakeholders to believe that their contributions are meaningless
- Managing the project with an unethical leadership style or interfacing with the stakeholders in an unethical manner

2.2 PROJECT AUDITS AND THE PMO

In recent years, the necessity for a structured independent review of various parts of a business, including projects, has taken on a more important role. Part of this can be attributed to the Sarbanes-Oxley Law compliance requirements. These audits are now part of the responsibility of the PMO.

These independent reviews are audits that focus on either discovery or decision making. They also can focus on determining the health of a project. The audits can be scheduled or random and can be performed by in-house personnel or external examiners.

SITUATION: You have been notified by the PMO that part of their new responsibility is to audit projects on a regular basis. In order to do this, the PMO has requested that all projects track certain metrics that are of interest to the PMO. The tracking of some of these will be costly and was not included when determining the original cost baseline.

There are several types of audits. Some common types include:

- **Performance Audits:** These audits are used to appraise the progress and performance of a given project. The project manager, project sponsor, or an executive steering committee can conduct this audit.
- **Compliance Audits:** These audits are usually performed by the project management office (PMO) to validate that the project is using the project management methodology properly. Usually the PMO has the authority to perform the audit but may not have the authority to enforce compliance.
- **Quality Audits:** These audits ensure that the planned project quality is being met and that all laws and regulations are being followed. The quality assurance group performs this audit.
- **Exit Audits:** These audits are usually for projects that are in trouble and may need to be terminated. Personnel external to the project, such as an exit champion or an executive steering committee, conduct the audits.
- **Best Practices Audits:** These audits can be conducted at the end of each life cycle phase or at the end of the project. Some companies have found that project managers may not be the best individuals to perform the audit. In such situations, the company may have professional facilitators trained in conducting best practices reviews.
- **Metric and KPI Audits:** These audits are similar to Best Practices Audits and used to establish a library for metrics.

2.3 INTRODUCTION TO SCOPE CREEP

There are three things that most project managers know will happen with almost certainty: death, taxes, and scope creep. Scope creep is the continuous enhancement of the project's requirements as the project's deliverables are being developed. Scope creep is viewed as the growth in the project's scope.

Although scope creep can occur in any project in any industry, it is most frequently associated with information systems development projects. Scope changes can occur during any project life cycle phase. Scope changes occur because it is the nature of humans not to be able to completely describe the project or the plan to execute the project at the start. This is particularly true on large, complex projects. As a result, we gain more knowledge as the project progresses, and this leads to creeping scope and scope changes.

Scope creep is a natural occurrence for project managers. We must accept the fact that this will happen. Some people believe that there are magical charms, potions, and rituals that can prevent scope creep. This is certainly not true. Perhaps the best we can do is to establish processes, such as configuration management systems or change control boards, to get some control over scope creep. However, these processes are not designed to prevent scope creep but more to prevent unwanted scope changes from taking place.

Therefore, we can argue that scope creep isn't just allowing the scope to change but an indication of how well we manage changes to the scope. If all of the parties agree that a scope change is needed, then perhaps we can argue that the scope simply changed rather than crept. Some people view scope creep as scope changes not approved by the sponsor or the change control board.

Scope creep is often viewed as being detrimental to the success of a project because it increases the cost and elongates the schedule. While this is true, scope creep can also produce favorable results such as add-ons that give your product a competitive advantage. Scope creep can also please the customer if the scope changes are seen as providing additional value for the final deliverable.

Defining Scope Creep

Perhaps the most critical step in the initiation phase of a project is the defining of the scope. The first attempt at scope definition may occur as early as the proposal or competitive bidding stage. At this point, sufficient time and effort may not be devoted to an accurate determination or understanding of the scope and customer requirements. To make matters worse, all of this may be done well before the project manager is brought on board.

TIP Metrics can be established for the tracking of scope creep. However, the usefulness of these metrics is questionable because of the many causes of scope creep.

Once the project manager is brought on board, he/she must either become familiar with and validate the scope requirements if they have already been prepared, or interview the various stakeholders and gather the necessary information for a clear understanding of the scope. In doing so, the project manager prepares a list of what is included and excluded from his/her understanding of the requirements. Yet no matter how meticulously the project manager attempts to do this, the scope is never known with 100 percent clarity. This is one of the primary reasons why metrics may need to change over the life of the project.

The project manager's goal is to establish the boundaries of the scope. To do this, the project manager's vision of the project and each stakeholder's vision of the project must be aligned. There must also be an alignment with corporate business objectives because there must be a valid business reason for undertaking this project. If the alignments do not occur, then the boundary for the project will become dynamic or constantly changing rather than remaining stationary. The same can be said to hold true when selecting metrics.

Figure 2–3 shows the boundaries of the project. The project's overall boundary is designed to satisfy both business objectives established by your company and technical/scope objectives established by your customer, assuming it is an external client. The project manager and the various stakeholders, including the customer, can have different interpretations of the scope boundary and the business boundary. Also, the project manager may focus heavily on the technology that the customer needs rather than business value that the project manager's company desires. Simply stated, the project manager may seek to exceed the specifications, whereas the stakeholders and your company want to meet the minimum specification levels in the shortest amount of time.

When scope creep occurs and scope changes are necessary, the scope boundary can move. However, the scope boundary may not be able to move if it alters the business boundary and corporate expectations. As an example, a scope change to add value to a product might not be approved if it extends the launch date of the product.

It is important to understand that the scope of the project is not what the customer asked for, but what we agree to deliver. What we agree to can have inclusions and exclusions from what the customer asked for.

Figure 2-3 **Project Boundaries**

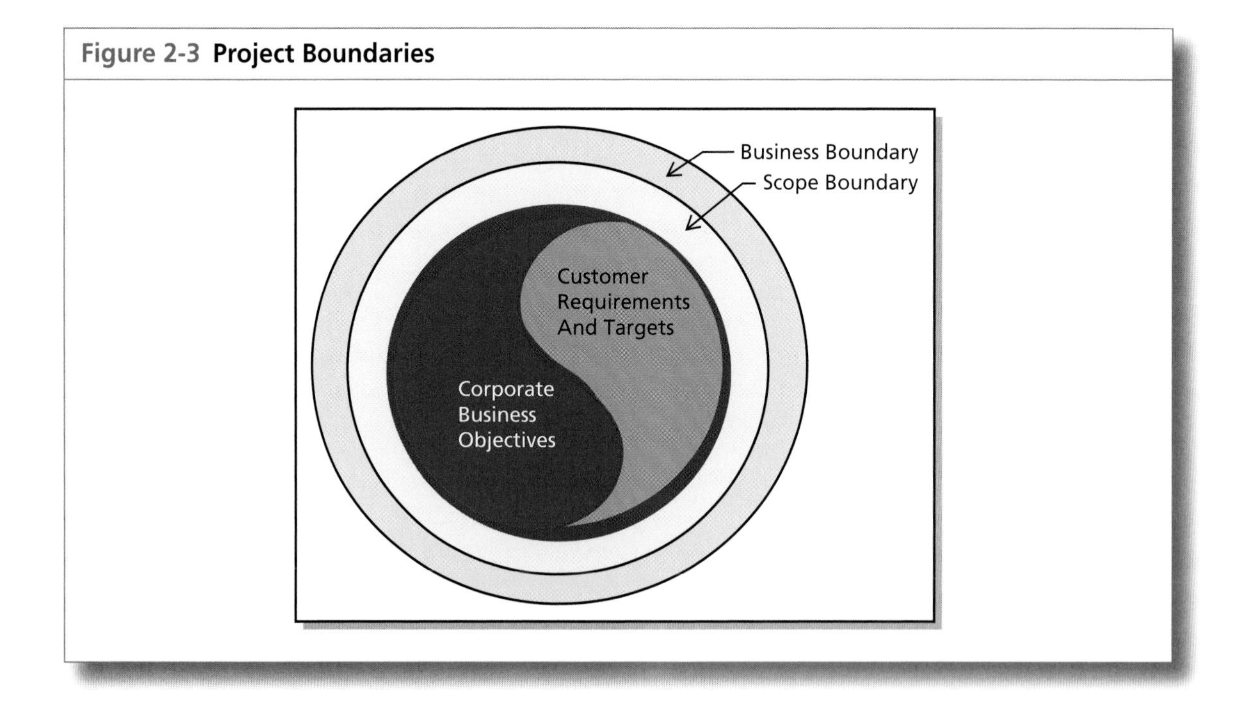

There are certain facts that we now know:

- The scope boundary is what the project manager commits to delivering.
- The boundary is usually never clearly defined at the start of the project.
- Sometimes the boundary may not be clearly defined until we are well into the project.
- We may need to use progressive or rolling wave planning to clearly articulate the scope.
- Sometimes the scope isn't fully known until the deliverables are completed and tested.
- Finally, even after stakeholders' acceptance of the deliverables, the interpretation of the scope boundary can still be up for debate.

The scope boundary can drift during the implementation of the project because, as we get further into the project and more knowledge is gained, we identify unplanned additions to the scope. This scope creep phenomenon is then accompanied by cost increases and schedule elongations. But is scope creep really evil? Perhaps not; it is something we must live with as project managers. Some projects may be fortunate enough to avoid scope creep. In general, the larger the project, the greater the likelihood that scope creep will occur.

The length of the project also has an impact on scope creep. If the business environment is highly dynamic and continuously changing, products and services must be developed to satisfy market needs. Therefore, on long-term projects, scope creep may be seen as a necessity for keeping up with customer demands, and project add-ons may be required to obtain customer acceptance.

TIP Metrics must be established to track alignment to both the business boundary and the scope boundary. However, not all of these metrics are KPIs that are reported to the stakeholders. Most of these metrics are for the project manager's use only.

Scope Creep Dependencies

Often, scope changes are approved without evaluating the downstream impact that the scope change can have on work packages that have not started yet. As an example, making a scope change early on in the project to change the design of a component may result in a significant cost overrun if long-lead raw materials that were ordered and paid for are no longer needed. Also, there could be other contractors who have begun working on their projects assuming that the original design was finalized. Now, a small scope change by one contractor could have a serious impact on other downstream contractors. Dependencies must be considered when approving a scope change because the cost of reversing a previous decision can have a severe financial impact on the project.

Causes of Scope Creep

In order to prevent scope creep from occurring, one must begin by understanding the causes of scope creep. The causes are numerous, and it is wishful thinking to believe that all of these causes can be prevented. Many of the causes are well beyond the control of the project manager, even though for some of these we can establish metrics that function as an early warning sign. Some causes are related to business scope creep, and others are part of technical scope creep.

- **Poor understanding of requirements:** This occurs when we accept or rush into a project without fully understanding what must be done.
- **Poorly defined requirements:** Sometimes the requirements are so poorly defined that we must make numerous assumptions, and as we get into the later stages of the project, we discover that some of the assumptions are no longer valid.
- **Complexity:** The more complex the project, the greater the impact of scope creep. Being too ambitious and believing that we can deliver more than we can offer on a complex project can be disastrous.
- **Failing to "drill down":** When a project is initiated using only high-level requirements, scope creep can be expected when we get involved in the detailed activities in the work breakdown structure.
- **Poor communications:** Poor communication between the project manager and the stakeholders can lead to ill-defined requirements and misinterpretation of the scope.
- **Misunderstanding expectations:** Regardless of how the scope is defined, stakeholders and customers have expectations of the outcome of the project. Failure to understand these expectations up front can lead to costly downstream changes.
- **Featuritis:** This is also called gold-plating a project and occurs when the project team adds in their own often unnecessary features and functionality in the form of "bells and whistles."
- **Perfectionism:** This occurs when the project team initiates scope changes in order to exceed the specifications and requirements rather than just meeting them. Project teams may see this as a chance for glory.
- **Career advancement:** Scope creep may require additional resources, perhaps making the project manager more powerful in the eyes of senior management. Scope creep also elongates projects and provides team members with a much longer tenure in a temporary home if they are unsure about their next assignment.
- **Time-to-market pressure:** Many projects start out with highly optimistic expectations. If the business exerts pressure on the project manager to commit to an unrealistic product launch date, then the project manager may need to reduce functionality. This could be less costly or even more costly based upon where the de-scoping takes place.

- **Government regulations:** Compliance with legislation and regulatory changes can cause costly scope creep.
- **Deception:** Sometimes we know well in advance that the customer's statement of work has "holes" in it. Rather than inform the customer about the additional work that will be required, we underbid the job based upon the original scope, and after contract award, we push through profitable scope changes.
- **Penalty clauses:** Some contracts have penalty clauses for late delivery. By pushing through (perhaps unnecessary) scope changes that will elongate the schedule, the project manager may be able to avoid penalty clauses.
- **Placating the customer:** Some customers will request "nice to have but not necessary" scope changes after the contract begins. While it may appear nice to placate the customer, always saying "yes" does not guarantee follow-on work.
- **Poor change control:** The purpose of a change control process is to prevent unnecessary changes. If the change control process is merely a rubber stamp that approves all of the project manager's requests, then continuous scope creep will occur.

The Need for Business Knowledge

Scope changes must be properly targeted prior to approval and implementation, and this is the weakest link because it requires business knowledge as well as technical knowledge. As an example, scope changes should not be implemented at the expense of risking exposure to product liability lawsuits or safety issues. Likewise, making scope changes exclusively for the sake of enhancing one's image or reputation should be avoided if it could result in an unhappy client. Also, scope changes should not be implemented if the payback period for the product is drastically extended in order to capture the recovery costs of the scope change.

Scope changes should be based upon a solid business foundation. For example, developing a very high-quality product may seem nice at the time, but there must be customers willing to pay the higher price. The result might be a product that nobody wants or can afford.

There must exist a valid business purpose for a scope change. This includes the following factors at a minimum:

- An assessment of the customers' needs and the added value that the scope change will provide
- An assessment of the market needs, including the time required to make the scope change, the payback period, return on investment, and whether the final product's selling price will be overpriced for the market.
- An assessment of the impact on the length of the project and product life cycle

- An assessment on the competition's ability to imitate the scope change
- An assessment on product liability associated with the scope change and the impact on the company's image

The Business Side of Scope Creep

In the eyes of the customer, scope creep is viewed as a detriment to success unless it provides add-ons or added value. For contractors, however, scope creep has long been viewed as a source of added profitability on projects. Years ago, it was common practice on some Department of Defense contracts to underbid the original contract during competitive bidding to ensure the award of the contract and then push through large quantities of lucrative scope changes. Scope creep was planned for.

Customers were rarely informed of gaps in their statements of work that could lead to scope creep. Even if the statement of work was clearly written, it was often intentionally or unintentionally misinterpreted for the benefit of seeking out profitable scope changes whether or not the scope changes were actually needed. For some companies, scope changes were the prime source of corporate profitability, more so than the initial contract. During competitive bidding, executives would ask the bidding team two critical questions before submitting a bid: (1) What is our cost of doing the work we are promising? and (2) How much money can we expect from scope changes once the contract is awarded to us? Often, the answer to the second question determined the magnitude of the initial bid. In other words, the contractor may plan for significant scope creep before the project even begins.

Ways to Minimize Scope Creep

Some people believe that scope creep should be prevented at all costs. However, not allowing necessary scope creep to occur can be dangerous and possibly detrimental to business objectives. Furthermore, it may be impossible to prevent scope creep. Perhaps the best we can do is to control scope creep by minimizing the amount and extent of it. Some of the activities that may be helpful include:

- **Realize that scope creep will happen:** Scope creep is almost impossible to prevent. Rather, attempts should be made to control scope creep.
- **Know the requirements:** You must fully understand the requirements of the project, and you must communicate with the stakeholders to make sure you both have the same understanding.
- **Know the client's expectations:** Your client and the stakeholders can have expectations that may not be in alignment with your interpretation of the requirements on scope. You must understand the expectations, and continuous communication is essential.
- **Eliminate the notion that the customer is always right:** Constantly saying "yes" to placate the customer can cause sufficient scope creep that a

good project becomes a distressed project. Some changes could probably be clustered together and accomplished later as an enhancement project.

- **Act as the devil's advocate:** Do not take for granted that all change requests are necessary even if they are internally generated by the project team. Question the necessity for the change. Make sure that there is sufficient justification for the change.
- **Determine the effect of the change:** Scope creep will affect the schedule, cost, scope/requirements, and resources. See whether some of the milestone dates can or cannot be moved. Some dates are hard to move, whereas others are easy. See if additional resources are needed to perform the scope change and if the resources will be available.
- **Get user involvement early:** Early user involvement may prevent some scope creep or at least identify the scope changes early enough such that the effects of the changes are minimal.
- **Add in flexibility:** It may be possible to add some flexibility into the budget and schedule if a large amount of scope creep is expected. This could appear as a management/contingency monetary reserve for cost issues and a "reserve" activity built into the project schedule for timing issues.
- **Know who has signature authority:** Not all members of the scope change control board possess signature authority to approve a scope change. You must know who possesses this authority.

TIP It is very difficult, if not impossible, to determine the real health of a project with the metrics that are in common use today.

In general, people who request scope changes do not do so in an attempt to make your life miserable. They do so because of a desire to "please," through a need for perfection, to add functionality, or to increase the value in the eyes of the client. Some scope changes are necessary for business reasons, such as add-ons for increased competitiveness. Scope creep is a necessity and cannot be eliminated, but it can be controlled.

2.4 PROJECT HEALTH CHECKS

Projects seem to progress quickly until they are about 60%–70% complete. During that time, everyone applauds that work is progressing as planned. Then, perhaps without warning, the truth comes out and we discover that the project is in trouble. This occurs because of:

- Our disbelief in the value of using project's metrics
- Selecting the wrong metrics
- Our fear of what project health checks may reveal

Some project managers have an incredible fixation with project metrics and numbers, believing that metrics are the Holy Grail in determining status. Most projects seem to focus on only two metrics: time and cost. These are the primary metrics in all earned value measurement systems (EVMS). While these two metrics "may" give you a reasonable representation of where you are today, using these two metrics to provide forecasts into the future produces "gray" areas and may not indicate future problem areas that could prevent a successful and timely completion of the project. At the other end of the spectrum, we have managers who have no faith in the metrics and, therefore, focus on vision, strategy, leadership, and prayers.

Rather than relying on metrics alone, the simplest solution might be to perform periodic health checks on the project. In doing this, three critical questions must be addressed:

- Who will perform the health check?
- Will the interviewees be honest in their responses?
- Will management and stakeholders overreact to the truth?

The surfacing of previously unknown or hidden issues could lead to loss of employment, demotions, or project cancellation. Yet project health checks offer the greatest opportunity for early corrective action to save a potentially failing project. Health checks can also discover future opportunities. It is essential to use the right metrics.

Understanding Project Health Checks

People tend to use audits and health checks synonymously. Both are designed to ensure successful repeatable project outcomes, and both must be performed on projects that appear to be heading for a successful outcome as well as those that seem destined to fail. There are lessons learned and best practices that can be discovered from both successes and failures. Also, detailed analysis of a project that appears to be successful at the moment might bring to the surface issues that show that the project is really in trouble.

Table 2–2 shows some of the differences between audits and health checks. Although some of the differences may be subtle, we will focus our attention on health checks.

SITUATION: During a team meeting, the project manager asks the team, "How's the work progressing?" The response is: "We're doing reasonably well. We're just a little bit over budget and a little behind schedule, but we think we've solved both issues by using lower-salaried resources for the next month and having them work overtime. According to our enterprise project management methodology, our unfavorable cost and schedule variances are still within the threshold limits and the generation of an exception report for management is not necessary. The customer should be happy with our results thus far."

TABLE 2-2 Audits versus Health Checks

VARIABLE	AUDIT	HEALTH CHECKS
Focus	On the present	On the future
Intent	Compliance	Execution effectiveness and deliverables
Timing	Generally scheduled and infrequent	Generally unscheduled and done when needed
Items to be searched	Best practices	Hidden, possible destructive issues and possible cures
Interviewer	Usually someone internal	External consultant
How interview is led	With entire team	One-on-one sessions
Time frame	Short term	Long term
Depth of analysis	Summary	Forensic review
Metrics	Use of existing or standard project metrics	Special health check metrics may be necessary

These comments are representative of a project team that has failed to acknowledge the true status of the project because they are too involved in the daily activities of the project. Likewise, we have project managers, sponsors, and executives who are caught up in their own daily activities and readily accept these comments with blind faith, thus failing to see the big picture. If an audit had been conducted, the conclusion might have been the same, namely that the project is successfully following the enterprise project management methodology and that the time and cost metrics are within the acceptable limits. A forensic project health check, on the other hand, might disclose the seriousness of the issues.

Just because a project is on time and/or within the allotted budget does not guarantee success. The end result could be that the deliverable has poor quality so that it is unacceptable to the customer. In addition to time and cost, project health checks focus on quality, resources, benefits, and requirements to name just a few factors. The need for more metrics than we now use should be apparent. The true measure of the project's future success is the value that the customers see at the completion of the project. Health checks must, therefore, be value-focused. Audits, on the other hand, usually do not focus on value.

Health checks can function as an ongoing tool, being performed randomly when needed or periodically throughout various life cycle stages. However, there are specific circumstances that indicate that a health check should be accomplished quickly. These include:

- Significant scope creep
- Escalating costs accompanied by a deterioration in value and benefits

- Schedule slippages that cannot be corrected
- Missed deadlines
- Poor morale accompanied by changes in key project personnel
- Metric measurements that fall below the threshold levels

Periodic health checks, if done correctly and using good metrics, eliminate ambiguity so that the true status can be determined. The benefits of health checks include:

- Determining the current status of the project
- Identifying problems early enough that sufficient time exists to take corrective action
- Identifying the critical success factors that will support a successful outcome or the critical issues that can prevent successful delivery
- Identifying lessons learned, best practices, and critical success factors that can be used on future projects
- Evaluating compliance with and improvements to the enterprise project management methodology
- Validate that the project's metrics are correct and provide meaningful data
- Identifying which activities may require or benefit from additional resources
- Identifying present and future risks as well as possible risk mitigation strategies
- Determining if the benefits and value will be there at completion
- Determining if euthanasia is required to put the project out of its misery
- The development of or recommendations for a fix-it plan

There are misconceptions about project health checks. Some of these are:

- The person doing the health check does not understand the project or the corporate culture, and is, thus, wasting time.
- The health check is too costly for the value we will get by performing it.
- The health check ties up critical resources in interviews.
- By the time we get the results from the health check, either it will be too late to make changes or the nature of the project may have changed.

Who Performs the Health Check?

One of the challenges facing companies is whether the health check should be conducted by internal personnel or by external consultants. The risk with using internal personnel is that they may have loyalties or relationships with people on the project team and, therefore, may not be totally honest in determining the true status of the project or in deciding who was at fault.

Using external consultants or facilitators is often the better choice. External facilitators can bring to the table:

- A multitude of forms, guidelines, templates, and checklists used in other companies and similar projects
- A promise of impartiality and confidentiality
- A focus on only the facts and hopefully free of politics
- An environment where people can speak freely and vent their personal feelings
- An environment that is relatively free from other day-to-day issues
- New ideas for project metrics

Life Cycle Phases

There are three life cycle phases for project health checks. These include:

- Review of the business case and the project's history
- Research and discovery of the facts
- Preparation of the health check report

Reviewing the business case and project's history may require the health check leader to have access to proprietary knowledge and financial information. The leader may have to sign nondisclosure agreements and also noncompete clauses before being allowed to perform the health check.

In the research and discovery phase, the leader prepares a list of questions that need to be answered. The list can be prepared from the *PMBOK® Guide's* domain areas or areas of knowledge. The questions can also come from the knowledge repository in the consultant's company and may also come from business case analysis templates, guidelines, checklists, or forms. The questions can change from project to project and industry to industry.

Some of the critical areas that must be investigated include:

- Performance against baselines
- Ability to meet forecasts
- Benefits and value analyses
- Governance
- Stakeholder involvement
- Risk mitigation
- Contingency planning

If the health check requires one-on-one interviews, the health check leader must be able to extract the truth from interviewees who have different interpretations or conclusions about the status of the project. Some people will be truthful, whereas others will either say what they believe the interviewer wants to hear or distort the truth as a means of self-protection.

The final phase is the preparation of the report. This should include:

- A listing of the issues
- Root cause analyses, possibly including identification of the individuals who created the problems
- Gap analysis
- Opportunities for corrective action
- A get-well or fix-it plan

Project health checks are not "Big Brother Is Watching You" activities. Rather, they are part of project oversight. Without these health checks, the chances for project failure significantly increase. Project health checks also provide us with insight on how to keep risks under control. Performing health checks and taking corrective action early is certainly better than having to manage a distressed project.

2.5 MANAGING DISTRESSED PROJECTS

Professional sports teams treat each new season as a project. For some teams, the only definition of success is winning the championship, whereas for others success is viewed as just a winning season. Not all teams can win the championship, but having a winning season is certainly within reach.

At the end of the season, perhaps half of the teams will have won more games than they lost. However, for the other half of the teams, who had losing records, the season (i.e., project) was a failure. When a project failure occurs in professional sports, managers and coaches are fired, there is a shakeup in executive leadership, some players are traded or sold to other teams, and new players are brought on board. These same tactics are used to recover failing projects in industry.

There are some general facts about troubled projects:

- Some projects are doomed to fail regardless of recovery attempts.
- The chances of failure on any given project may be greater than the chances of success.
- Failure can occur in any life cycle phase; success occurs at the end of the project.
- Troubled projects do not go from "green" to "red" overnight.
- There are early warning signs, but they are often overlooked or misunderstood.
- Most companies have a poor understanding of how to manage troubled projects.
- Not all project managers possess the skills to manage a troubled project.

Not all projects will be successful. Companies that have a very high degree of project success probably are not working on enough projects and certainly are not taking on very much risk. These types of companies eventually become followers rather than leaders. For companies that desire to be leaders, knowledge on how to turn around a failing or troubled project is essential.

Projects do not get into trouble overnight. There are early warning signs, but most companies seem to overlook them or misunderstand them. Some companies simply ignore the tell-tale signs and continue on, hoping for a miracle. Failure to recognize these signs early can make the downstream corrections a very costly endeavor. Also, the longer you wait to make the corrections, the more costly the changes become.

Some companies perform periodic project health checks. These health checks, when applied to healthy looking projects, can lead to the discovery that the project may be in trouble even though on the surface the project looks healthy. Outside consultants are often hired for the health checks in order to get an impartial assessment. The consultant rarely takes over the project once the health check is completed but may have made recommendations for recovery.

When a project gets way off track, the cost of recovery is huge, and vast or even new resources may be required for corrections. The ultimate goal for recovery is no longer to finish on time, but to finish with reasonable benefits and value for the customer and the stakeholders. The project's requirements may change during recovery to meet the new goals if they have changed. Regardless of what you do, however, not all troubled projects can be recovered.

"Root" Causes of Failure

There are numerous causes of project failure. Some causes are quite common in specific industries, such as information technology, whereas others can appear across all industries. Here is a generic list of common causes of failure:

- End user stakeholders not involved throughout the project.
- Minimal or no stakeholder backing; lack of ownership.
- Weak business case.
- Corporate goals not understood at the lower organizational levels.
- Plan asks for too much in too little time.
- Poor estimates, especially financial.
- Unclear stakeholder requirements.
- Passive user stakeholder involvement after handoff.
- Unclear expectations.
- Assumptions, if they exist at all, are unrealistic.
- Plans are based upon insufficient data.
- No systemization of the planning process.
- Planning is performed by a planning group.

- Inadequate or incomplete requirements.
- Lack of resources.
- Assigned resources lack experience.
- Staffing requirements are not fully known.
- Constantly changing resources.
- Poor overall project planning.
- Enterprise environmental factors have changed, causing outdated scope.
- Missed deadlines and no recovery plan.
- Budgets are exceeded and out of control.
- Lack of replanning on a regular basis.
- Lack of attention provided to the human and organizational aspects of the project.
- Project estimates are best guesses and not based upon history or standards.
- Not enough time provided for proper estimating.
- No one knows the exact major milestone dates or due dates for reporting.
- Team members working with conflicting requirements.
- People are shuffled in and out of the project with little regard for the schedule.
- Poor or fragmented cost control.
- Each stakeholder uses different organizational process assets, which may be incompatible with the assets of project partners.
- Weak project and stakeholder communications.
- Poor assessment of risks if done at all.
- Wrong type of contract.
- Poor project management; team members possess a poor understanding of project management, especially virtual team members.
- Technical objectives are more important than business objectives.

These causes of project failure can be sorted into three broad categories:

- **Management mistakes:** These are the result of a failure in stakeholder management perhaps by allowing too many unnecessary scope changes, failing to provide proper governance, refusing to make decisions in a timely manner, and ignoring the project manager's requests for help. This can also be the result of wanting to gold-plate the project. This is the result of not performing project health checks.
- **Planning mistakes:** These are the result of poor project management, perhaps not following the principles stated in the *PMBOK® Guide*, not having a timely "kill switch" in the plan, not planning for project audits or health checks, and not selecting the proper tracking metrics.
- **External influences:** These are normally failures in assessing the environmental input factors correctly. This includes the timing for getting approvals and authorization from third parties, and a poor understanding of the host country's culture and politics.

The Definition of Failure

Historically, the definition of success on a project was viewed as accomplishing the work within the triple constraints and obtaining customer acceptance. Today, the triple constraints are still important, but they have taken a "back seat" to the business and value components of success. In today's definition, success is when the planned business value is achieved within the imposed constraints and assumptions, and the customer receives the desired value.

While we seem to have a reasonably good understanding of project success, we have a poor understanding of project failure. The project manager and the stakeholders can have different definitions of project failure. The project manager's definition might just be not meeting the competing constraints criteria. Stakeholders, on the other hand, might seem more interested in business value than the competing constraints once the project actually begins. Stakeholders' perception of failure might be:

- The project has become too costly for the expected benefits or value.
- The project will be completed too late.
- The project will not achieve its targeted benefits or value.
- The project no longer satisfies the stakeholders' needs

Early Warning Signs of Trouble

Projects do not become distressed overnight. They normally go from "green" to "yellow" to "red," and along the way are early warning signs or metrics indicating that failure may be imminent or that immediate changes may be necessary.

Typical early warning signs include:

- Business case deterioration.
- Different opinions on project's purpose and objectives.
- Unhappy/disinterested stakeholders and steering committee members.
- Continuous criticism by stakeholders.
- Changes in stakeholders without any warning.
- No longer a demand for the deliverables or the product.
- Invisible sponsorship.
- Delayed decisions resulting in missed deadlines.
- High-tension meetings with team and stakeholders.
- Finger-pointing and poor acceptance of responsibility.
- Lack of organizational process assets.
- Failing to close life cycle phases properly.
- High turnover of personnel, especially critical workers.
- Unrealistic expectations.
- Failure in progress reporting.
- Technical failure.

- Having to work excessive hours and with heavy workloads.
- Unclear milestones and other requirements.
- Poor morale.
- Everything is a crisis.
- Poor attendance at team meetings.
- Surprises, slow identification of problems, and constant rework.
- Poor change control process.

The earlier the warning signs are discovered, the more opportunities exist for recovery. This is the time when a project health check should be conducted. Successful identification and evaluation of the early warning signs can tell us that the distressed project:

- Can succeed according to the original requirements, but some minor changes are needed
- Can be repaired, but major changes may be necessary
- Cannot succeed and should be killed

There are three possible outcomes when managing a troubled project:

- The project must be completed; that is, it is required by law.
- The project can be completed, but with major costly changes to the requirements.
- The project should be canceled.
  - Costs and benefits are no longer aligned.
  - What was once a good idea no longer has merit.

Some projects cannot be canceled because they are required by law. These include those necessary for compliance with government laws on environmental issues, health, safety, pollution, and the like. For these projects, failure is not an option. The hardest decision to make is obviously to hit the "kill switch" and cancel the project. Companies that have a good grasp on project management establish processes to make it easy to kill a project that cannot be saved. There is often a great deal of political and cultural resistance to killing a project. Stakeholder management and project governance play a serious role in the ease by which a project can be terminated.

Selecting the Recovery Project Manager (RPM)

Companies often hire outside consultants to perform a health check on a project. If the health check report indicates that an attempt should be made to recover the troubled project, then perhaps a new project manager should be brought on board with skills in project recovery. Outside consultants normally do not take over the troubled project because they may not have a good grasp of the company's culture, business and project management processes, politics, and employee working relationships. Not all project managers possess the skills to be an effective RPM. In addition to possessing project management knowledge, typical skills needed include:

- Strong political courage and political savvy
- A willingness to be totally honest when attacking and reporting the critical issues
- Tenacity to succeed even if it requires a change in resources
- Understanding that effective recovery is based upon information, not emotions
- Ability to deal with stress, personally and with the team

Recovering a failing project is like winning the World Series of Poker. In addition to having the right poker skills, some degree of luck is also required.

Taking over a troubled project is not the same as starting up a new project. Recovery project managers must have a good understanding of what they are about to inherit, including high levels of stress. This includes:

- A burned out team
- An emotionally drained team
- Poor morale
- An exodus of the talented team members, who are always in high demand elsewhere
- A team that may have a lack of faith in the recovery process
- Furious customers
- Nervous management
- Invisible sponsorship and governance
- Either invisible or highly active stakeholders

Project managers who do not understand what is involved in the recovery of a troubled project can make matters worse by hoping for a miracle and allowing the "death spiral" to continue to a point where recovery is no longer possible. The death spiral continues if we:

- Force employees to work excessive hours unnecessarily
- Create unnecessary additional work
- Replace team members at an inappropriate time
- Increase team stress and pressure without understanding the ramifications
- Search for new "miracle" tools to resolve some of the issues
- Hire consultants who cannot help or make matters worse by taking too long to understand the issues

Recovery Life Cycle Phases

A company's existing enterprise project management methodology may not be able to help recover a failing project. After all, the company's standard enterprise project management methodology (EPM), which may not have been appropriate for this project, may have been a contributing factor to the project's decline. It is a mistake to believe that any methodology is the miracle cure. Projects are "management by people," not tools or methodologies. A different approach may be necessary for the recovery project to succeed.

Figure 2–4 shows the typical life cycle phases for a recovery project. These phases can significantly differ from the company's standard methodology life cycle phases. The first four phases in Figure 2–4 are used for problem assessment and to evaluate and, hopefully, verify that the project may be able to be saved. The last two phases are where the actual recovery takes place.

The Understanding Phase

The purpose of the understanding phase is for the newly assigned RPM to review the project and its history. To do this, the RPM will need some form of mandate or a project charter that may be different from that of his predecessor. This must be done as quickly as possible because time is a constraint rather than a luxury. Typical questions that may be addressed in the mandate include:

- What authority will you have to access proprietary or confidential information? This includes information that may not have been available to your predecessor, such as contractual agreements and actual salaries.

Figure 2-4 Recovery Life Cycle Phases

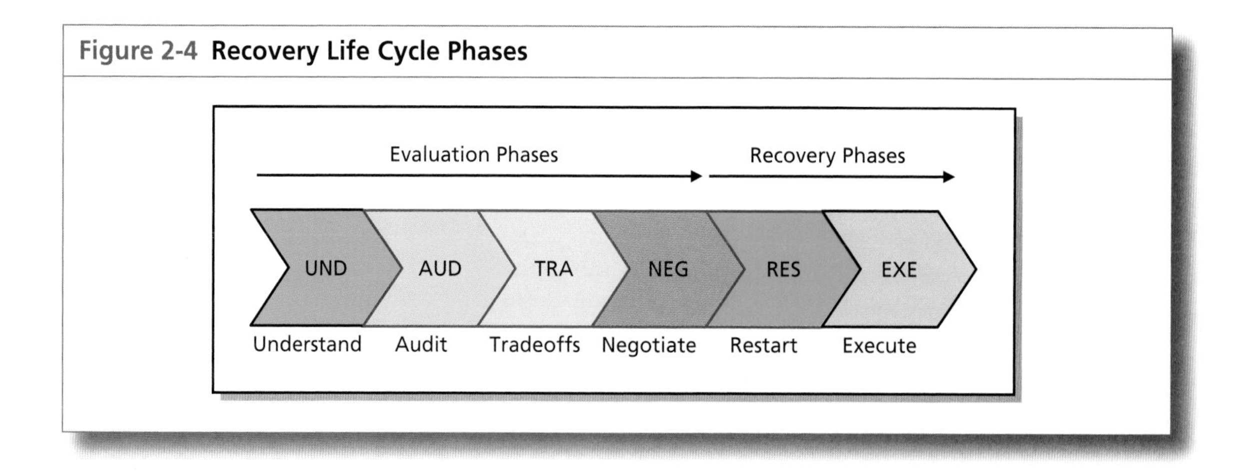

- What support will you be given from the sponsor and the stakeholders? Are there any indications that they will accept less than optimal performance and a descoping of the original requirements?
- Will you be allowed to interview the team members in confidence?
- Will the stakeholders overreact to brutally honest findings even if the problems were caused by the stakeholders and governance groups?

Included in this phase are the following:

- Understanding of the project's history
- Reviewing the business case, expected benefits, and targeted value
- Reviewing the project's objectives
- Reviewing the project's assumptions
- Familiarizing yourself with the stakeholders, and their needs and sensitivities
- Seeing if the enterprise environmental factors and organizational process assets are still valid

The Audit Phase

Now that we have an understanding of the project's history, we enter the audit phase, which is a critical assessment of the project's existing status. The following is part of the audit phase:

- Assessing the actual performance to date
- Identifying the flaws
- Performing a root cause analysis
- Looking for surface (or easy to identify) failure points
  - Looking for hidden failure points
  - Determining what are the "must have," "nice to have," "can wait," and "not needed" activities or deliverables
- Looking at the issues log and seeing if the issues are people issues. If there are people issues, can people be removed or replaced?

The audit phase also includes the validation that the objectives are still correct, the benefits and value can be met but perhaps to a lesser degree, the assigned resources possess the proper skills, the roles and responsibilities are assigned to the correct team members, the project's priority is correct and will support the recovery efforts, and executive support is in place. The recovery of a failing project cannot be done in isolation. It requires a recovery team and strong support/sponsorship.

The timing and quality of the executive support needed for recovery is most often based upon the perception of the value of the project. Five important questions that need to be considered as part of value determination are:

- Is the project still of value to the client?
- Is the project still aligned to your company's corporate objectives and strategy?

- Is your company still committed to the project?
- Are the stakeholders still committed?
- Is there overall motivation for rescue?

Since recovery cannot be accomplished in isolation, it is important to interview the team members as part of the audit phase. This may very well be accomplished at the beginning of the audit phase to answer the previous questions. The team members may have strong opinions on what went wrong as well as good ideas for a quick and successful recovery. You must obtain support from the team if recovery is to be successful. This includes:

- Analyzing the culture
- Data gathering and assessment involving the full team
- Making it easy for the team to discuss problems without finger-pointing or the laying of blame
- Interviewing the team members perhaps on a one-on-one basis
- Reestablishing work-life balance
- Reestablishing incentives, if possible

It can be difficult to interview people and get their opinion on where we are, what went wrong, and how to correct it. This is especially true if the people have hidden agendas. If you have a close friend associated with the project, how will you react if they are found guilty of being part of the problem? This is referred to as an emotional cost.

Another problem is that people may want to hide critical information if something went wrong and they could be identified with it. They might view the truth as affecting their chances for career advancement. You may need a comprehensive list of questions to ask to extract the right information.

When a project gets into trouble, people tend to play the "Blame Game" trying to make it appear that someone else is at fault. This may be an attempt to muddy the water and detract the interviewer from the real issues. It is done as part of a person's sense of self-preservation. It may be difficult to decide who is telling the truth and who is fabricating information.

You may conclude that certain people must be removed from the project if it is to have a chance for recovery. Regardless what the people did, you should allow them to leave the project with dignity. You might say, "Annie is being reassigned to another project that needs her skills. We thank her for the valuable contribution she has made to this project."

Perhaps the worst situation is when you discover that the real problems were with the project's governance. Telling stakeholders and governance groups that they were part of the problem may not be received well. The author's preference is always to be honest in defining the problems even if it hurts. This response must be handled with tact and diplomacy.

You must also assess the team's morale. This includes:

- Looking at the good things first to build morale
- Determining if the original plan was overly ambitious
- Determining if there were political problems that led to active or passive resistance by the team
- Determining if the work hours and workloads were demoralizing

The Tradeoff Phase

Hopefully, by this point you have the necessary information for decision making as well as the team's support for the recovery. It may be highly unlikely that the original requirements can still be met without some serious tradeoffs. You must now work with the team and determine the tradeoff options that you will present to the stakeholders.

When the project first began, the triple constraints most likely looked like what was shown previously in Figure 1–3. Time, cost, and scope were the primary constraints and tradeoffs would have been made on the secondary constraints of quality, risk, value, and image/reputation. When a project becomes distressed, stakeholders know that the original budget and schedule may no longer be valid. The project may take longer and may cost significantly more money than originally thought. Therefore, the primary concerns for the stakeholders as to whether or not to support the project further may change to value, quality and image/reputation, as shown in right side of Figure 1–3. The tradeoffs that the team will present to the customer and stakeholders will then be tradeoffs on time, cost, scope, and possibly risk.

One way of looking at tradeoffs is to review the detailed Work Breakdown Structure (WBS) and identify all activities remaining to be accomplished. The activities are then placed on the grid in Figure 2–5. The "must have" and "nice to have" work packages or deliverables are often the most costly and the hardest to use for tradeoffs. If vendors are required to provide work package support, then we must perform vendor tradeoffs as well, which include:

- Assessing vendor contractual agreements
- Determining if the vendor can fix the problems
- Determining if vendor concessions and tradeoffs are possible
- Establishing new vendor schedules and pricing

Once all of the elements are placed on the grid in Figure 2–5, the team will assist the RPM with tradeoffs by answering the following questions:

- Where are the tradeoffs?
- What are the expected casualties?
- What can and cannot be done?
- What must be fixed first?
- Can we stop the bleeding?

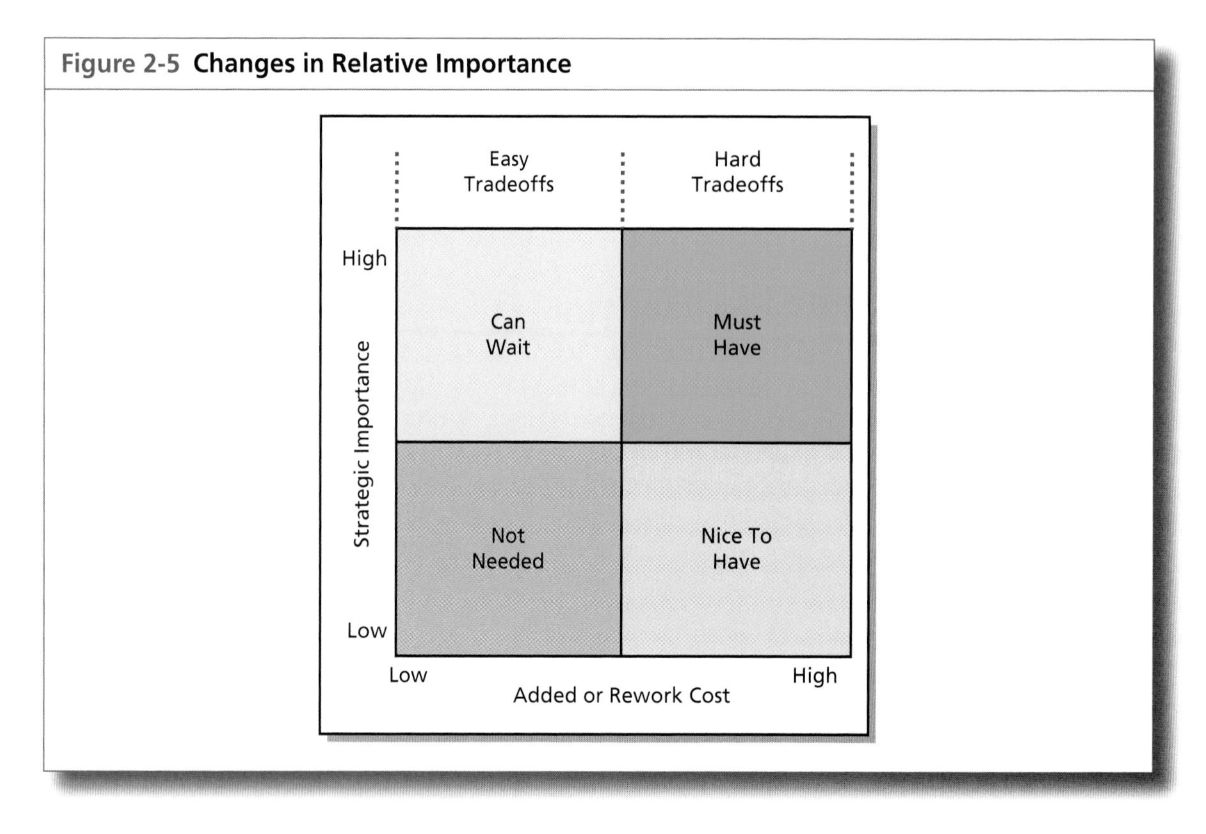

- Have the priorities of the competing constraints changed?
- Have the features changed?
- What are the risks?

Once the tradeoffs have been discovered, the RPM and the team must prepare a presentation for the stakeholders. There are two primary questions that the RPM will need to discuss with the stakeholders:

- Is the project worth saving? If the project is not worth saving, then you must have the courage to say so. Unless a valid business reason exists for continuation, you must recommend cancellation.
- If the project is worth saving, can the stakeholders expect a full or partial recovery, and by when?

There are also other factors that most likely are concerns of the stakeholders and must be addressed. These factors include:

- Changes In the political environment
- Existing or potential lawsuits

TIP When things go bad and you are trying to recover a potentially failing project, concessions may have to be made by allowing additional metrics and KPIs to be introduced into project. This may be the only way the project can be saved.

- Changes in the enterprise environmental factors
- Changes in the organizational process assets
- Changes in the business case
- Changes in the assumptions
- Changes in the expected benefits and final value

The Negotiation Phase

At this point, the RPM is ready for stakeholder negotiations. Items that must be addressed as part of stakeholder negotiations include:

- What items are important to the stakeholders? (i.e., time, cost, value, etc.)
- Prioritization of the tradeoffs
- Honesty in your beliefs for recovery
- Not giving them unrealistic expectations
- Getting their buy-in
- Negotiating for the needed sponsorship and stakeholder support

SITUATION: You have been placed in charge of a distressed project. You have done your homework correctly and are ready for negotiations with the stakeholders and the client. They inform you that they now wish to be more actively involved in the project and want additional metrics to be included, especially metrics directly related to the success and/or failure of the project. Inserting these metrics may be costly and were not priced out as part of the original cost baseline.

The Restart Phase

Assuming the stakeholders have agreed to a recovery process, you are now ready to restart the project. This includes:

- Briefing the team on stakeholder negotiations
- Making sure the team learns from past mistakes
- Introducing the team to the stakeholders' agreed-upon recovery plan, including the agreed upon milestones
- Identifying any changes to the way the project will be managed
- Fully engaging the project sponsor as well as the key stakeholders for their support
- Identifying any changes to the roles and responsibilities of the team members

There are three restarting options. These include:

- **Full anesthetic:** Bring all work to a standstill until the recovery plan is finalized.

- **Partial anesthetic:** Bring some work to a standstill until the scope is stabilized.
- **Scope modification:** Continue work but with modifications as necessary.

Albert Einstein once said: "We cannot solve our problems with the same thinking we used when we created them." It may be necessary to bring on board new people with new ideas. However, there are risks. You may want these people full-time on your project, but retaining highly qualified workers that may be in high demand elsewhere could be difficult. Since your project most likely will slip, some of your team members may be committed to other projects about to begin. However, you may be lucky enough to have strong executive-level sponsorship and retain these people. This could allow you to use a co-located team organization.

The Execution Phase

During the execution phase, the project manager must focus upon certain back-to-work implementation factors. These include:

- Learning from past mistakes
- Stabilizing scope
- Rigidly enforcing the scope change control process
- Performing periodic critical health checks and using earned value measurement reporting
- Providing effective and essential communications
- Maintaining positive morale
- Adopting proactive stakeholder relations management
- Not relying upon or expecting the company's EPM system to save you
- Not allowing unwanted stakeholder intervention, which increases pressure
- Carefully managing stakeholder expectations
- Insulating the team from politics

Recovery project management is not easy, and there is no guarantee you can or will succeed. You will be under close supervision and scrutinized by superiors and stakeholders. You may even be required to explain all of your actions, but saving a potentially troubled project from disaster is certain worth the added effort.

3 METRICS

CHAPTER OVERVIEW

This chapter describes the characteristics of a metric. Not all metrics are equal. The value of a metric must be well understood in order for it to be used correctly and provide the necessary information for informed decision making.

CHAPTER OBJECTIVES

- To understand the complexities in determining project status
- To understand the meaning and use of a metric
- To understand the benefits of using metrics
- To understand the components and types of metrics
- To understand the resistance to using metrics

KEY WORDS

- Information systems
- Measurement
- Metrics

3.0 INTRODUCTION

Metrics keep stakeholders informed as to the status of the project. Stakeholders must be confident that the correct metrics are used and that the measurement portrays a clear and truthful representation of the status. Metrics may determine if it is feasible to take on a certain project or if a certain course of action should be taken. Metrics can be developed to track organizational maturity in project management as well as innovation progress.

The project manager and the appropriate stakeholders must come to an agreement on which metrics to be used and how measurements will be made. There must also be agreement on which metrics will be part of the dashboard reporting system and how the metric measurement will be interpreted. Metrics management is taking on a much higher level of importance, so a metrics management expert may be part of the PMO.

Figure 3-1 Determining Project Status

3.1 PROJECT MANAGEMENT METRICS: THE EARLY YEARS

In the early years of project management, the United States government discovered that the project managers in the contractors' firms were functioning more as project monitors than as project managers. Monitors would simply record information and then pass it along to higher levels of management for consideration. Project managers were reluctant to take any action as a result of negative information because they either lacked sufficient authority to implement change or did not know what actions to take. The result was often customer micromanagement of projects. Unfortunately, as projects became larger and more complex, government micromanagement became exceedingly difficult.

Determining the true project status became difficult, as shown humorously in Figure 3–1. On some larger projects, it became questionable as to who was controlling costs. This is shown in Figure 3–2.

The solution was to get the contractors to learn and implement project management rather than project monitoring. Project managers were now expected to perform the following:

- Establish boundaries, baselines, and targets for performance
- Measure the performance

Figure 3-2 **Who Controls Costs?**

Do You Control Costs? OR Do Costs Control You?

- Determine the variances from the baselines or targets
- Develop contingency plans to reduce or eliminate unfavorable variances
- Obtain approval of the contingency plans
- Implement the contingency plans
- Measure the new variances
- Repeat the process when necessary

For this to work, metrics would be needed. The government then created the Cost/Schedule Control Systems and later the Earned Value Measurement System (EVMS).[1] The metrics that were established as part of these systems allowed us to determine project status, at least what we thought was the status. As an example:

- The project
  - A budget of $1.2 million.
  - Time duration is 12 months.
  - Production requirement is 10 deliverables.
- Time line
  - Elapsed time is six months.
  - Money spent to date is $700,000.
  - Deliverables produced: 4 complete and 2 partial.

1. In 1967, DOD Instruction 7000.2 identified 35 Cost/Schedule Control System Criteria (C/SCSC). In 1997, DOD Regulation 5000.2-R identified 32 Earned Value Measurement System (EVMS) criteria.

With the EVMS metrics, we were able to reasonably determine status. The metrics helped us determine both present and an often questionable prediction of the future. The metrics provided an early warning system that allowed the project managers sufficient time to make course corrections in small increments. The metrics emphasized prevention over cures by identifying and resolving problems early. Metrics can function as risk triggers such that the impact of downstream risks can be minimized.

The benefits of using these metrics now became abundantly clear:

- Accurate displaying of project status
- Early and accurate identification of trends
- Early and accurate identification of problems
- Reasonable determination of the project's health
- A source of critical information for controlling projects
- Basis for course corrections

The prolonged use of these metrics led us to the identification of several best practices that had to take place:

- Thorough planning of the work to be performed to complete the project
- Good estimating of time, labor, and costs
- Clear communication of the scope of the required tasks
- Disciplined budgeting and authorization of expenditures
- Timely accounting of physical progress and cost expenditures
- Frequent, periodic comparison of actual progress and expenditures to schedules and budgets, both at the time of comparison and at project completion
- Periodic reestimation of time and cost to complete the remaining work

For more than 40 years, these metrics have been treated as the Gospel. There are limitations to the use of the metrics in EVMS, however:

- Time and cost are basically the only two metrics. Most of the other metrics being reported are derivatives of time and cost.
- Measurements of time and cost can be inaccurate, thus leading to faulty status reports.
- The quality and value of the project cannot be calculated using time and cost metrics alone.
- Completing a project within time and cost does not imply that the project is a success.
- Time and cost information can be fudged.
- Unfavorable metrics do not necessarily mean that we are in trouble.
- Unfavorable metrics do not provide us with information for corrective action.

- Customers and stakeholders do not always understand the meaning of these metrics.
- Others.

The conclusion is clear. In today's project management environment where projects are becoming more complex, the metrics of EVMS, by themselves, may not be sufficient for managing projects. EVMS does not address some fundamental issues that can lead to project failure such as unrealistic planning, poor governance, low quality of the resources, and poor estimates. We are not claiming that EVMS does not work, rather that additional metrics will be needed. We need to establish metrics that cover the big picture, namely business value to be delivered, quality of the results, effort, productivity, and team performance to name just a few. In another example, in IT projects today there is a push for the establishment of code-related metrics. Also a grouping of metrics rather than a single metric may be needed to accurately determine performance.

3.2 PROJECT MANAGEMENT METRICS: CURRENT VIEW

One of the reasons that the government created the EVMS was for standardization in status reporting. Companies then followed the government's lead, using time and cost as the primary metrics. If other metrics were considered, they were for internal use only and not often shared with customers.

Today, part of the project manager's new role is to understand what critical metrics need to be identified and managed for the project to be viewed as a success by all of the stakeholders. Project managers have come to the realization that defining project-specific metrics and key performance indicators are joint ventures between the project manager, client, and stakeholders. Getting stakeholders to agree on the metrics is difficult, but it must be done as early as possible in the project.

Unlike financial metrics used for the Balanced Scorecard, project-based metrics can change during each life cycle phase, as well as from project to project. This can be seen in Figure 3–3. Therefore, the establishment and measurement of metrics may be an expensive necessity to validate the critical success factors (CSFs) and maintain customer satisfaction. Many people believe that the future will be metric-driven project management.

There are several reasons for the growth in project management metrics:

- There is a need for "paperless" project management. The cost of paperwork has become quite expensive.
- New techniques, such as Agile project management, are pressuring us to reduce costs and eliminate unnecessary waste.
- The growth in complex projects requires more frequent health checks which, in turn, require a better understanding of metrics.

Figure 3-3 Selecting Metrics

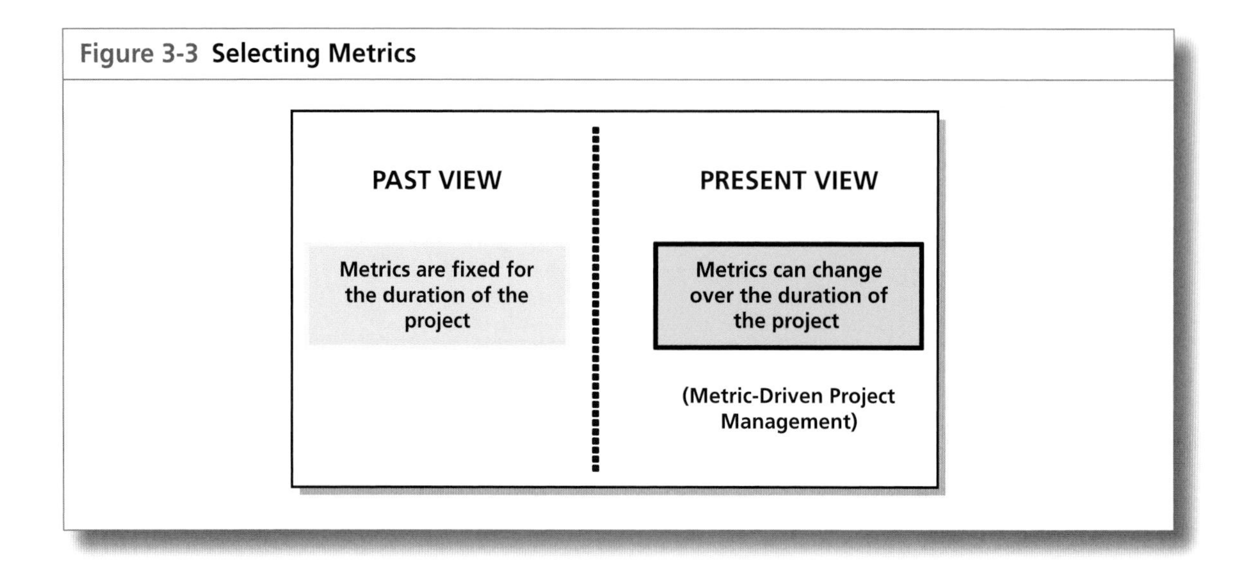

- Project success is now defined as the delivery of both business value and project value. Therefore, we need metrics that can track and report business value as well as the traditional project value.
- Collecting metrics may be the only way that we can validate performance and success.

Metrics and Small Companies

The implementation of metrics management and the use of dashboards is relatively inexpensive. This makes it attractive to small companies. Many small companies are afraid that this will entail a large cost expenditure, and that certainly is not the case. Successful metrics management builds customer loyalty, and this in turn generates new business. The only major expense might be the hiring of someone familiar with dashboard design.[2]

3.3 METRICS MANAGEMENT MYTHS

There are several myths concerning the use of metrics for project management. Several of the myths are listed below and will be discussed in later sections.

- Metric management is easy.
- Dashboard viewers always seem to understand the metrics they are viewing.

2. For an excellent paper on the application of metrics to small businesses, see "5 KPI Metrics Every Small Business Owner Should Monitor," posted by DashboardSpy around November 1, 2011 (www.DashboardSpy.com)

- Dashboards can be designed with minimal effort.
- With a little effort, we can select a perfect set of metrics for each project.
- Once metrics are selected, anyone can perform the measurements.
- You can never have too many metrics.
- Project management metrics cannot change over the duration of the project.
- Not all metrics can be measured.
- Metrics tell you what steps to take to remedy an unfortunate situation.
- Good project metrics should be tied to employee performance reviews.

3.4 SELLING EXECUTIVES ON A METRICS MANAGEMENT PROGRAM

Selling metrics management to executives, and being able to receive a timely response, requires showing both the benefits and the cost savings. There are several approaches:

- If the company is already using financial metrics for business strategy, and having reasonable success, relate the implementation of project metrics to financial metrics.
- Demonstrate quantitatively the cost savings from fewer meetings and less paperwork as a result of using metrics management.
- Perform benchmarking studies to show how many other companies are using metrics management and obtaining favorable results.
- Show the executives how it will work using a pilot project.

A Detroit-based company went to metrics management to reduce the amount of paperwork appearing on projects. Previously, executive project review meetings were preceded by the preparation of reports that included the following steps for each page in the report:

- Organizing
- Writing
- Typing
- Proofing
- Editing
- Retyping
- Graphic arts
- Approvals
- Reproduction
- Classification
- Distribution
- Security
- Storage
- Disposal

The company eliminated all of these reports and instead conducted the meetings with dashboards and traffic lights for each major work package in the work breakdown structure. In the first year alone, the PMO estimated the cost savings to be more than $1 million resulting from fewer meetings and a reduction in paperwork.

While it is true that there will be a startup cost for implementing a metrics management program, the downstream rewards and financial savings can be orders of magnitude greater than the startup costs. Convincing an organization of this is often difficult, however, because senior managers are often resistant to challenging the status quo.

3.5 UNDERSTANDING METRICS

TIP Be prepared for new or changing metrics as the project progresses.

TIP Metrics may not provide any real value unless they can be measured.

Although most companies use some type of metrics for measurement, they seem to have a poor understanding of what constitutes a metric, at least for use in project management. You cannot effectively manage a project without having metrics and accompanying measurement capable of providing you with complete or almost complete information. Therefore, the simplest definition of a metric is something that is measured. Consider the following:

- If it cannot be measured, then it cannot be managed.
- What gets measured gets done.
- You never really understand anything fully unless it can be measured.

Metrics can be measured and recorded as:

TIP When performance is measured, performance generally improves.

TIP You may never know with any reasonable degree of precision the exact performance of the project. However, good metrics can provide you with a close estimate.

TIP Despite having a variety of measurement techniques, be careful when promising that you can measure efficiency, effectiveness, and productivity. Some goals are difficult to measure accurately.

- Observations
- Ordinal (i.e., four or five stars) and nominal (i.e., male or female) data tables
- Ranges/sets of value
- Simulation
- Statistics
- Calibration estimates and confidence limits
- Decision models (EV, EVPI, etc.)
- Sampling techniques
- Decomposition techniques
- Human judgment

If you cannot offer a stakeholder something that can be measured, then how can you promise that his/her expectations will be met? You cannot control what you cannot measure. Good metrics lead to proactive project management rather than reactive project management, if the metrics are timely and informative. Likewise, some desired metrics simply may not work and should be abandoned.

For years, measurement itself was not well understood. We avoided metrics management because we did not understand it, but authors such as Douglas Hubbard have helped to resolve the problem:[3]

- Your problem is not as unique as you think.
- You have more data than you think.
- You need less data than you think.
- There is useful measurement that is much simpler than you think.

Not all metrics have the same time frames for measurements and life expectancies. This will impact the frequency when we will perform metric measurements. Some metrics may be measured and reported in real time, whereas others may be looked at weekly or monthly. For simplicity's sake, project-based metrics can be broken down according to the following time frames for measurement purposes:

Metrics with full project duration measurements: These are metrics, such as cost and schedule variances, that are used for the entire duration of the project and measured either weekly or monthly.

Metrics with life cycle phase measurements: These are metrics that are in existence just during a particular life cycle phase. As an example, metrics that track the amount or percent of direct labor dollars used for project planning would probably be measure just in the project planning phase.

Metrics with limited life measurements: These are metrics that are in existence for the life of an element of work or work package. As an example, we could track the manpower staffing rate for specific work packages or the number of deliverables produced in a specific month.

Metrics that use rolling-wave or moving-window measurements: These are metrics where the starting and finishing measurement dates can change as the project progresses. As an example, calculations for the cost performance index (CPI) and schedule performance index (SPI) are used to measure trends for forecasting. On long-term projects, we may establish a moving window of the most recent six data points (monthly measurements) to be used to obtain a linear curve fit for the trend line.

Alert metrics and measurements: These metrics are used to indicate that an out-of-tolerance condition exists. The metric may be in existence just

3. Douglas W. Hubbard, *How to Measure Anything; Finding the Value of Intangibles in Business*, Hoboken, NJ: John Wiley & Sons, 2007, p. 31.

until the out-of-tolerance condition is corrected, but they may appear later on in the project if the situation appears again. Alert metrics could also be metrics that are used continuously but are highlighted differently when an out-of-tolerance condition exists.

Over the years, numerous benefits have surfaced from the use of metrics management. Some benefits are:

- Metrics tell us if we are hitting the targets/milestones, getting better or getting worse.
- Metrics allow us to catch mistakes before they lead to other mistakes; early identification of issues.
- Good metrics lead to informed decision making, whereas poor or inaccurate metrics lead to bad management decisions.
- Good metrics can assess performance accurately.
- Metrics allow for proactive management in a timely manner.
- Metrics improve future estimating.
- Metrics improve performance in the future.
- Metrics make it easier to validate baselines and maintain the baselines with minimal disruptions.
- Metrics can more accurately assess success and failure.
- Metrics can improve client satisfaction.
- Metrics are a means of assessing the project's health.
- Metrics track the ability to meet the project's critical success factors.
- Good metrics allow the definition of project success to be made in terms of factors other than the traditional triple constraints.
- Metrics help in resolving crises.
- Metrics allow you to identify and mitigate risks.

It is important to remember that metrics are measurements and, therefore, provide us with opportunities for continuous improvements to the project management processes. Selecting metrics without considering a plan for future action is a waste of time and money. If a measurement indicates that the metric is significantly far away from the target, then the team must investigate the root cause of the deviation, determine what can be done to correct the deviation, get the plan to correct the deviation approved, and then implement the new plan. Metrics also allow us to create a database of historical information from which to analyze trends and improve future estimating.

However, there is a dark side to using metrics. Metric management requires an understanding of human behavior. Care must be taken that the use of metrics and measurement techniques does not encourage unintended behavior.

While metrics are most frequently used to validate the health of a project, they can also be used to discover best practices in the processes.

Capturing best practices and lessons learned are necessities for long-term continuous improvement. Without effective use of metrics, companies could spend years trying to achieve sustained improvements. In this regard, metrics are a necessity because:

- Project approvals are often based upon insufficient information and poor estimating.
- Project approvals are based upon unrealistic return on investment (ROI), net present value (NPV) and payback period calculations.
- Project approvals are often based upon a best-case scenario.
- The true time and cost requirements may be either hidden or not fully understood during the project approval process.

Metrics require:

- A need or purpose
- A target, baseline, or reference point
- A means of measurement
- A means of interpretation
- A reporting structure

Even with good metrics, metrics management can fail. The most common causes of failure are:

- Poor governance, especially by stakeholders
- Slow decision-making processes
- Overly optimistic project plans
- Trying to accomplish too much in too little time
- Poor project management practices and/or methodology
- Poor understanding of how the metrics will be used

Sometimes the failure of metrics management is the result of poor stakeholder relations management. Typical issues that can lead to failure include:

- Failing to resolve disagreements among the stakeholders
- Failure to resolve mistrust among the stakeholders
- Failing to define critical success factors
- Failing to get an agreement on the definition of project success
- Failing to get an agreement on the metrics needed to support the CSFs and the definition of success
- Failing to see if the critical success factors are being met
- Failing to get an agreement on how to measure the metrics
- Failing to understand the metrics
- Failing to use the metrics correctly

Unless there is stakeholder agreement on how the metrics will be used to define or predict success and failure, all you can hope for are just best guesses. Disagreements on the use of the metrics and their meaning can result in a loss of credibility for the use of metrics.

3.6 CAUSES FOR LACK OF SUPPORT FOR METRICS MANAGEMENT

During the past few years, one of the drivers for effective metrics management has been the growth in complex projects. The larger and more complex the project is, the greater the difficulty in measuring and determining success. Therefore, the larger and more complex the project is, the greater the need for metrics.

Determining the metrics requires answering certain critical questions, however:

- Measurements
  - What should be measured?
  - When should it be measured?
  - How should it be measured?
  - Who will perform the measurement?
- Collecting information and reporting
  - Who will collect the information?
  - When will the information be collected?
  - When and how will the information be reported?

For many companies, answering these questions especially on complex projects was a challenge. As a result, metrics were often ignored because they were hard to define and collect.

Other reasons for the lack of support included:

- Metrics management was viewed as extra work and a waste of productive time.
- There was no guarantee that the correct metrics would be selected.
- If the wrong metrics are selected, then we are wasting time collecting the wrong data.
- Metrics management is costly and the benefits do not justify the cost.
- Metrics are expensive and useless.
- Metrics require change and people often dislike changes to their work habits.
- Metrics encourage unintended and/or unwanted behavior.

Metric management is often seen as an add-on to the existing work of the project team, but without these metrics, we often focus on reactive rather than proactive management. The result is a focus on the completion of individual work packages rather than a focus on completion of the business solution for the client.

Everyone understands the value in using metrics, but there is still the inherent fear among team members that metrics will be seen as "Big brother is watching you!" Employees will not support a metrics management effort that looks like a spying machine. Metrics should track the performance of a project rather than the performance of individuals. Metrics should never be used as justification for punishment.

3.7 USING METRICS IN EMPLOYEE PERFORMANCE REVIEWS

Job descriptions for project managers are now including a requirement for some knowledge about metrics and metrics management. However, using the value of the project-based metrics, whether favorable or unfavorable, as part of employee performance reviews may be a bad idea. Some reasons for this include:

- Metrics are usually the results of more than one person's contribution.
- Unfavorable metrics may be the result of circumstances beyond the employee's control.
- The employee may fudge the numbers in the metrics to look good during performance reviews, and the stakeholders may not get a true representation of the project's status.
- Real problems may be buried so that they do not appear in performance reviews.
- The person doing the performance review may not understand that the true value of the metric may not be known until sometime in the future.
- Employees working on the same project may end up competing with one another rather than collaborating, and the project's results could end up being suboptimal.

3.8 CHARACTERISTICS OF A METRIC

There are certain basic characteristics that a metric should possess. These include:

- Has a need or a purpose
- Provides useful information
- Focuses toward a target
- Can be measured with reasonable accuracy
- Reflects the true status of the project
- Supports proactive management

- Assists in assessing the likelihood of success or failure
- Accepted by the stakeholders as a tool for informed decision making

There are several types of metrics. For simplicity's sake, they can be broadly identified as:

- **Results indicators (RIs):** These tell us what we have accomplished.
- **Performance indicators (KPIs):** These metrics are the critical performance indicators that can drastically increase performance or accomplishment of the project's objectives.

Most companies use an inappropriate mix of these two and label them all as KPIs. However, there is a difference between a metric and a KPI.

- Metrics generally focus on the accomplishment of performance objectives, focusing on "Where are we today?"
- KPIs focus on future outcomes and address "Where will we end up?"

For simplicity's sake, we will consider only metrics in this chapter. KPIs will be discussed in more depth in Chapter 4.

Financial metrics have been used for more than a decade for analyzing business strategies. Financial metrics are business-related metrics and are based upon measurements of how well business goals are being met as part of a corporate strategy. Even with the long-term use of business and financial metrics, limitations still exist:

- Metrics such as profitability tell us if things look good or bad, but they do not necessarily provide meaningful information on what we must do to improve performance.
- Business-based or financial metrics are usually the result of many factors, and it is therefore difficult to isolate what must be done to implement change.
- Business-based or financial metrics are linked to long-term strategic objectives and usually do not change much.
- Words such as "customer satisfaction" and "reputation" have no real use as a metric unless they can be measured with some precision.
- Some business-based metrics cannot be measured until well into the future because it is the beneficial use of the deliverable that determines success.

Over the years, we have overcome most of these limitations by using KPIs, which are specialized metrics. While business metrics work well when focusing on a business strategy, there are significant differences between business and project metrics. This is shown in Table 3–1.

In contrast to business environments, which are long term, project environments are much shorter and, therefore, more susceptible to changing

TABLE 3-1 Business versus Project Management Metrics

VARIABLE	BUSINESS/FINANCIAL	PROJECT
Focus	Financial measurement	Project performance
Intent	Meeting strategic goals	Meeting project objectives, milestones, and deliverables
Reporting	Monthly or quarterly	Real-time data
Items to be looked at	Profitability, market share, repeat business, number of new customers, etc.	Adherence to competing constraints, validation and verification of performance
Length of use	Decades or longer	Life of the project
Use of the data	Information flow and changes to the strategy	Corrective action to maintain baselines
Target audience	Executive management	Stakeholders and working levels

metrics. In a project environment, metrics can change from project to project, during each life cycle phase, and at any time because of:

- The way the company defines value internally
- The way the customer and the contractor jointly define success and value at project's initiation
- The way the customer and contractor come to an agreement at project's initiation as to what metrics should be used on a given project
- New or updated versions of tracking software
- Improvements to the enterprise project management methodology and accompanying project management information system
- Changes in the enterprise environmental factors
- Changes in the project's business case assumptions

3.9 METRIC CATEGORIES AND TYPES

In the previous sections, we defined metrics as being either business metrics or project management metrics. We can expand this list even further to include four broad categories:

- Business-based or financial metrics
- Success-based metrics
- Project-based metrics
- Project management process metrics

Historically, metrics were most commonly used to evaluate a business strategy. Thus, typical business-based metrics included:

- ROI
- NPV
- Payback period
- Cost reduction
- Improved efficiency
- Paperwork reduction
- Future opportunities
- Accuracy and timing of information
- Profitability
- Market share
- Sales growth rate
- Number of new customers
- Amount of repeat business

Another category includes those metrics directly related to the success of the project. Examples of these are:

- Benefits achieved
- Value achieved
- Goals/milestones achieved
- Stakeholder satisfaction
- User satisfaction

Project-based metrics can be large in number. They will be discussed in more depth in Chapter 4. However, for simplicity's sake, they might include:

- Time
- Cost
- Scope and the number of scope changes
- Rate of change in the requirements (i.e., requirements growth over time)
- Quality
- Customer satisfaction with project performance
- Safety considerations
- Risk mitigation

Regardless of how we select metrics, there must exist a tracking metric for each constraint on the project. Since the number and criticality of the constraints can change from project to project, pressure is being placed upon the project team to establish more sophisticated metrics for the tracking and reporting of each constraint. You cannot control a constraint that cannot be monitored, measured, and reported.

Project management process metrics are directly related to lessons learned and best practices. Included in this category might be metrics related to:

- Continuous improvements
- Benchmarking
- Accuracy of the estimates
- Accuracy of the measurements
- Accuracy of the targets for the metrics and the KPIs

Thus far, we have identified the four broad categories of metrics. Within each category there are and can be subcategories or types of metrics based upon how the metric will be used. As an example, the following are seven types of metrics or metric indicators that could appear in each major category:

- Quantitative metrics (planning dollars or hours as a percentage of total labor)
- Practical metrics (improved efficiencies)
- Directional metrics (risk ratings getting better or worse)
- Actionable metrics (affect change as the number of unstaffed hours)
- Financial metrics (profit margins, ROI, etc.)
- Milestone metrics (number of work packages on time)
- End result or success metrics (customer satisfaction)

As mentioned previously, and as will be shown in later chapters, there are almost endless metrics that can be established for project management applications. Only a few of these, however, can be classified as a KPI. Typical metrics that may be established as KPIs, depending on the use, include:

- Cost variance
- Schedule variance
- Cost performance index
- Schedule performance index
- Resource utilization
- Number of unstaffed hours
- Percent of milestones missed
- Management support hours as a percent of labor
- Planning cost as a percentage of labor
- Percent of assumptions that have changed
- Customer loyalty
- Percent turnover of key workers
- Percent of labor hours spent on overtime
- Cost per page for customer reporting

Care must be taken when setting up a classification system for metrics. Some important factors include:

- It doesn't matter which classification system of metrics you use as long as you use some system.
- Do not get stuck in a "metrics mania" mode, where you must create significantly more metrics than you need just for metrics' sake.
- The metrics identified must be used. If you cannot continue using a metric, then the metric may be flawed.
- Consideration should be given to combining metrics that provide little or no value.

3.10 SELECTING THE METRICS

Quite often, the wrong project metrics are selected. The reason is that the selection is based upon who is doing the asking. Selecting a commonly used metric is easy, but it may be inappropriate for the project at hand. The result will be useless data. Another reason why the wrong metrics are selected is the law of least resistance, whereby metrics are selected based upon the ease and speed with which they can be measured.

According to Owen Head:[4]

> When establishing the processes to be used in gathering metrics, it is important to prioritize the list in order of importance, and avoid processing any that aren't truly needed. No metric should be included, for example, unless the team will actually be able to take the time to react to it. If the team won't be able to take action in response to an undesirable metric, then it would be a waste of time to gather it. It's always possible to increase the number of metrics as the project moves forward, but attempting to take on too much at the beginning can steal time and attention from other critical project activities.

Sometimes, the people doing the selection do not understand the use of the metric. Metrics do not tell us what action to take or whether the success or recovery of a failing project is possible.

Figure 3–4 shows the metrics/KPI spectrum. On the left-hand side, you have people who want an abundance of metrics regardless of the value of the metric's information. On the right-hand side you have people who ignore the use of metrics because they are hard to define and collect. Finding a compromise is not easy, but we must determine how many metrics are needed.

- With too many metrics:
  - Metric management steals time from other work.

4. http://www.pmhut.com/a-minimalists-approach-to-project-metrics.

Figure 3-4 **The Metrics Value Spectrum**

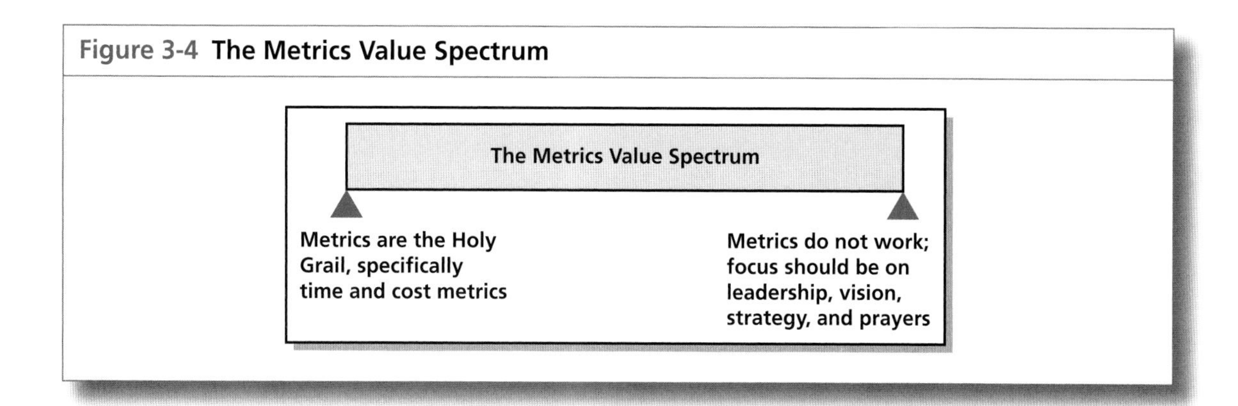

- We end up providing too much information to stakeholders so that they cannot determine what information is critical.
- We end up providing information that has limited value.
- With too few metrics:
  - Not enough critical information is provided.
  - Informed decision making becomes difficult.

Project teams and stakeholders tend to select too many rather than too few metrics. According to Douglas Hubbard:[5]

> In business, only a few key variables merit deliberate measurement efforts. The rest of the variables have an "information value" at or near zero.

There are certain ground rules we can establish as part of the metric selection process:

- Make sure that the metrics are worth collecting.
- Make sure that we use what we collect.
- Make sure that the metrics are informative.
- Train the team in the use and value of metrics.

Selecting metrics is a lot easier when you have competent baselines from which to make measurements. It is very difficult or even impossible to use metrics management effectively when the baselines undergo continuous transformation. For work that has not been planned yet, benchmarks and standards can be used instead of baselines.

TIP Effective selection of the metrics cannot be done in a vacuum.

5. Douglas W. Hubbard, *How to Measure Anything: Finding the Value of Intangibles in Business*, Hoboken, NJ: John Wiley & Sons, 2007, p. 33.

Metrics by themselves are just numbers or trends resulting from measurements. Metrics have no real value unless they can be properly interpreted by the stakeholders or subject matter experts and a corrective plan, if necessary, can be developed. It is important to know who will benefit from each metric. The level of importance can vary from stakeholder to stakeholder.

There is always the risk that the metric information presented on the dashboard will be misunderstood and the wrong conclusions will be drawn. Some people argue that metrics should not exist without context. It is possible to use information drill down buttons beside certain important metrics. There are two purposes for the drill-down buttons:

- The information is optional and is used to reassure the project manager that the stakeholders understand the information that is being provided. This may be necessary in the early stages of using dashboards. The information can be in just cursory format.
- The information is mandatory and must be provided to explain the meaning of the metric. As an example, assume a metric shows that there are action items that have been in the system for more than three months. This may not mean that the project team is derelict in their responsibilities to resolve issues in a timely manner. The unresolved action items may be the result of special circumstances such as waiting for the results of a certain test, the need for additional funding, and other such arguments.

While information drill-down buttons may be necessary, having too many or unnecessary drill-down buttons may end up creating rather than reducing paperwork. The intent of dashboards is to minimize or eliminate paperwork rather than increase it.

There are several questions that can be addressed during metric selection:

- How knowledgeable are the stakeholders in project management?
- How knowledgeable are the stakeholders in metrics management?
- Do we have the necessary organizational process assets for metric measurements?
- Will the baselines and standards undergo transformations during the project?

There are two additional factors that must be considered when selecting metrics. First, there is a cost involved in performing the measurements, and based upon the frequency of the measurements, the costs can be quite large. Second, we must recognize that metrics need to be updated. Metrics are like best practices; they age and may no longer provide the value or information that was expected. There are several reasons, therefore, for periodically reviewing the metrics:

- Customers may desire real-time reporting rather than periodic reporting, thus making some metrics inappropriate.
- The cost and complexity of the measurement may make a metric inappropriate for use.
- The metric does not fit well with the organizational process assets available for an accurate measurement.
- Project funding limits may restrict the number of metrics that can be used.

In reviewing the metrics, there are three possible outcomes:

- Update the metric
- Leave the metric as is but possibly put it on hold
- Retire the metric from use

Finally, metrics should be determined after the project is selected and approval is obtained. Selecting a project based upon available or easy-to-use metrics often results in either the selection of the wrong project or metrics that provide useless data.

3.11 SELECTING A METRIC/KPI OWNER

Many of the companies that have recognized the importance of maintaining a best practices library have also created the position of a best practice owner. The concept of a KPI owner has been used for financial metrics but only recently has it been adopted for project management metrics. Since companies are expected to maintain a metric/KPI library, it is only customary that they also maintain the position of a metric/KPI owner. Each project team must accept ownership for the KPIs they use. Based upon the number of KPIs, it may be advisable for each project team to assign a KPI owner to each KPI. The other choice, which is more common, is for one person in the company to be assigned as the KPI owner.

The metric/KPI owner:

- Must understand the company's culture
- Must have the respect of the labor force
- Must be able to foster support for the use of metrics
- May serve as a mentor for people using the metric
- Must perform continuous improvements on the metric, including improvements in metric measurement techniques
- Must support the PMO in determining whether the metric/KPI is still valid or has aged

3.12 METRICS AND INFORMATION SYSTEMS

It is possible on a given project to have several different project management information and reporting systems. As an example, on the same project, we can have an information system for:

- The project manager's personal use
- The project manager's parent company
- The client and the stakeholders

There can be different metrics and KPIs in each of the information systems. The greatest number of metrics will appear in the project manager's information system. These metrics, which would be for the project manager's personal use, could include resource utilization, details related to work packages, risk-related activities, and cost-estimating accuracies. Executives in the parent company might focus on the project's profit margins, project headcount, customer satisfaction, and the potential for future business. The information presented to the stakeholders is usually the KPIs, which are the critical metrics for informed decision making and can include metrics on cost, schedule, value, and other such factors.

3.13 CRITICAL SUCCESS FACTORS

The ultimate purpose for working on projects, whether they are for internal or external clients, is to support some type of business strategy. This is shown in Figure 3–5. Once the project is selected as part of the portfolio selection of projects, a project strategy is developed around the project's objectives, criteria for success, and metrics/KPIs. The greater the agreement between the customer and the contractor on the project's strategy, the greater the chances for success.

The same holds true for stakeholders. Simply stated, it is important that everyone who has a stake in the project communicate what they believe to be success on the project.

As mentioned in Chapter 1, one of the first steps in the development of a customer-contractor relationship on a project, perhaps beginning even during the initial engagement with the client, is to come to an agreed-upon definition of success. Some companies first define success in terms of critical success factors (CSFs) and then establish metrics and KPIs to determine whether these CSFs are being met. Critical success factors identify those activities necessary to meet the desired deliverables of the customer and maintain effective stakeholder relations management. Typical CSFs include:

- Adherence to schedules
- Adherence to budgets
- Adherence to quality
- Appropriateness and timing of signoffs
- Adherence to the change control process

Figure 3-5 Establishing the Project's Strategy

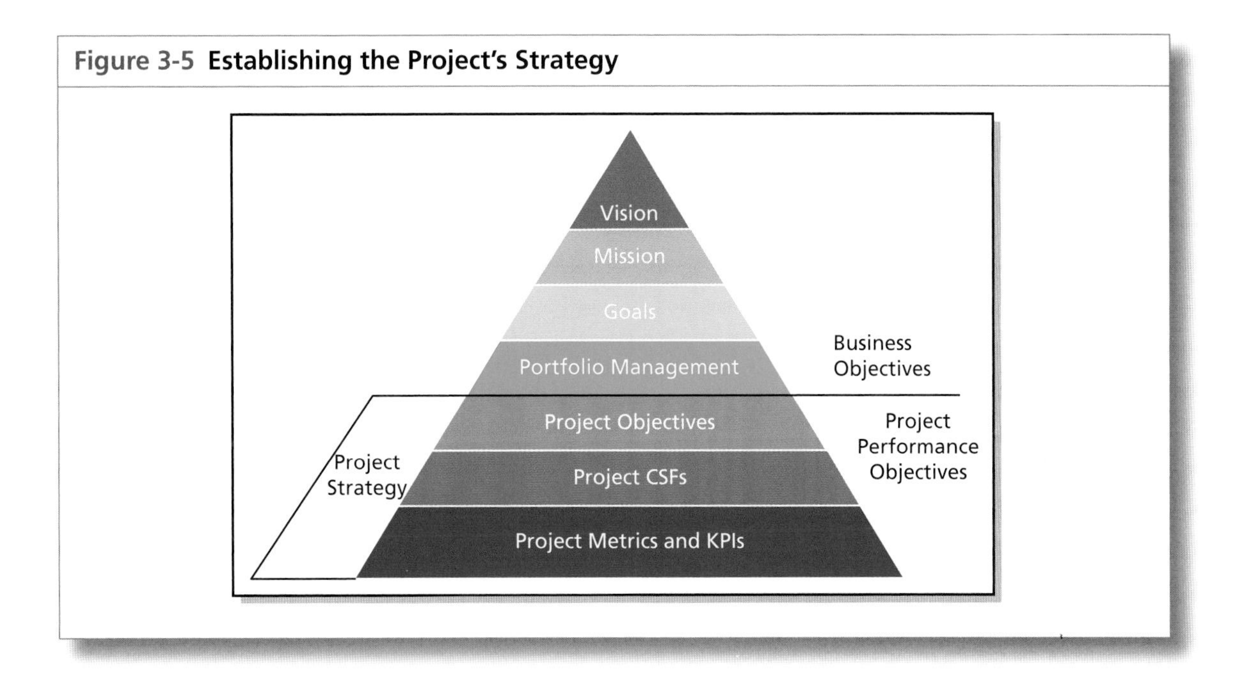

- Add-ons to the contract
- Proper scoping of the existing environment
- Understanding the customer's requirements
- Early involvement by the customer and stakeholders
- Agreement upon and documenting of project objectives
- Provisioning all of the required resources
- Managing expectations
- Identifying all project risks and planned handling—with the customer
- Effective exception-handling process
- Control of requirements; preventing scope creep
- Explicit and specific communications with customer
- Defined processes and formalized gate reviews

Every company has its own definition of success because they have different clients, different requirements with these clients, and different stakeholders. Some CSFs may be heavily oriented toward an internal definition of success rather than a customer's definition, although the ideal situation would a compromise. At Convergent Computing, some of the CSFs include:[6]

- Have experienced and well-rounded technical resources. These resources need to not only have outstanding technical skills, but also be good

6. Kerzner, H., *Project Management Best Practices; Achieving Global Excellence*, Hoboken, NJ: John Wiley & Sons, 2006, pp.26–27.

communicators, work well in challenging environments, and thrive in a team environment.
- Make sure we understand the full range of the clients' needs, including both technical and business needs, and document a plan of action (the scope of work) for meeting these needs.
- Have well-defined policies and processes for delivering technology services that leverage "best-practice" project management concepts and practices.
- Have carefully crafted teams, with well-defined roles and responsibilities for the team members, designed to suit the specific needs of the client.
- Enhance collaborations and communications both internally (within the team and from the team to CCO) and externally with our clients.
- Leverage our experience and knowledge base as much as possible to enhance our efficiency and the quality of our deliverables.

According to Bill Cattey,[7]

> Stated simply and generically, it's important that everyone who has a stake in a project communicate about what is believed to be critical to the success of the project. Projects succeed better when there is broad agreement on what's needed for the project to be a success, and measurement is made to see if there is a gap between what is needed, and what's actually happening, and corrective action is taken to close the gap.

CSFs measure the end result, usually as seen through the eyes of both the customer and the contractor. KPIs and metrics generally measure the quality of the processes used to achieve the end results and accomplish the CSFs. KPIs and metrics are internal measures and can be reviewed on a periodic basis throughout the life cycle of a project. Some people believe that CSFs are the same as metrics and KPIs and in confusion try to track them. CSFs are usually broad categories and difficult to track, whereas metrics and KPIs are more specific and are, therefore, more appropriate for measurement and then reporting through means such as dashboards. CSFs are often interim steps between the definition of success and the establishment of metrics.

Metric measurements can be too costly. Even if all of the stakeholders are in agreement on the CSFs, the cost of measuring the metrics to support the CSFs can be prohibitive and the benefits achieved may not support the cost. The measurements should pay for themselves in supporting the CSFs; overly expensive or useless measurements should not be made. Stakeholders

7. Bill Cattey, "Project Management Metrics," http://web.mit.edu/wdc/www/project-metrics.html.

must believe that the correct metrics were chosen and that the measurements accurately portray the true status. It is important to understand that some metrics for success cannot instantaneously determine the success of a project. True measurements may not be able to be made until well after the project is completed.

Projects often fail because the project manager and the stakeholders cannot agree on the CSFs and then may end up selecting useless metrics that cannot provide meaningful data. It is not uncommon for the stakeholders to believe that having the fewest scope changes is a CSF, whereas the project manager believes that following the change control process rigidly is the true CSF regardless of the number of scope changes. Some causes of failure include:

- Improper understanding of the meaning of the CSFs.
- Failing to believe in the value of the CSFs.
- Each stakeholder is working toward his/her own definitions of the CSFs.
- Stakeholders refuse to come to an agreement on the correct CSFs for the project at hand. This can occur during the project as well as at the onset.
- Failure to understand the gap between the actual performance and the CSFs.
- Believing that the measurement costs for the metrics to support the CSFs are too great.
- Believing that it is a waste of productive time measuring the metrics needed to support the CSFs.

3.14 METRICS AND THE PMO

Previously we stated that you cannot effectively manage a project without having measurement and metrics capable of providing you with complete information. Project managers do not always possess expertise in selecting the correct metrics, KPIs, and CSFs. This is where the PMO can be of assistance. With regard to metrics, the PMO can:

- Assist each project team with the development of project-based metrics
- Recognize that some metrics may be project or client specific
- Maintain a metrics library
- Develop a metric/KPI template for the library
- Recognize that metric and KPI improvements must evolve over time and that updating is necessary

As part of the PMO's responsibility as the guardian of all project management intellectual property, the PMO will coordinate the efforts for the updating of the metrics. Metric owners usually report dotted to the PMO. A periodic review of the effectiveness of each metric/KPI is essential because metrics have life cycles. As stated previously, the metrics can remain as is, be updated, or be retired from service. Metrics can become irrelevant without supporting data. Overseeing metrics and KPIs can be expensive and difficult. Creating unnecessary metrics should be avoided.

The PMO may have the responsibility of debriefing the project teams in order to capture lessons learned and best practices. A typical debriefing pyramid is shown in Figure 3–6. The PMO will look at metrics related to the project, use of organizational process assets such the EPM systems, the business units, and perhaps even corporate strategic metrics.

At the same time, the PMO may evaluate all of the metrics used to see if the metrics should be part of the metrics library and if the cost of using the metric (i.e., measurement) was worth the effort. Figure 3–7 show a typical way that the metrics may be evaluated. Below the risk boundary or the parity line, the value exceeds the cost and the risks of using the metric are acceptable. Above the parity line, the measurement cost may exceed the value of using the metric and careful risk analysis is necessary.

If the metrics are deemed to be of value, then the metrics can be classified in the metrics library the same way that we often classify best practices. A typical example is shown in Figure 3–8. For companies that are immature in project management, the focus is on promoting the use of metrics but perhaps only a few. Companies that are reasonably mature in project management build and maintain metric and KPI libraries.

Figure 3-6 The Postmortem Pyramid

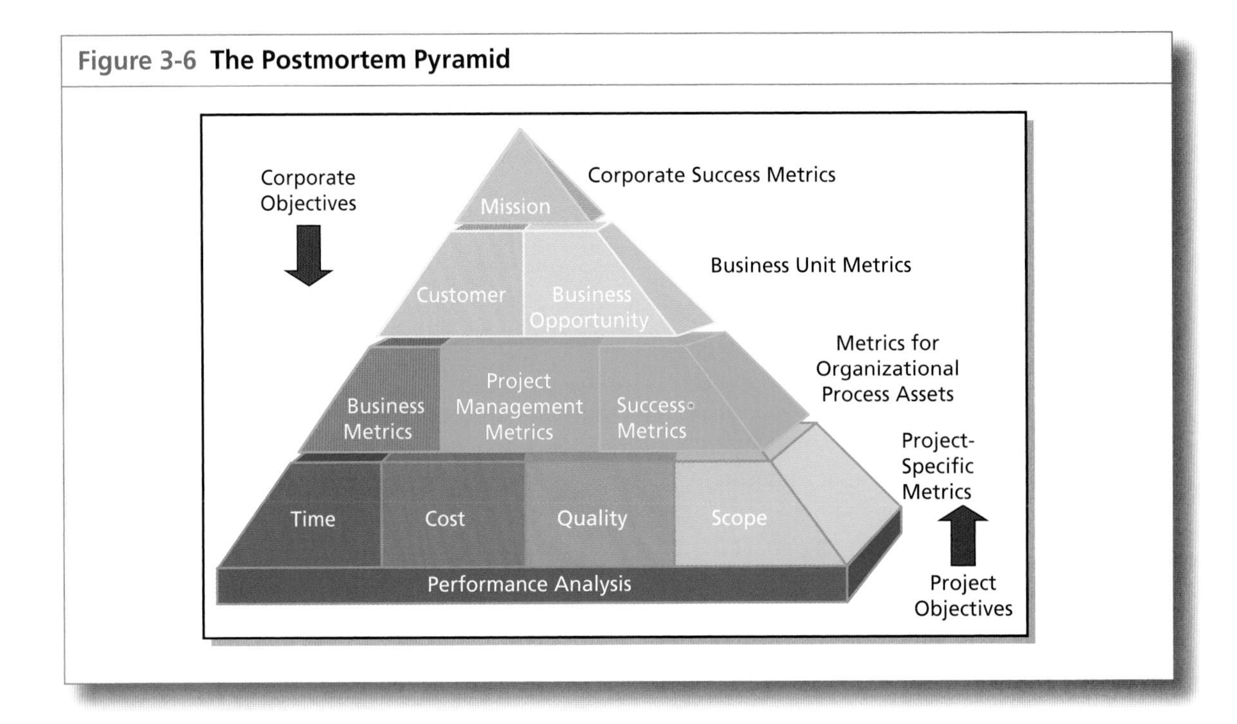

Figure 3-7 **Metric Cost versus Value**

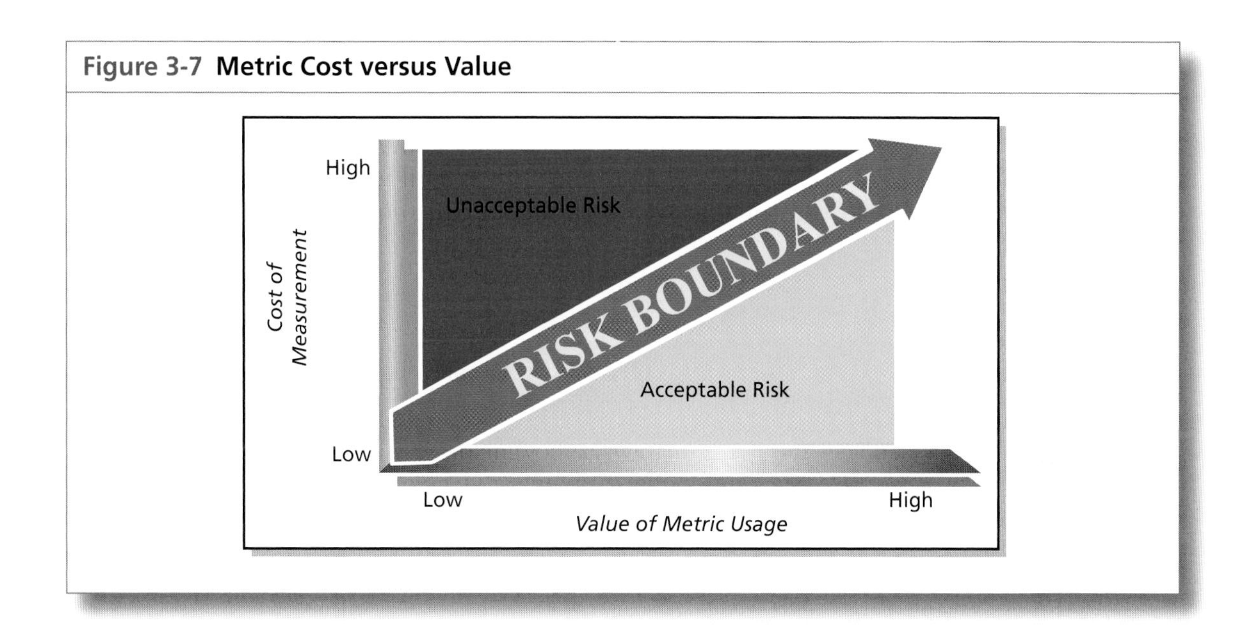

Figure 3-8 **Best-Practices Classification**

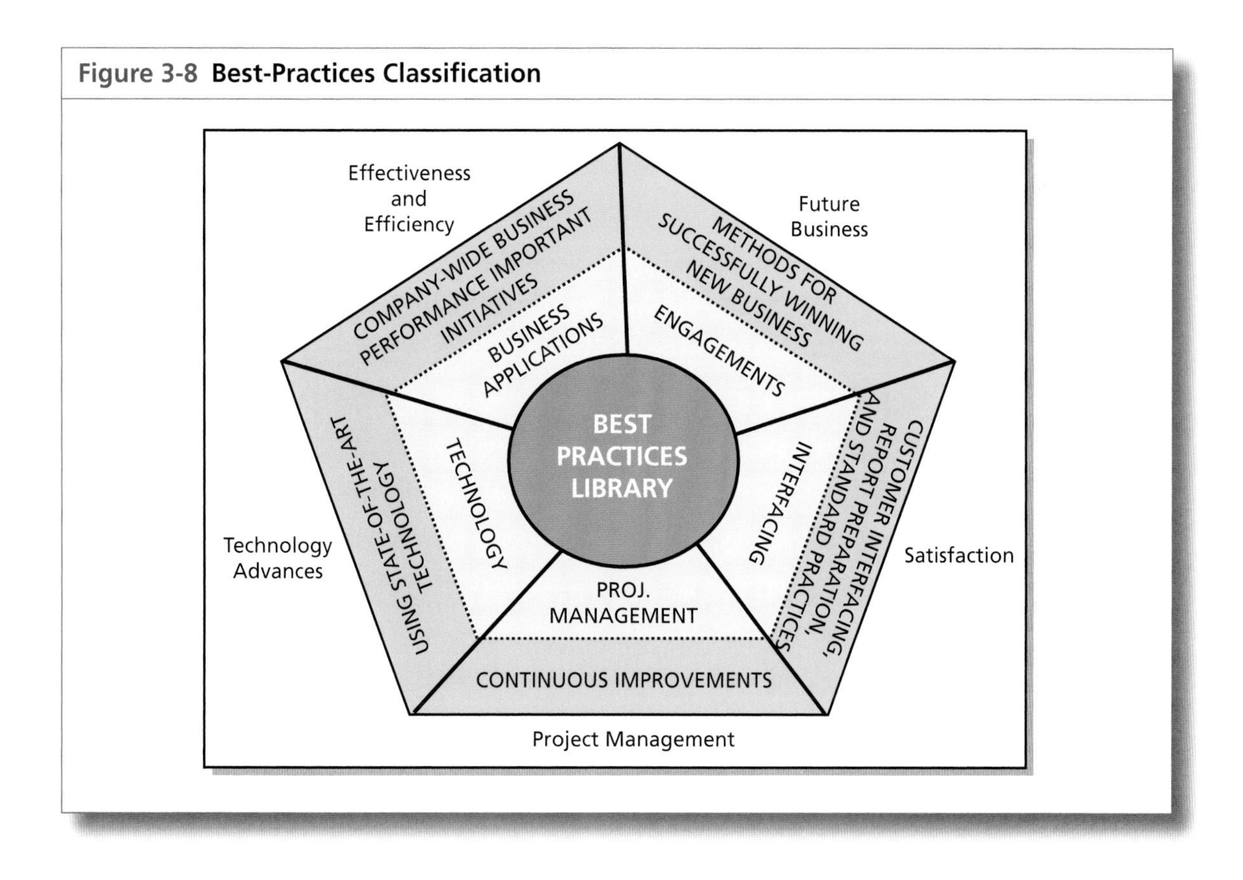

In determining the best possible metrics for a project, the PMO may find it necessary to perform metrics benchmarking. Two critical factors must be considered:

- The project management maturity level of your organization as well as the company against which you are benchmarking
- The project management maturity level of the stakeholders

There are also misconceptions that must be considered such as:

- Metrics that work well for one company may not work well for another company, if the use of the metrics is based upon in-house practices.
- Identifying the metrics is easy. Using them is difficult.
- Some metrics may be more a rough guide than a precise benchmark.

Since most PMOs are overhead rather than direct labor charges, it may be necessary for the PMO to establish their own metrics to show the PMO's contribution to the success of the company. Typical ROI metrics that the PMO use include:

- Percent of projects using/following the EPM system/framework
- Ratio of project manager to total project staff
- Customer satisfaction ratings
- Year-over-year throughput
- Percent of projects at risk or in trouble
- Number of projects per headcount (staffing tolerance for projects)
- Ways to improve faster closure
- Percent of scope changes per project
- Percentage of projects completed on time
- Percentage of projects completed within budget

It is important to understand that metrics management is an essential component of knowledge management and involvement by the PMO is essential. It is very difficult to improve processes and work flow without gathering metrics and storing the results for traceability.

3.15 METRICS AND PROJECT OVERSIGHT/GOVERNANCE

Every project governance group has their own distinguishing characteristics on their role, the types of decisions they are expected to make and how they should interface with the project. In general, the governance group must be able to balance the risks of the project against the benefits or value of the final outcome. To do this, metrics are needed. It is important to understand what type of decision rights should be delegated to an executive oversight or governance group and what rights should go to the program team before selecting the metrics:

- Program team: authority to maintain baselines
- Oversight group: scope changes above a certain dollar level, additional funding, alignment to business objectives, health checks and project/program termination (exit champion)

While standard high level metrics can be used for oversight groups, additional metrics may be necessary when a crisis occurs. The metrics needed to resolve a crisis are different from the traditional metrics that are used to monitor performance.

3.16 METRIC TRAPS

Throughout this chapter, we have discussed ways that we can get in trouble using the wrong metrics. These were examples of metric traps. Fancy colors and charts displaying the metrics can often lead to a metrics trap. Some other common mistakes that people often make with metrics include:

- Believing that several fancy images are needed to display the same information that can be shown with one image
- Spending a great deal of time using trial-and-error solutions to determine which image is best
- Selecting metrics that cannot be measured effectively
- Promising stakeholders metrics before knowing how to perform the measurement
- Making stakeholders believe that metrics alone can predict the success or failure of a project

3.17 PROMOTING THE METRICS

Most people that work on projects are motivated by seeing the results of their efforts. Promoting metrics creates awareness and support for the project, keeps people informed and motivates workers. Historically, metrics were displayed only in the project's war room or command center. War rooms usually have one door and no windows. All of the walls are covered with charts and displays showing the health of the project. On large projects, it was common to have an assistant project manager whose job includes updating all of the metrics and charts on a regular basis.

Some companies are creating a "wall of metrics" that everyone can see, including vendors and clients. Publishing these metrics outside of the traditional war room does not require additional effort and the rewards can significantly outweigh the small cost.

3.18 CHURCHILL DOWNS INCORPORATED'S PROJECT PERFORMANCE MEASUREMENT APPROACHES

Since a PMO is the guardian of the company's project management intellectual property as well as the organization responsible for the project management methodology, the PMO can make it relatively painless to improve project management performance measurements. Centralizing these continuous improvement efforts in the PMO can accelerate the implementation and usage of metrics and KPIs. This can also be accomplished in a simple manner rather than by using massive sophistication that could scare away possible users.

The remainder of this section has graciously been provided by Chuck Millhollan, director of program management, Churchill Downs Incorporated.[8]

The Churchill Downs Incorporated Project Management Office (PMO) has two basic premises that helped form our approach to project related measurements. The first, and most important, is that the dashboard and report content is not as important as the discussion generated from the information. The second, with a genesis in lessons learned through our experiences, is that key performance indicators that are not defined, documented, and tracked have a much higher potential for being missed. Since most of the factors that influence what, when, and how we deliver are beyond the project manager's control, we focus on proactively managing expectations to ensure that those expectations match the delivered reality.

Since our stakeholder's define our success, we do not use project management "process" indicators to define project success. While schedule and budget targets are part of the criteria, sponsor acceptance, project completion, and ultimately project success, are based on meeting defined business objectives. To enhance the value of project performance measurement at Churchill Downs Incorporated (CDI), we purposefully separate project management–related measurements and reporting from the benefit measurement associated with the delivered product or service. The remainder of this paper references to our project management related metrics.

To understand our approach to defining critical-to-quality parameters and reporting processes, it is important to understand how CDI defines project success. When our PMO was chartered in April 2007, we developed a definition for project success with input from our executive leadership team. CDI considers a project as a success when the following are true:

1. Predefined business objectives and project goals were achieved or exceeded.
2. A high-quality product is fully implemented and utilized.
3. Project delivery meets or beats schedule and budget targets.

8. Reproduced by Permission of CDI. For information on the CDI approach, contact Chuck Millhollan, MBA, MPM, PMP, PgMP, IIBA Certified Business Analysis Professional (CBAP), ASQ Certified Six Sigma Black Belt, ASQ Certified Manager of Quality/Organizational Excellence, and ASQ Certified Software Quality Engineer. (Email: chuck.millhollan@kyderby.com).

4. There are multiple winners:
   a. Project participants have pride of ownership and feel good about their work.
   b. The customer's (internal and/or external) expectations are met.
   c. Management has met its goals.
5. Project results helped build a good reputation for the project team and the product or service.
6. Methods are in place for continual monitoring and evaluation (benefit realization).

Another key consideration when evaluating CDI's project performance reporting process is that we made a conscious decision not to invest in complex portfolio reporting tools that required enterprise acceptance and adoption. Instead, our modus operandi is to ensure that the heavy project management lifting is done behind the scenes and the information distributed leverages common desktop applications available and understood by a vast majority of our stakeholders. We are not opposed to using more sophisticated reporting tools that generate graphs, bar charts, etc.; however, we must first identify both the need and value. Finally, it is important to note that we are quick to evaluate and modify our reporting processes if we discover that senior/executive leadership is not leveraging the information to support their decision making. Our goal is to ensure that the PMO is not perceived as a score keeper, but instead that performance measurement and report is providing a defined benefit.

Toll Gates (Project Management–Related Progress and Performance Reporting)

Since PMO inception (April 2007), our project tracking process has evolved from a basic one-page document that displayed key milestones (such as Investment Council approval, completion, project health indicators, projected completion dates, etc.) to a toll-gate process and associated graphic we refer to as the project quad (See Figure 3–9).

As our processes matured, so did the desire for and use of project-specific progress information. Since the initial Green/Yellow/Red project health indicators were subjective (however, valuable at the time), we had a need for more quantifiable performance and progress metrics. The project quad is divided into four intuitive sections that provide an executive summary level of information.

Quad sections:

I. **Project Overview:** This section provides project charter level information on the quantifiable business objectives, work authorization date, approved capital budget, and the standard project health indicator (now less subjective).

II. **Current State:** This section provides a quick reference for project progress through the defined Toll Gates and progress according to plan. The

Figure 3-9 Project Quad

Project Title

CHURCHILL DOWNS INCORPORATED

Sponsor:______________ **PM:**______________

Report Out Date:________

Project Overview:
Business Goal(s):
Start Date:
Approved Budget:
Current Status (R, Y, G):

Current State:			
Project Phase:			
% Complete:			
Dates:			
Toll Gate	Plan	Estimate	Actual
TG-1			
TG-2			
TG-3			
TG-4			

Issues:	
Schedule:	Budget:
Scope:	Objectives:

Next Steps:
1. 2. 3. 4.

CDI PMO ▶ 1

variances are immediately obvious and frequently initiate the discussion necessary to remove barriers to progress. Figure 3–10 is a training document we use to provide a high-level overview of the Toll Gates.

III. **Issues:** The project manager uses this section to provide bulletized statements about barriers to either progress or meeting the approved business objectives.

IV. **Next Steps:** This section is used to either identify the next milestones or for recommendations to bring performance back in line with the plan.

In order to support the Toll Gate process and facilitate the discussions necessary for accurate reporting and decision making, we use a standard list of questions that project managers must address as they move through the gates. Naturally, each question does not apply to each project and there is

Figure 3-10 Toll Gate Overview

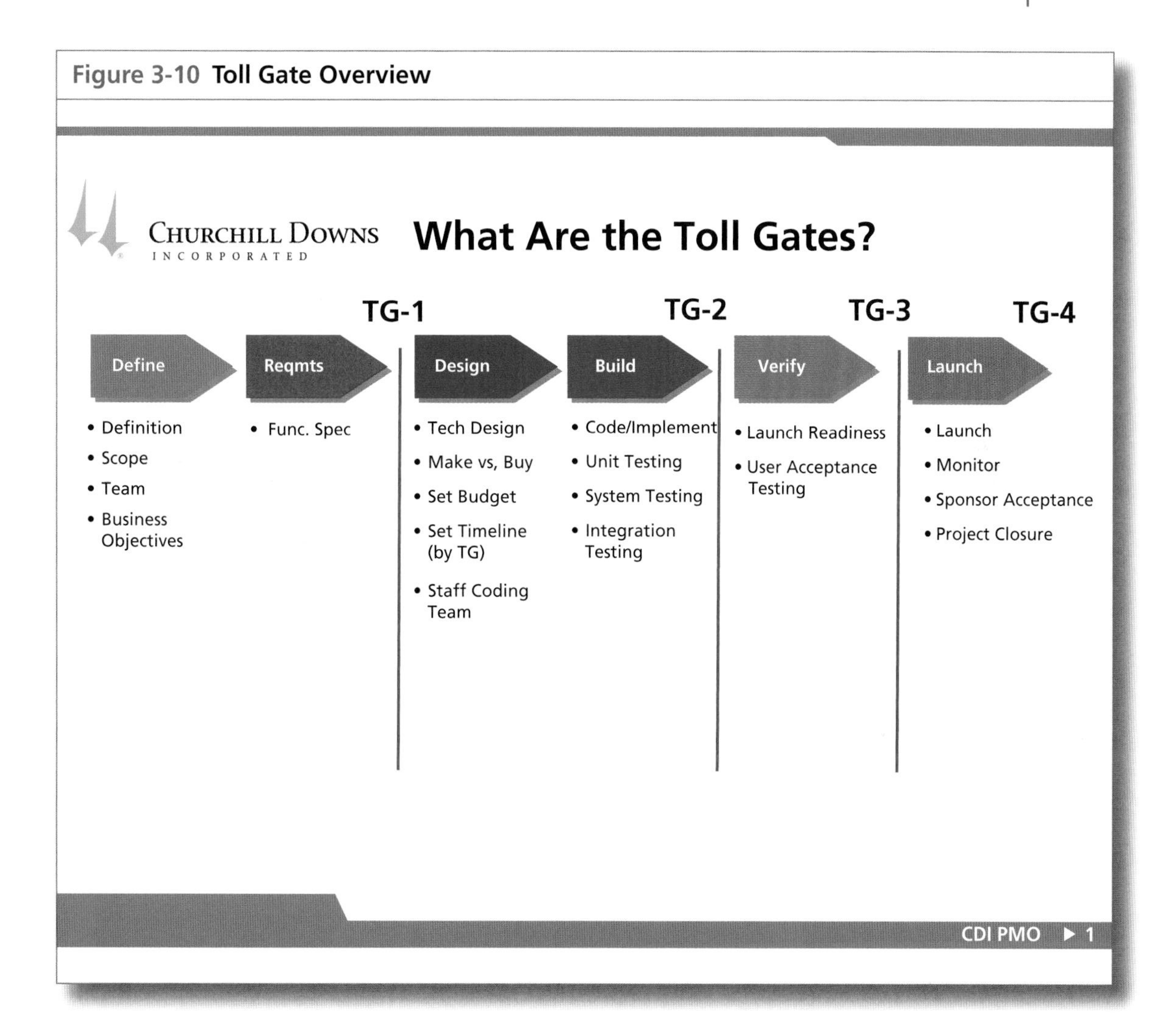

an iterative nature to many of the questions; however, the overall structure has proven effective for ushering projects from initiation to implementation. Figure 3–11 is the Toll Gate 2 checklist.

The nature of CDI's business makes delivery dates the primary constraint for the majority of our projects because of race meet openings dates for each track, marquee racing events (such as the Kentucky Derby), concerts, etc. During our biweekly project portfolio reviews, we use a one-page portfolio dashboard with summary-level data consisting primarily of Toll Gate delivery dates. This dashboard is used to drill down into the quads for specific projects that require additional discussion to facilitate decision making. Figure 3–12 illustrates the project dashboard used to facilitate the project reviews with the leadership team.

Figure 3-11 Toll Gate 2 Checklist

CHURCHILL DOWNS
INCORPORATED

PROJECT NAME:
PROJECT NUMBER:
EXECUTIVE SPONSOR:
BUSINESS SPONSOR:
IT SPONSOR:
PM:

DESIGN	ANSWER	BUILD	ANSWER
Can we design & develop a solution to meet the requirements?		Are we ready for business user acceptance testing?	
What is the IT solution and it's ling term supportability?		Do we have the resources to get the job done?	
How will we address any functionality gaps?		Do we have a well defined testing strategy/plans including	
Did we translate all the business requirements into detailed technical specifications?		Code Reviews & unit testing?	
Do we still have business buy in?		Performance and load testing?	
Do we have a firm project plan with commitment from all parties?		Integration testing	
What about scope, budget, CTQs, and Benefits?		User acceptance testing?	
Does the design ensure we will meet out CTQs?		Piolet testing?	
		Did the testing show that we can meet our CTQs?	
		How will we commercialize?	

Approvals

Project Manager	IT Sponsor	Business Sponsor
Date	Date	Date

Figure 3-12 **Project Toll Gate Dashboard**

CHURCHILL DOWNS
INCORPORATED

PROJECT NAME	STATUS	TOLL GATE 1			TOLL GATE 2			TOLL GATE 3			TOLL GATE 4			NOTES / CRITICAL ISSUES
		Plan	Estimate	Actual	Plan	Estimate	Actual	Plan	Estimate	Actual	Plan	Estimate	Actual	
Enterprise Ticketing Solution	G													
Enterprise Vault	Hold													
Fair Grounds OTB Failover Testing	G													
Network Refresh	Hold													
Epiphany Service	R													
CDI Video Teleconferencing	G													
Youbet Merger Program	G													
Enterprise Printing Analysis	Y													
CDI Webex Services	G													
Twinspires Call Center Expansion	G													
Hullabalou & Ticketmaster Interface	G													
Peak-10 Clean-up	G													
Hullabalou Mobile App	G													
Digital Asset Management	G													
FGRC Oasis Upgrade	G													
SABO Derby Invitation Process Enhancements	G													
Avamar System Training	G													

4 KEY PERFORMANCE INDICATORS

CHAPTER OVERVIEW

People tend to use the words *metrics* and *key performance indicators* interchangeably. Unfortunately, they are different. In this chapter, we will discuss the differences as well the role of KPIs in project management. At the end of the chapter is a white paper that provides a good summary of the information in the chapter.

CHAPTER OBJECTIVES

- To understand the differences between metrics and KPIs
- To understand that KPIs are controllable factors
- To understand how to use KPIs correctly
- To understand the characteristics of a KPI
- To understand the components of a KPI
- To understand the categories of KPIs
- To understand the issues with KPI selection

KEY WORDS

- Actionable characteristics
- Critical success factors
- KPI owner
- Lagging indicators
- Leading indicators
- Relevant characteristics
- SMART Rule
- Stakeholder classification

4.0 INTRODUCTION

As stated in previous chapters, part of the project manager's role is to understand what the critical metrics are that need to be identified, measured, reported, and managed so that the project will be viewed as a success by all of the stakeholders, if possible. The term "metric" is generic, whereas a KPI is specific. KPIs serve as early warning signs that, if an unfavorable condition exists and is not addressed, the results could be poor. KPIs and metrics can be displayed in dashboards, scorecards and reports.

Defining the correct metrics or key performance indicators is a joint venture of the project manager, client, and stakeholders, and is a necessity in order to get stakeholder agreement. KPIs give everyone a clear picture of what is important on the project. One of the keys to a successful project is the effective and timely management of information. This includes the KPIs. KPIs give us information to make informed decisions and reduce uncertainty by managing risks.

Getting stakeholders' agreement on the KPIs is difficult. If you provide the stakeholders with 50 metrics to select from, they will somehow justify the need for all 50 of them. If you show them 100 metrics, they will find a reason why all 100 should be reported. The hard part is to select from the metrics library those critical metrics that can function as key performance indicators.

For years, metrics and KPIs were used primarily as part of business intelligence techniques. When applied to projects, KPIs answer the question, "What is really important for different stakeholders to monitor on the project?" In business, once a KPI is established, it becomes difficult to change as enterprise environmental factors change for fear that historical comparison data will be lost. Benchmarking industry KPIs, however, is still possible because the KPIs are long term. In project management, given of the uniqueness of projects, benchmarking is more complex because of the relatively short life span of the KPIs. In both business applications and projects, however, we assume that, if the KPI targets are being met or exceeded favorably, we are adding value to the business or the project.

4.1 THE NEED FOR KPIs

Most often, the items that appear in the dashboards are elements that both customers and project managers track. These items are referred to as key performance indicators. According to Eckerson:[1]

> A KPI is a metric measuring how well the organization or an individual performs an operational, tactical or strategic activity that is critical for the current and future success of the organization.

Although Eckerson's comment is more appropriate for business-oriented rather than project-oriented metrics, the application to a project environment still exists. KPIs are high-level snapshots of how a project is progressing toward predefined targets. Some people confuse a KPI with a leading indicator. A leading indicator is actually a KPI that measures how

1. Wayne W. Eckerson, *Performance Dashboards: Measuring, Monitoring and Managing Your Business*, Hoboken, NJ: John Wiley and Sons, 2006, pp. 294.

the work you are doing now will affect the future. KPIs can be treated as indicators but not necessarily leading indicators.

> **SITUATION:** By the end of the second month of a 12- month project, the cost variance indicated that the project was over budget by $40,000. The client then believed that if this continued until the end of the project, the final result would be a cost overrun of $240,000. The client became irate and called for a clear explanation as to why we were heading for a $240,000 cost overrun.

While some metrics may appear to be leading indicators, care must be taken as to how they are interpreted. The misinterpretation of a metric or the mistaken belief that a metric is a leading indicator can lead to faulty conclusions.

KPIs are critical components of all earned value measurement systems. Terms such as cost variance, schedule variance, schedule performance index, cost performance index, and time/cost at completion are actually KPIs if used correctly but not always referred to as such. The need for these KPIs is simple: What gets measured gets done! If the goal of a performance measurement system is to improve efficiency and effectiveness, then the KPI must reflect controllable factors. There is no point in measuring an activity if the users cannot change the outcome.

For more than four decades, the only KPIs we looked at were time and cost or derivatives of time and cost. Today, we realize that true project status cannot be measured from just time and cost alone. Therefore, the need for additional KPIs has grown. Sometimes, several metrics and KPIs can be rolled up into a single KPI. As an example, a customer satisfaction KPI can be a composite of time, cost, quality, and effective customer communications. KPIs are often derived from formulas on how to combine these other metrics into a single KPI that may be specific and/or beneficial to a particular company.

What is and is not a KPI must be defined by individual decisions unique to that project. Project managers must explain to the stakeholders the differences between metrics and KPIs and why only the KPIs should be reported on dashboards. As an example, metrics focus on the completion of work packages, achievement of milestones, and accomplishment of performance objectives. KPIs focus on future outcomes, and this is the information stakeholders need for decision making. Simply stated, with metrics we often get bogged down looking at what happened in the past. With KPIs, we figure out how to use this data for decision making in the future. Neither metrics nor KPIs can truly predict that the project will be successful, but KPIs provide more accurate information on what might happen in the future if the existing trends continue. Both metrics and KPIs provide useful information, but neither can tell you what action to take or whether a distressed project can be recovered.

KPIs have been used in a variety of industries, as shown in the following material, and for specialized purposes such as:

TIP The project manager must explain to the dashboard users what is and is not a leading indicator, and how the metrics should be interpreted.

TIP Although KPIs reflect controllable factors, not all unfavorable situations can be completely corrected. Stakeholders must be made aware of this fact.

- Construction
- Maintenance
- Risk management
- Safety
- Quality
- Sales
- Marketing
- IT
- Supply chain management
- Nonprofit organizations

The fastest growing area for research in metrics appears to be IT. The risk of project failure seems to grow exponentially with the increase in the size of the IT project. Capers Jones, in his book *Assessment and Control of Software Risks* (Yourdon Press, 1994) stated that the majority of project failures in IT seem to be the result of inaccurate metrics and inadequate measurement techniques. More people today, this author included, agree with his observations. Sometimes, the only difference between a complete success and a complete failure is whether or not we correctly recognize the early warning signs in the metrics.

In IT, commonly used KPIs include:

- **Code:** Number of lines of code
- **Language understandability:** Language and/or code is easy to understand and read
- **Movability/immovability:** The ease by which information can be moved
- **Complexity:** Loops, conditional statements, etc. . .
- **Math complexity:** Time and money needed to execute algorithms
- **Input/output understandability:** How difficult it is to understand the program
- **Antivirus and spyware:** Percent of the systems with the latest updates
- **Repairs:** Mean time to make system repairs

Once the stakeholders understand the need for correct KPIs, other questions must be discussed, including:

- How many KPIs are needed?
- How often should they be measured?
- What should be measured?
- How complex will the KPI become?
- Who will be accountable for the KPI (i.e., the KPI's owner)?
- Will the KPI serve as a benchmark?

Simply because stakeholders are interested in metrics/KPIs and dashboards does not mean that they will understand what they are viewing. It is imperative that stakeholders understand the information on the dashboards and draw the correct conclusions. The risk is that stakeholders may not understand the metric, draw the wrong conclusion, and lose faith in the metrics concept. Getting back lost faith in a concept may take a great deal of time and become costly. Therefore, during the first few months of using metrics and dashboards, it is imperative that the project manager periodically debriefs each stakeholder to make sure they understand what they are viewing and that they are arriving at right conclusions.

We stated previously that what gets measured gets done and that it is through measurement that a true understanding of the information is obtained. If the goal of a metric measurement system is to improve efficiency and effectiveness, then the KPI must reflect controllable factors. There is no point in measuring an activity or a KPI if the users cannot change the outcome. Such KPIs would not be acceptable to stakeholders.

Working with stakeholders is challenging. There are complexities that must be overcome such as:

- Getting stakeholders to agree on the KPIs maybe difficult even if the stakeholders understand KPIs and possess a reasonable level of maturity in project management.
- Before agreeing to provide KPI data to a stakeholder, we must determine if the KPI data is in the system or needs to be collected.
- We must determine the cost, complexity, and timing for obtaining the data.
- We may have to consider the risks of information system changes and/or obsolescence in some of the organizational process assets that can affect the KPI data collection over the life of the project.
- We must consider that some KPIs may not appear until well into the project and that, over time, the stakeholders may request that additional KPIs be included in the system.

There are two other critical issues that project managers need to consider. First, if the project manager maintains multiple information systems, a measurement can appear and be treated as a KPI in one information system but be recognized as a simple metric in another. As an example, maintaining the project's profit margins might be a simple metric in the project manager's information system but a KPI in the corporate information system. With regard to stakeholders, this metric would not be provided to them.

TIP KPIs in one industry may not be transferable to another industry. Even in the same industry, KPIs may be used differently in each company.

TIP KPI measurement techniques must be explained to the stakeholders to get their buy-in and approval.

The second issue involves the contractors you hire. If you hire consultants and contractors to assist you in the management of the project, they may bring with them their own project management methodology, metrics and KPIs. You must make sure that the information they report to you is compatible with your business needs, especially if this information will be presented to the stakeholders as well. The contractor's definition of a KPI may not be the same as your definition.

4.2 USING THE KPIs

Although most companies use metrics and perform measurements, they seem to have a poor understanding of what constitutes a KPI for projects and how they should be used. Some general principles include:

- KPIs are agreed to beforehand and reflect the critical success factors on the project.
- KPIs indicate how much progress has been made toward the achievement of the project's targets, goals, and objectives.
- KPIs are not performance targets.
- The ultimate purposes of a KPI are the measurement of items directly relevant to performance and the provision of information on controllable factors appropriate for decision making that will lead to positive outcomes.
- Good KPIs drive change but do not prescribe a course of action. They indicate how close you are to a target but do not tell you what must be done to correct deviations from the target.
- KPIs assist in the establishing of objectives to be targeted with the ultimate purpose of either adding value to the project or achieving the prescribed value.
- KPIs force us to look at the future, whereas metrics alone may allow us to get bogged down looking at history.

Some people argue that the high-level purposes of a KPI are to encourage effective measurement. In this regard, the three high-level purposes are:

- Measurements that leads to motivation of the team
- Measurements that leads to compliance with use of organizational process assets and alignment to business objectives
- Measurements that lead to performance improvements and the capturing of lessons learned and best practices

Some companies post KPI information on bulletin boards, in the company cafeteria, on the walls of conference rooms, or in company newsletters as a means of motivating the organization by showing progress toward that target. However, unfavorable KPIs can have an adverse effect on morale.

4.3 THE ANATOMY OF A KPI

Some metrics, such as project profitability, can tell us if things look good or bad, but do not necessarily provide meaningful information on what we must do to improve performance. Therefore, a typical KPI must do more than just function as a metric. If we dissect the KPIs, we will see the following:

- **KEY** = A major contributor to the success or failure of the project. A KPI metric is therefore only "key" when it can make or break the project.
- **PERFORMANCE** = A metric that can be measured, quantified, adjusted and controlled. The metric must be controllable to improve performance.
- **INDICATOR** = Reasonable representation of present and future performance.

A KPI is part of a measurable objective. Defining and selecting the KPIs are much easier if you define the critical success factors first. KPIs should not be confused with CSFs. CSFs are things that must be in place to achieve an objective. A KPI is not a CSF but may provide a leading indication that the CSF can be met.

Selecting the right KPIs and the right number of KPIs will:

- Allow for better decision making
- Improve performance on the project
- Help identify problem areas faster
- Improve customer–contractor–stakeholder relations

David Parmenter defines three categories of metrics:[2]

- **Results indicators (RIs):** What have we accomplished?
- **Performance indicators (PIs):** What must we do to increase or meet performance?
- **Key performance indicators (KPIs):** What are the critical performance indicators that can drastically increase performance or accomplishment of the objectives?

TIP KPIs can change over the life of a project, but CSFs usually remain the same. Changing CSFs in midstream can be devastating.

Most companies use an inappropriate mix of these three and label them as KPIs. Having too many KPIs can slow down projects because of excessive measurements and reporting requirements. Too many can also blur one's vision of actual performance. Too few can likewise cause delays because of the lack of critical information. Typically, we end up with too many rather than too few KPIs.

2. David Parmenter, *Key Performance Indicators*, Hoboken, NJ: John Wiley & Sons, 2007, p. 1.

The number of KPIs can vary from project to project, and they may be affected by the number of stakeholders. Some people select the number of KPIs based upon the Pareto principle, which states that 20 percent of the total indicators will have an impact on 80 percent of the project. David Parmenter states that the 10/80/10 rule is usually applied when selecting the number of KPIs:[3]

- **RIs:** 10
- **PIs:** 80
- **KPIs:** 10

Typically, between six and ten KPIs are standard. Factors influencing the number of KPIs include:

- The number of information systems that the project manager uses (i.e., 1, 2, or 3)
- The number of stakeholders and their reporting requirements
- The ability to measure the information
- The organizational process assets available to collect the information
- The cost of measurement and collection
- Dashboard reporting limitations

4.4 KPI CHARACTERISTICS

The literature abounds with articles defining the characteristics of metrics and KPIs. All too often, authors use the "SMART" rule as a means of identifying the characteristics:

- **S** = Specific: The KPI is clear and focused toward performance targets or a business purpose.
- **M** = Measurable: The KPI can be expressed quantitatively.
- **A** = Attainable: The targets are reasonable and achievable.
- **R** = Realistic or relevant: The KPI is directly pertinent to the work done on the project.
- **T** = Time-Based: The KPI is measurable within a given time period

TIP Try to educate the stakeholders that there are limits to the number of KPIs that will be reported. This should happen prior to the actual selection process.

The SMART rule was originally developed for establishing meaningful objectives for projects and later adapted to the identification of metrics and KPIs. While the use of the SMART rule does have some merit, its applicability to KPIs is questionable because the relationship to CSFs is limited.

3. Ibid, p. 9.

TABLE 4-1 Twelve Characteristics of Effective KPIs

1. **Aligned**. KPIs are always aligned with corporate strategy and objectives.
2. **Owned**. Every KPI is "owned" by an individual or group on the business side who is accountable for its outcome.
3. **Predictive**. KPIs measure drivers of business value. Thus, they are "leading" indicators of performance desired by the organization.
4. **Actionable**. KPIs are populated with timely, actionable data so users can intervene to improve performance before it is too late.
5. **Few in number**. KPIs should focus users on a few high-value tasks, not scatter their attention and energy on too many things.
6. **Easy to understand**. KPIs should be straightforward and easy to understand, not based on complex indexes that users do not know how to influence directly.
7. **Balanced and linked**. KPIs should balance and reinforce each other, not undermine each other and suboptimize processes.
8. **Trigger changes**. The act of measuring a KPI should trigger a chain reaction of positive changes in the organization, especially when it is monitored by the CEO.
9. **Standardized**. KPIs are based on standard definitions, rules, and calculations so they can be integrated across dashboards throughout the organization.
10. **Context driven**. KPIs put performance in context by applying targets and thresholds to performance so users can gauge their progress over time.
11. **Reinforced with incentives**. Organizations can magnify the impact of KPIs by attaching compensation or incentives to them. However, they should do this cautiously, applying incentives only to well-understood and stable KPIs.
12. **Relevant**. KPIs gradually lose their impact over time, so they must be periodically reviewed and refreshed.

The most important attribute of a KPI may be that it is actionable. If the trend of the metric is unfavorable, then the users should know what action is necessary to correct the unfavorable trend. The user must be able to control the outcome. This is a weakness when using the SMART rule to select KPIs.

Wayne Eckerson has developed a more sophisticated set of characteristics for KPIs. The list is more appropriate for business-oriented KPIs than project-oriented KPIs, but it can be adapted for project management usage. Table 4–1 shows Eckerson's Twelve Characteristics.[4]

Eckerson then goes on to explain the characteristics in Table 4–1:[5]

Accountability

An actionable KPI implies that an individual or group exists that "owns" the KPI is held accountable for its results, and knows what to do when performance declines. Without accountability, measures are meaningless. Thus, it is critical to assign a single business owner to each KPI and make it part of

4. Wayne W. Eckerson, *Performance Dashboards: Measuring, Monitoring and Managing Your Business*, Hoboken, NJ: John Wiley & Sons, 2006, p. 201.
5. Ibid., adapted from pp. 201–204.

his or her job description and performance review. It is also important to train users to interpret the KPIs and how to respond. Often, this training is best done "on the job" by having veterans transfer their knowledge to newcomers. Some companies attach incentives to metrics, which always underscores the importance of the metric in the minds of individuals. However, just publishing performance scores among peer groups is enough to get most people's competitive juices flowing. It is best to assign accountability to an individual or small group rather than a large group, in which the sense of ownership and accountability for the metric become diffused.

Empowered

Companies also need to empower individuals to act on the information in a performance dashboard. This seems obvious, but many organizations that deploy performance dashboards hamstring workers by circumscribing the actions they can take to meet goals. Companies with hierarchical cultures often have difficulty here, especially when dealing with front-line workers whose actions they have historically scripted. Performance dashboards require companies to replace scripts with guidelines that give users more leeway to make the right decisions.

Timely

Actionable KPIs require right-time data. The KPI must be updated frequently enough so the responsible individual or group can intervene to improve performance before it is too late. Operational dashboards usually do this by default, but many tactical and strategic dashboards do not. Many of these latter systems contain only lagging indicators of performance and are only updated weekly or monthly. These types of performance management systems are merely electronic versions of monthly operational review meetings, not powerful tools of organizational change.

Some people argue that executives do not need actionable information because they primarily make strategic decisions for which monthly updates are good enough. However, the most powerful change agent in an organization is a top executive armed with an actionable KPI.

Trigger Points

Effective KPIs sit at the nexus of multiple interrelated processes that drive the organization. When activated, these KPIs create a ripple effect throughout the organization and produce stunning gains in performance.

For instance, late planes affect many core metrics and processes at airlines. Costs increase because airlines have to accommodate passengers who miss connecting flights; customer satisfaction declines because customers dislike missing flights, worker morale slips because they have to deal with unruly customers, and supplier relationships are strained because missed flights disrupt service schedules and lowers quality.

When an executive focuses on a single, powerful KPI, it creates a ripple effect throughout the organization and substantially changes the way an organization carries out its core operations. Managers and staff figure out ways to change business processes and behaviors so that they do not receive a career-limiting memo from the CEO.

Easy to Understand

KPIs must be understandable. Employees must know what is being measured, how it is being calculated, and, more, importantly, what they should do (and should not do) to affect the KPI positively. Complex KPIs that consist of indexes, rations, or multiple calculations are difficult to understand and, more importantly, not clearly actionable.

However, even with straightforward KPIs, many users struggle to understand what the KPIs really mean and how to respond appropriately. It is critical to train individuals whose performance is being tracked and follow up with regular reviews to ensure they understand what the KPIs mean and know the appropriate actions to take. This level of supervision also helps spot individuals who may be cheating the system by exploiting unforeseen loopholes.

It is also important to train people on the targets applied to metrics. For instance, is a high score good or bad? If the metric is customer loyalty, a high score is good, but if the metric is customer churn, a high score is bad. Sometimes a metric can have dual polarity, that is, a high score is good until a certain point and then it turns bad. For instance, a telemarketer who makes 20 calls per hour may be doing exceptionally well, but one who makes 30 calls per hour is cycling through clients too rapidly and possibly failing to establish good rapport with callers.

Accurate

It is difficult to create KPIs that accurately measure an activity. Sometimes, unforeseen variables influence measures. For example, a company may see a jump in worker productivity, but the increase is due more to an uptick in inflation than internal performance improvements. This is because the company calculates worker productivity by dividing revenues by the total number of workers it employs. Thus, a rise in the inflation rate artificially boosts revenues—the numerator in the metric—and increases the worker productivity score even though workers did not become more efficient during this period.

Also, it is easy to create metrics that do not accurately measure the intended objective. For example, many organizations struggle to find a metric to measure employee satisfaction or dissatisfaction. Some use surveys, but some employees do not answer the questions honestly. Others use absenteeism as a sign of dissatisfaction, but these numbers are skewed significantly by employees who miss work to attend a funeral, care for a sick family member, or stay home when daycare is unavailable. Some experts suggest that a better metric, although not a perfect one, might be the number of sick days since unhappy employees often take more sick days than satisfied employees.

Relevant

A KPI has a natural life cycle. When first introduced, the KPI energizes the work-force and performance improves. Over time, the KPI loses its impact and must be refreshed, revised, or discarded. Thus, it is imperative that organizations continually review KPI usage.

Performance dashboard teams should track KPI usage automatically, using system logs that capture the number of users and queries for each metric in the system. The team should then present this information to the performance dashboard steering committee, which needs to decide what to do about underused metrics.

Business or financial metrics are usually the results of many factors and it, therefore, may be difficult to isolate what must be done to implement change. For project-oriented KPIs, the follow six characteristics, which will be discussed in more depth in Section 4.6, may very well be sufficient:

- **Predictive:** The KPI is able to predict the future of this trend.
- **Measurable:** The KPI can be expressed quantitatively.
- **Actionable:** The KPI triggers changes that may be necessary for corrective action.
- **Relevant:** The KPI is directly related to the success or failure of the project.
- **Automated:** Reporting minimizes the chance of human error.
- **Few in number:** Only what is necessary.

All of these characteristics are not equal. It may be necessary to prioritize these characteristics. Aaron Hursman has written an interesting article entitled, "How Do You Spell KPI?"[6] The contents of the paper follow:

Seven Strategies for Selecting Relevant Key Performance Indicators

Conventional wisdom tells us a few things about establishing key performance indicators (KPIs). It goes something like this: Determine your corporate goals. Identify metrics to grade progress against those goals. Capture

6. The paper is reproduced with permission of Slalom Consulting and Aaron Hursman. Aaron Hursman, User Experience Lead for nGame, http://aaron.hursman.com.

actual data for those metrics. Jam metrics into scorecards. Jam scorecards down the throats of employees. Cross fingers. Hope for the best.

Remember the episode of *Undercover Boss* that aired after this year's Super Bowl? Waste Management COO, Larry O'Donnell, walked in the shoes of his employees for a few days (under the guise of an alternative identity). He discovered the effects his KPIs had on employees, first-hand. Specifically, a productivity and efficiency KPI convinced one of his "co-workers for a day" that she needed to urinate in a coffee cup to satisfy her production quota. As a truck operator, stopping to find and use the restroom adversely affected her performance grades. Therefore, she decided it was more efficient to use a coffee cup she kept with her in the vehicle. He later acknowledged that this was not exactly what he had in mind when he selected KPI.

Something strange happened here but not uncommon. Well-intentioned executives attempted to establish goals and track their progress. This is perfectly reasonable. In fact, the intent is downright reasonable. Unfortunately, the events that follow frequently turn into a twisted game of "telephone." Many would argue the cause for this scenario was a failure in communication. Maybe the communication plan was ineffective or maybe the organization was just incapable of supporting the specifications of the plan. Worse yet, maybe there was no plan at all.

Although a well-defined and executed communication plan is essential, that alone does not solve execution problems related to establishing KPIs. In reality, communication problems are merely friction. Although that friction can be strong enough to prevent an intended execution, reducing or clearing that friction alone does not guarantee success.

Effective KPIs share some core attributes. Many organizations have adopted a specific approach for establishing KPIs. It is called the SMART Criteria technique, and, in a nutshell, it requires that a KPI must satisfy these five criteria: (S)pecific, (M)easurable, (A)ttainable, (R)elevant, and (T)ime-bound. "S-M-A-R-T" is a fine way to spell KPI as this a solid framework for making decisions about KPIs. Unfortunately, organizations still find themselves unsatisfied with the results due to a misinterpretation of the term "relevant." Usually, this is narrowly defined as "relevant to *company* goals," but what about the individual? If KPIs only become effective when individuals throughout the organization are aware of them and working toward improving them, they will only achieve widespread adoption when the metrics are made relevant to the *individual*. Without relevancy, organizations are left to bet on communication alone to convince, persuade, and cajole others into acceptance.

Putting the R in KPI

By making KPIs individually relevant, you can begin to reach individuals capable of having a positive impact on those KPIs; keeping them motivated to perform well against specific metrics. Fortunately, the journey to pervasive adoption is straightforward. Leverage these s simple strategies to put the (R)elevancy back into your KPIs.

1. **Identify target audiences:** How can we select KPIs that are meaningful to others if we know nothing about these people are or even who they are? It is especially important to identify teams and individuals across the organization that [have] the ability to directly affect the health of the business. These are usually not the leaders and strategists, but delivery folks executing on and managing the front lines. To gain initial momentum, it can be helpful to first identify specific individuals and then extrapolate this list into cross-functional audience types.
2. **Ethnography:** Take a holistic approach to studying your people – observing them in their actual work environment to better appreciate their needs, motivations, goals, desires, constraints and obstacles. Use research methods like participant observation and contextual inquiry to gain these insights. If these methods are not feasible, interviews and questionnaires can suffice. Focus your research on answer questions like: Are they driven by financial, intellectual, and/or emotional goals? Are they motivated by fear? This information can then be used to establish tangible personas that synthesize these attributes. Personas can serve as powerful communication tools and grounding mechanisms that aid critical business decisions like selecting KPIs.
3. **Identify business rhythms:** People and businesses have their schedules and routines. Once key individuals and teams have been identified, determine the patterns and frequency of their activities. The SMART Criteria tell us that good KPIs are also (T)ime-bound, so select metrics that align with these business rhythms.
4. **Perform affinity diagramming:** An important part of selecting KPIs is understanding where individual goals and activities are not aligned with corporate goals and strategy. The prework necessary for this sort of gap analysis exercise can be accomplished through Affinity Diagramming. Affinity Diagramming (also known as the KJ Method) is an effective technique for efficiently making sense of large quantities of qualitative data and unstructured content, and is even more effective when executed as a team. Just write any extracted insights (focus on individual motivations, goals, and activities) about personas on sticky notes. Then, group similar sticky notes together into a number of physical groups/piles. Next, create groups of groups if possible. Finally, label the groups (both the original groups and the new super groups) with meaningful 1–3 word phrases. Capture this information electronically, preferably arranged in a spreadsheet file. Place the super-group labels across the top of the spreadsheet in the first row. Place the group labels in separate columns in the second row under the super-group headings. Transfer the words from the sticky notes to cells under the coordinating group label. Repeat this exercise, but instead create sticky notes that describe the corporate goals, strategy,

and initiatives. Also, reuse the same group and super-group labels that were just created instead of creating new labels. Affinity Diagramming for both scenarios is critical to the next strategy, gap analysis.

5. **Conduct gap analysis:** To move forward, it is crucial that we understand the current state. Misalignment between company and personal goals can impact effective KPI selections. Use a gap analysis to uncover any misalignments. Another organization technique, Mental Modeling, builds upon the Affinity Diagramming strategy and clearly identifies gaps with to a visual representation of its inputs. To build the Mental Model, merge the results of the two Affinity Diagramming scenarios (individual vs. corporate) by copying the results of corporate-focused scenario and pasting them below the results of the first scenario. Since the labels were reused for the second scenario, the data should align. Now, although the data points align, the values indicate visually where the individual and the company deviate. Each group for each scenario creates a virtual tower of varying height. Analyze the results of the illustration simply by identifying where the tower sizes are relatively and significantly unequal. The Mental Model clearly illustrates when the individual is focused and/or motivated to affect tasks/metrics that are not consistent with corporate strategy (and vice versa). Knowing this information is extremely advantageous, as it gives your organization a blueprint of areas to (1) address from a business process/organization standpoint or (2) consider and/or target when selecting KPIs.
6. **Consider the domain of control:** Select KPIs that fall within the actionable domain of these very key personnel. For example, a large retail client once described a series of periodic reports that were packed with pages of metrics, to which the store managers were held accountable. They were affectionately labeled as "Worry Reports," because the reports contained too many metrics that the managers had no ability to influence. This is where knowing the intimate work–life details for these individuals [is] crucial for selecting the right KPIs. These KPIs should be easy to calculate, clearly defined, and focused in purpose. Also, they tell should tell a very "rich" story in that they take into consideration a comparable entity (versus budget, forecast, last year, variance to average, etc.). Selecting KPIs using this criteria increases clarity, focus, determination, and motivation in the individual.
7. **Compensation alignment:** This is the ultimate strategy to make KPIs relevant for an individual. This one is very simple but effective. Identify metrics that are tied to the compensation (bonus or base) for an individual. If those metrics are not aligned with corporate goals or strategy, assess and adjust the compensation model as necessary. Ultimately, it can be very difficult to consistently motivate individuals to work to improve KPIs when they are not rewarded for doing so, even if the improvement directly benefits the health of the benefits. In the eyes of the individual, they perceive any such benefit to be extremely indirect, if any.

8. **Follow up:** Executing the previous strategies has the beneficial side effect of a creating a personal relationship. Seeking out someone's perspective does wonders for initiative adoption programs, especially at the delivery levels of the organization. Take advantage of that momentum and follow up with these individuals on a recurring basis to continue to fortify those relationships and strengthen the purpose behind establishing KPIs in the first place.

Take First Prize

Engage. Understand. Empathize. Show Compassion. Achieve corporate and individual alignment by selecting KPIs that are personally relevant. Next time you begin selecting KPIs, remember to spell KPI with a capital R. You may not win any spelling bees, but you will be better positioned to effectively monitor and improve business performance.

4.5 CATEGORIES OF KPIs

KPIs can be segmented or clustered per industry. They can also be reported as a group. This is common for business or financial KPIs. Project-based metrics are treated differently because of their inherent differences from financial KPIs, as shown previously in Table 3–1. Unlike financial metrics used for the Balanced Scorecard, project-based metrics can change during each life cycle phase as well as from project to project. Project-based metrics may be highly specific for each project, even in similar industries, and reported individually rather than as a group. Not all KPIs can be grouped. As an example, the KPIs shown below are not easily grouped.

- Percent of work packages adhering to the schedule
- Percent of work packages adhering to the budget
- Number of assigned resources versus planned resources
- Percent of actual versus planned baselines completed to date
- Percent of actual versus planned best practices used
- Project complexity factor
- Customer satisfaction ratings
- Number of critical assumptions made
- Percent of critical assumptions that have changed
- Number of cost revisions
- Number of schedule revisions
- Number of scope change review meetings
- Number of critical constraints
- Percent of work packages with a critical risk designation
- Net operating margins

Sometimes KPIs are categorized according to what they are intended to indicate, similar to the metrics categories discussed in Chapter 3:

- **Quantitative KPIs:** Numerical values
- **Practical KPIs:** Interfacing with company processes
- **Directional KPIs:** Getting better or worse
- **Actionable KPIs:** Effect change
- **Financial KPIs:** Performance measurements

Another means of classification might be leading, lagging, or diagnostic indicators or KPIs:

- Lagging KPIs measure past performance.
- Diagnostic KPIs measure current performance.
- Leading KPIs measure drivers for future performance.

Most dashboards have a combination of leading, diagnostic and lagging metrics.

4.6 KPI SELECTION

Identifying KPIs or even establishing a KPI library is easy, but selecting the right KPIs can be difficult. Sometimes we select a KPI that at first appears to be the perfect metric. Later we find out that it is actually a terrible measurement and leads to faulty conclusions by the stakeholders.

KPIs should provide some meaningful information for the following four questions usually asked by executives and stakeholders:

- Where are we today?
- Where will we end up?
- Where were we supposed to end up?
- If necessary, how can we get there in a cost-effective manner without any degradation in the quality of the deliverables or major scope changes?

KPIs are used for information dissemination and must be compatible with dashboard requirements. Some critical factors that can influence the selection process are:

- Size of the dashboard
- Number of dashboards
- Number of KPIs
- Type of audience
- Audience requirements
- Audience project management maturity level

Not all team members may understand the need for KPIs. This is particularly true when using virtual teams that are unfamiliar with KPI measurement practices. Understanding the importance of a KPI is an essential part of the selection process:

- Many things are measurable but not key to the project's success. KPIs are key metrics rather than merely metrics.
- It is important that the number of KPIs be limited so that everyone is focused on the same KPIs and understands them.
- Too many KPIs may distract the project team from what is really important.
- Good metrics are essential for tracking performance toward goals. Poor or inaccurate metrics and indicators lead to bad management decisions.

Without a good understanding of a KPI, we may end up with an improper selection process that works as follows:

- We identify everything that is easy to measure and count.
- We then develop sophisticated dashboards and reporting techniques for everything easy to measure and count.
- We then struggle trying to determine what to do with the information given its questionable use.

Sometimes the selection process may be hindered by factors beyond the control of the project manager. Reasons for this might happen if the project:

- Was bid on at a loss for political reasons
- Was bid on at a loss with the hope of winning future contracts
- Had its estimated budget slashed by management to win the contract
- Had a statement of work that was ill-defined
- Had a statement of work that was highly optimistic

TIP Selecting KPIs is easy. Selecting the right KPIs is difficult.

TIP Anything can be measured but perfect measurements may be unrealistic. Therefore, it may be impossible to select a perfect set of KPIs.

TIP Buy-in by the stakeholders and the team is significantly more important than trying to select the perfect set of KPIs.

The nature of the project together with the agreed-upon definition of success and the CSFs determine which KPIs to use. Given the potential number of stakeholders, as shown in Figure 4–1, problems can occur if each stakeholder has different needs:

- It may be difficult to get customer and stakeholder agreement on the KPIs.
- We must determine if the KPI data is in the system or needs to be collected.
- We must determine the cost, complexity, and timing for obtaining the data.
- We may have to consider the risks of information system changes and/or obsolescence that can have an impact on KPI data collection over the life of the project.

Figure 4-1 Typical Stakeholder Classification System

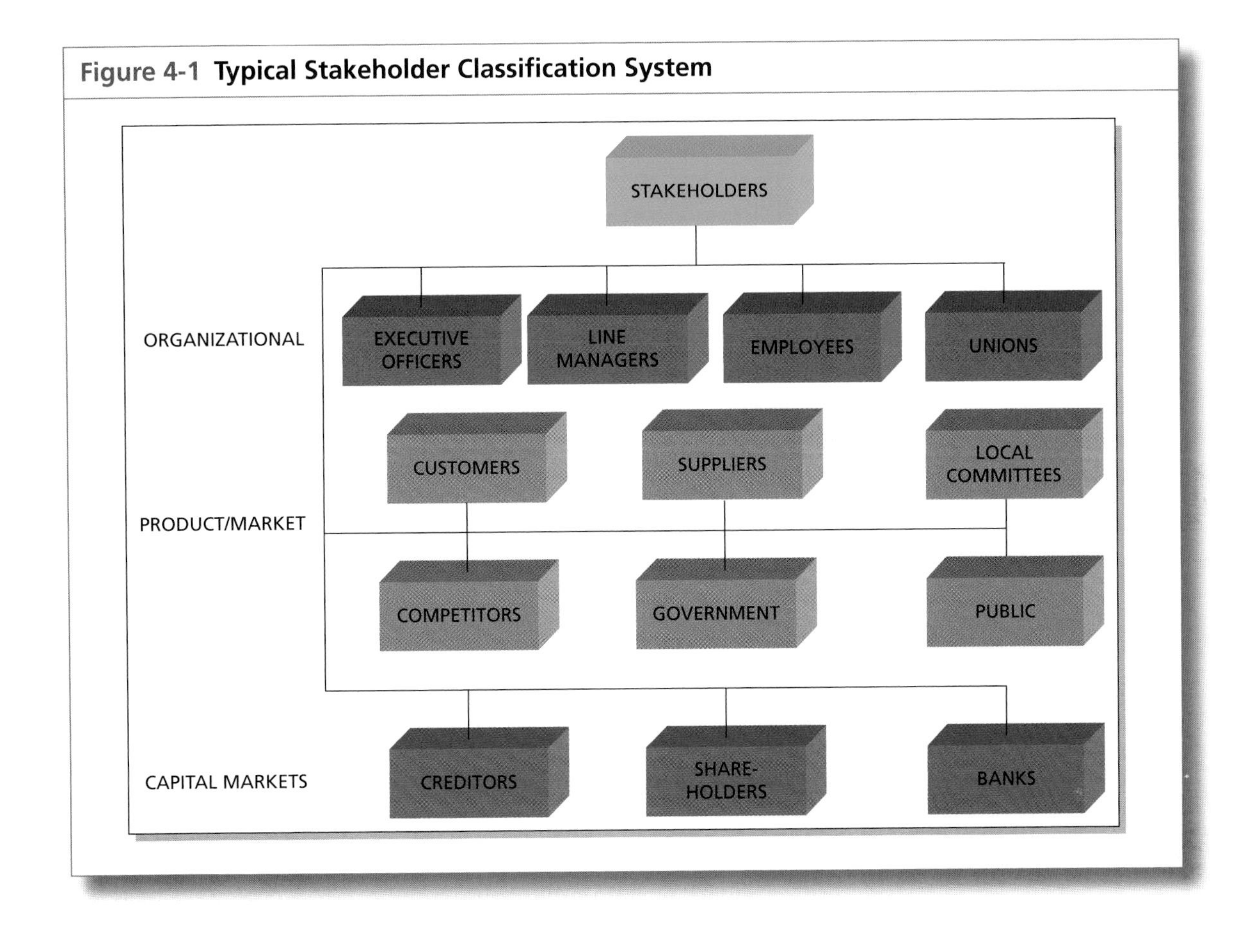

TIP Not all stakeholders will consider each metric to be a KPI.

In a project context, any single metric can be selected as a KPI for a given project because of the relative importance of that KPI to the project manager, client, or stakeholder. For example, the following four metrics can be viewed as KPIs according to who is doing the viewing:

- Project team morale
- Customer satisfaction
- Project profitability
- Performance trends such as CPI and SPI

It is possible that a given metric will function as a KPI for one stakeholder but serve as just a simple metric for another. As an example, look at Table 4.2.

The columns in Table 4–2 reflect five of the six criteria for a KPI as was discussed previously, namely predictive, quantifiable, actionable, relevant, and automated. The "yes" entries in Table 4–2 are subjective entries made by the project manager and possibly the team. As stated previously, the "yes" entries can vary from project to project and for each stakeholder.

The "yes" entries in the table are metrics that may have some of the characteristics of a KPI, but perhaps not all of the characteristics. For example:

- The number of milestones missed may not be actionable because the project manager may not be able to control this.
- Likewise, the same holds true for the number or percent of work packages on budget, and it is unlikely that this can be used as a predictive tool for future work packages.
- Customer loyalty falls into all categories as long as all of the stakeholders are in agreement and a viable measurement approach is undertaken.
- SV and CV are reasonably good indicators of the present but not as good as CPI and SPI for predicting the future.
- Changes in the risk profile can vary for each project. For a project with a reasonably low risk, this metric may not be used at all.

TABLE 4-2 Converting a Metric to a KPI

METRIC	PREDICTIVE	QUANTIFIABLE	ACTIONABLE	RELEVANT	AUTOMATED
Number of unstaffed hours	Yes	Yes	Yes	Yes	Yes
Number or % of milestones missed		Yes		Yes	Yes
Management support hrs as % of total labor	Yes	Yes			Yes
% of work packages on budget		Yes		Yes	Yes
# of scope changes		Yes		Yes	Yes
Changes in the risk profile	Yes	Yes	Yes	Yes	Yes
# or % of assumptions that have changed		Yes		Yes	
Customer loyalty	Yes	Yes	Yes	Yes	Yes
Turnover of key personnel, # or %		Yes		Yes	
% of labor hrs on overtime		Yes	Yes		Yes
SV		Yes			Yes
CV		Yes			Yes
SPI	Yes	Yes	Yes	Yes	Yes
CPI	Yes	Yes	Yes	Yes	Yes

- Turnover of key personnel is certainly of interest to the project manager but may or may not be of interest to all of the stakeholders. This metric may function as a KPI but only for a selected number of stakeholders.

Therefore, using the criteria for a KPI stated previously, only five of the fourteen metrics would be treated as true KPIs and would appear in the dashboard. The other metrics may still be reported but not necessarily through a dashboard reporting system.

Unlike business KPIs, which may remain the same for years, project-based metrics and KPIs can change for a variety of reasons and can have a short life expectancy. Metrics may be treated as KPIs at various stages of a project and replace certain KPIs that may no longer be needed or are treated as simple metrics for the remainder of the project. When a crisis occurs, the shifting of a metric to a KPI and back may happen. This can also happen when there are changes in the stakeholders.

Previously, we stated that the project manager may be working with three different information systems. Some of the 14 metrics in Table 4–2 may be treated as KPIs in only one of the information systems. This is shown in Table 4–3. All of the metrics, whether or not they are treated as a

TABLE 4-3 Possible Viewers for Each Metric

METRIC	PROJECT MANAGER	PROJECT SPONSOR	STAKEHOLDERS
Number of unstaffed hours	Yes	Yes	Yes
Number or % of milestones missed	Yes	Yes	Yes
Management support hrs as % of total labor	Yes	Yes	Yes
% of work packages on budget	Yes	Yes	Yes
# of scope changes	Yes		Yes
Changes in the risk profile	Yes	Yes	Yes
# or % of assumptions that have changed	Yes		Yes
Customer loyalty	Yes	Yes	Yes
Turnover of key personnel, # or %	Yes		
% of labor hrs on overtime	Yes		
SV	Yes		
CV	Yes		
SPI	Yes	Yes	Yes
CPI	Yes	Yes	Yes

KPI, are of interest to the project manager. Sponsors and stakeholders may be selective in the metrics that they wish to see in their information system.

Once the KPIs are selected, certain team members must accept ownership for the KPIs they use. Depending upon the number of KPIs, it may be advisable to assign a KPI owner to each KPI. However, certain KPIs, such as customer satisfaction, may not be able to be assigned to a single KPI owner.

Many companies today are maintaining KPI libraries. The KPI libraries must take into account the fact that KPIs must evolve over time. If a KPI library exists, then we must ask:

- Should there be a single owner in the organization for each KPI?
- Who, in addition to the KPI's owner, should attend the KPI review meetings?

4.7 KPI MEASUREMENT

KPIs serve no real value if they cannot be measured with any "reasonable" degree of accuracy. As stated by Warren Buffett, "It is better to be approximately right than to be precisely wrong." Anything can be measured, but perfect measurements may be unrealistic. Therefore, it may be impossible to select a perfect set of KPIs. KPIs function as a rough guide rather than as a precise value.

The relatively slow growth in the acceptance of metrics management for projects can be directly related to a poor understanding of metric measurement. We always knew what to measure but not how to measure. Sometimes, we even allow the team to invent their own measurement techniques, and the result is usually chaos. In the past, some of the most importance metrics needed for effective governance were considered as intangibles. We then refused to look for ways to measure the intangibles, believing that they were immeasurable, and the results were the selection of the wrong projects, assigning the wrong resources, misleading status reporting and poor decision making. Refusing to measure intangibles can give us a false impression of the true value of the project.

Metrics are needed so that we can better understand and hopefully reduce the uncertainties involved in the project. The word "uncertainty" has slightly different meanings according to the field of study. In a project management environment, uncertainty is a state of having limited knowledge such that you may not be able to exactly describe the present or future health of the project. There can be several possible outcomes based upon how work is progressing. Effective use of metrics can provide a clearer picture of the project's health and reduce the number of possible outcomes. Associated with each outcome is a risk, and each risk can have a favorable or unfavorable consequence. Metrics cannot reduce the risks but instead provide the project team, the stakeholders and the governance group with sufficient information such that the right decisions can be made to reduce or mitigate the risks. Therefore, the better the measurement of the metrics, the more informed the decision makers will be to reduce the possible outcomes and the associated risks.

Measurement can be defined as the quantification of the uncertainties based upon some type of observation. Measurement can never totally eliminate uncertainly. The person doing the observation must find the proper balance between overconfidence and underconfidence when using the data, especially when reporting the measurements to the stakeholders. There can be a different interpretation in the minds of the stakeholders as to whether the glass is half full or half empty. Stating that the glass is half full may lead stakeholders to believe that the glass may eventually get full and the liquid level is rising. Stating that the glass is half empty may create the illusion that we have lost half of the liquid in the glass.

Even with sophisticated measurement techniques, uncertainties will still exist, and it is unrealistic to believe that we will be able to have perfect information and complete certainty for decision making. Even if techniques existed by which we could obtain perfect information, the cost of the measurement would probably be prohibitive. Reality forces us to live with partial information obtained in a cost-effective manner.

The organizational process assets must be capable of capturing the data necessary to make the measurement. Sometimes, the method needed to capture the data has to be developed as the project progresses. In this case, all efforts should be made to get a process in place as quickly as possible.

Douglas Hubbard believes that five questions should be asked before we establish KPIs for measurement:[7]

- What is the decision this [KPI] is supposed to support?
- What really is the thing being measured [by the KPI]?
- Why does this thing [and the KPI] matter to the decision being asked?
- What do you know about it now?
- What is the value to measuring it further?

Hubbard also identifies four useful measurement assumptions that should be considered when selecting KPIs:[8]

- Your problem [in selecting a KPI] is not as unique as you think.
- You have more data than you think.
- You need less data than you think.
- There is a useful measurement that is much simpler than you think.

Selecting the right KPIs is essential. On most projects, only a few KPIs are needed. Sometimes we seem to select too many KPIs and end up with some

7. Douglas W. Hubbard, *How To Measure Anything*; Hoboken, NJ: John Wiley and Sons, 2007, p.43.

8. Ibid, p.31.

KPIs that provide us with little or no information value, and the KPI ends up being unnecessary or useless in assisting us in making project decisions.

KPIs are generally defined beforehand but may have to evolve as the project progresses if there are no methods or processes in place to capture the required data initially. When this happens, the result is usually the measurement inversion impact on the KPI selection process:

- The KPI with the highest information value, especially for decision making, will be avoided or never measured because of the difficulty of data collection.
- KPIs like time and cost, which are the easiest to measure will be selected, and we often spend too much time on these variables, which may have the least impact on decision making and the project's final value.

Today, numerous measurement techniques exist. A typical list would include:

- Observations
- Ordinal (i.e., four or five stars) and nominal (i.e., male or female) data tables
- Ranges/sets of value
- Simulation
- Statistics
- Calibration estimates and confidence limits
- Decision models (EV, EVPI, etc.)
- Sampling techniques
- Decomposition techniques
- Human judgment
- Rules and formulas (i.e., 50/50 Rule, 80/20 Rule, 0/100 Rule, % complete, etc.)

Regardless of which measurement technique is selected, arguments will always exist over the sample size, timing of the measurements, duration of the measurements, accuracy and precision, who is best qualified to perform the measurements, and other such factors.

The results of KPI measurements can create conflict if employees find loopholes believing that the information will be:[9]

- Collected on individuals and used against them, (e.g., for disciplinary purposes)
- Controlled by management
- Filtered in both content and distribution (e.g., "They show us information only when it suits their purposes.")
- Used to allocate blame for performance problems

9. David Parmenter, *Key Performance Indicators*, Hoboken, NJ: John Wiley & Sons, 2007, p 64.

4.8 KPI INTERDEPENDENCIES

It is almost impossible to determine the status of a project from a single metric or KPI. As shown in Figure 4–2, metrics are interlocked or related. For example, let's assume that metric #1 is the quality or availability of critical resources, metric #2 is time and metric #3 is cost. Small changes in the number of qualified resources can have a significant effect on the project's budget and schedule. The effect can be favorable or unfavorable depending upon whether we add or remove qualified resources.

Another important factor is the rate of change of the metrics. This can be seen from the size of the metrics (i.e., wheels) in Figure 4–2. A small change in metric #1 will cause metric #2 to move faster just to keep up. Likewise, metric #3 must move faster than metric #1 and #2 to keep up. In other words, if we lack some critically skilled resources, then the schedule may slip quickly and overtime may be needed at once to produce the deliverables.

KPIs are a set of interrelated performance measures that are necessary to meet the project's critical success factors. Looking at the 14 metrics in Table 4–2, you may not be able to determine the actual cause of poor performance or the necessary action to correct the problem. It may be necessary to look at several interrelated metrics. As an example, consider the following two possibilities where a "+" sign is favorable and a "–" sign is unfavorable:

- SV = + and CV = –
- SV = – and CV = +

Figure 4-2 **Metrics Are Related**

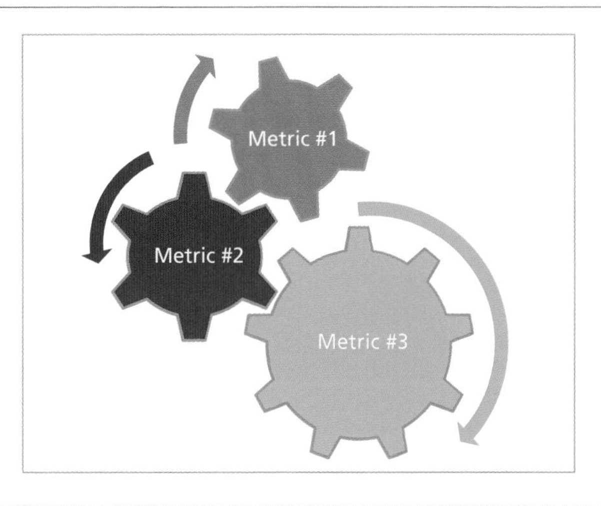

In the first bullet, you may have worked overtime, used higher salaried workers, or accelerated the schedule. In the second bullet, you may have insufficient resources on the project. In either case, it is hard to tell if performance is good or bad.

Now let's consider another situation:

- SV = −$100,000
- CV = −$250,000

It appears that the situation is bad, but if we look at two additional metrics, we may see a different picture:

- Number of approved scope changes = 34
- Turnover of critical skilled workers = 9

Combining all four of these KPIs we could argue that the project might not be in that much trouble at least for now.

As another example, consider what happens if we try to determine the status of the project from just one KPI, namely, CV:

- June: CV = −$10,000
- July: CV = −$20,000

It looks like the situation has gotten worse, since the unfavorable variance has doubled from −$10,000 to −$20,000 in one month. On the surface, this may look bad. But let's assume that in June, EV = $100,000 and in July, EV = $400,000.

If we convert the CV from dollars to percent using the formula:

$$CV(\%) = CV(\$)/EV(\$)$$

then CV(%) for June was −10%, but CV(%) for July is −5%. In other words, the situation has actually improved even though the magnitude of the variance has increased. It may take the integration of several KPIs to get an accurate picture of the project's real status.

4.9 KPIs AND TRAINING

Project managers and team members may need to attend training sessions on KPI identification, measurement, control, and reporting. Training sessions must include:

- A comprehensive understanding of KPIs
- How to identify KPIs

- How to select the right display for reporting each KPI
- How to design a project KPI database
- How to measure each KPI
- How to decide upon the necessary action, if appropriate, to correct performance
- How to update the corporate KPI library

The training should take place prior to the launch of the project. At the project's kick-off meeting, the team will then be briefed on:[10]

- Why KPIs are being introduced
- How KPIs will be developed
- How KPIs will be used
- What KPIs will not be used for

Care must be taken when launching a training course that the organization is ready for such a change. The organization must recognize:

- The value that it bring the organization
- The need for learning effecting measurement techniques
- The need for metrics to improve performance

If the company already has some form of metrics management in place, people may be willing to accept and support the training. Otherwise, the course may be seen as a threat and do more harm than good. Simply stated, you must know your organization's maturity level in project management and metrics management before beginning.

4.10 KPI TARGETS

Words such as "customer satisfaction" and "reputation" have no real use as metrics unless they can be measured with some precision. Therefore, we must establish KPI targets, thresholds, and baselines. Targets have the following properties:

- Targets represent a set of values against which measurements will be made.
- Targets must be realistic and not necessarily challenging. Otherwise, workers might try to circumvent the targets.
- Targets may require trial-and-error solutions.
- Targets must not be established in a vacuum.

10. Ibid, p. 129.

It must be understood that KPIs are not targets. KPIs represent how far an important metric is above or below a predefined target. Typical targets for a KPI might be:

- Simple quantitative targets
- Time-based targets: measurements made monthly or during a certain time interval
- At completion targets: measurements made at the completion of work packages or project completion
- Stretch targets: become best in class or a target which is greater than specification requirements
- Visionary targets well into the future: more repeat business from this client

Stretch targets and visionary targets are often a mix of possible and impossible outcomes, with the impossible outcomes providing encouragement. However, stretch targets can lead to the misinterpretation of information. For example, the customer might be happy with delivery of 10 units per month. However, in this case, the project manager sets up a stretch target of 12 units per month. If actual delivery is 10 units for the month, the metric may be misinterpreted as a sign of poor performance because we are below the established stretch target.

Examples of simple quantitative targets include:

- A single value (i.e., completion of 20 tests)
- An upper limit (i.e., ≤ $200,000)
- A lower limit (i.e., ≥ $100,000)
- A range of values (i.e., $400,000 ± 10%)
- A percentage of a specific quantity that may be fixed for the project (i.e., scrap is less than 5% of material costs)
- A percentage of a specific quantity that may change (i.e., planning dollars are less than 35% of total labor dollars)
- Accomplished milestones and deliverables (i.e., must produce and ship at least 10 deliverable each month)
- A percentage of a specific activity that may change over the duration of the project (i.e., planning dollars are not more than 35% of total labor dollars estimated)
- Accomplished milestones and deliverables (i.e., must produce and ship at least 10 deliverables each month)

Figure 4–3 represents a KPI target or boundary box. Normal performance is meeting the target ±10%. If you were more than 20% below the target, urgent attention would be required.

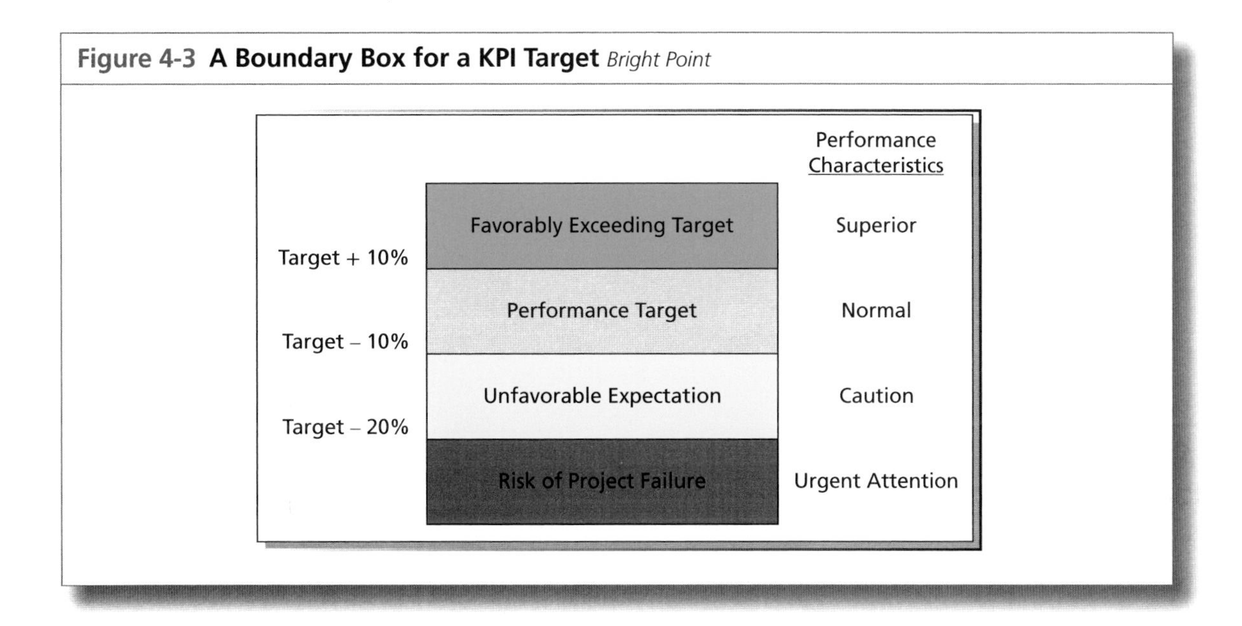

Figure 4-3 **A Boundary Box for a KPI Target** *Bright Point*

It is important to ask the stakeholders how much integrity is acceptable for reporting purposes. In Figure 4–3, the threshold or integrity of ±10% may result from a joint agreement with the client. Typical questions might be:

- Is the target ± 5% acceptable?
- Is the target ± 10% acceptable?
- Are integrity guidelines established in the project's business case?
- Are integrity guidelines established as part of the EPM system?

Some targets are very difficult to establish such as value targets. Establishing KPIs to identify present and future value is difficult but not impossible. We can select value-driven project KPIs by addressing the following questions:

- How can I show that the project is creating value for the client?
- How will the client and the stakeholders perceive the value measurements?
- Can I show that the project will also create value for my parent company?

This will be discussed in more detail in Chapter 5. Some people argue that customer satisfaction surveys are value-reflective KPIs. Figure 4–4 represents a simple customer satisfaction instrument. Mahindra Satyam refers to it as the Customer Delight Index. Companies use this with the belief that customer satisfaction may lead to repeat business.

Figure 4-4 **Mahindra Satyam Customer Delight Index**

SYMBOL	MEANING
○ (white circle)	Data not entered
● (dark circle)	Dissatisfied
△ (triangle)	Satisfied
■ (square)	Delighted

The colors are important, as will be discussed in Chapter 6. The color green represents something favorable, whereas red indicates something not so good.

When using these types of KPIs, it is the trend that is important rather than a single data point. If the trend shows that the customer satisfaction index is getting worse, then templates or checklist may exist to show what actions the project manager may take to change the trend to a more favorable indication. The problem is in determining who or which department has the lead role in the improvement of customer satisfaction.

4.11 KPI FAILURES

There are several reasons why the use of KPIs often fails on projects. Some of the reasons include:

- People believe that the tracking of a KPI ends at the first line manager level.
- The actions needed to regulate unfavorable indications are beyond the control of the employees doing the monitoring or tracking.
- The KPIs are not related to the actions or work of the employees doing the monitoring.
- The rate of change of the KPIs is too slow, thus making them unsuitable for managing the daily work of the employees.
- Actions needed to correct unfavorable KPIs take too long.

- Measurement of the KPIs does not provide enough meaning or data to make them useful.
- The company identifies too many KPIs, to the point where confusion reigns among the people doing the measurements.

Years ago, the only metrics that some companies used were those identified as part of the earned value measurement system. The metrics generally focused only on time and cost and neglected metrics related to business success as opposed to project success. Therefore, the measurement metrics were the same on each project and the same for each life cycle phase. Today, metrics can change from phase to phase and from project to project. The hard part is obviously deciding upon which metrics to use. Care must be taken that whatever metrics are established does not end up comparing apples and oranges. Fortunately, there are several good books in the marketplace that can assist in identifying proper or meaningful metrics.[11]

Selecting the right KPIs is critical. Since a KPI is a form of measurement, some people believe that KPIs should be assigned only to those elements that are tangible. Therefore, many intangible elements that should be tracked by KPIs never get looked at because someone believes that measurement is impossible. Anything can be measure regardless of what some people think. According to Douglas Hubbard:[12]

- Measurement is a set of observations that reduces uncertainty where the results are expressed as a quantity.
- A mere reduction, not necessarily elimination, of uncertainty will suffice for a measurement.

Therefore, KPIs can be established even for intangibles like those that will be discussed in Chapter 5 of this book.

4.12 KPIs AND INTELLECTUAL CAPITAL

The growth in information technology and the use of a PMO has made it apparent that metrics and KPIs are now being treated as intellectual capital. The curved arrows in Figure 4–5 represent the ten knowledge areas in the

11. Three books that provide examples of metric identification are Parviz F. Rad and Ginger Levin, *Metrics for Project Management*, Vienna, VA: Management Concepts, 2006; Mel Schnapper and Steven Rollins, *Value-Based Metrics for Improving Results*, Ft. Lauderdale, FL: J. Ross Publishing, 2006; and Douglas W. Hubbard, *How To Measure Anything*; Hoboken, NJ: John Wiley and Sons, 2007.

12. Douglas W. Hubbard, *How to Measure Anything*, Hoboken, NJ: John Wiley and Sons, 2007, p. 21.

Figure 4-5 The *PMBOK®* *Guide* and Metrics

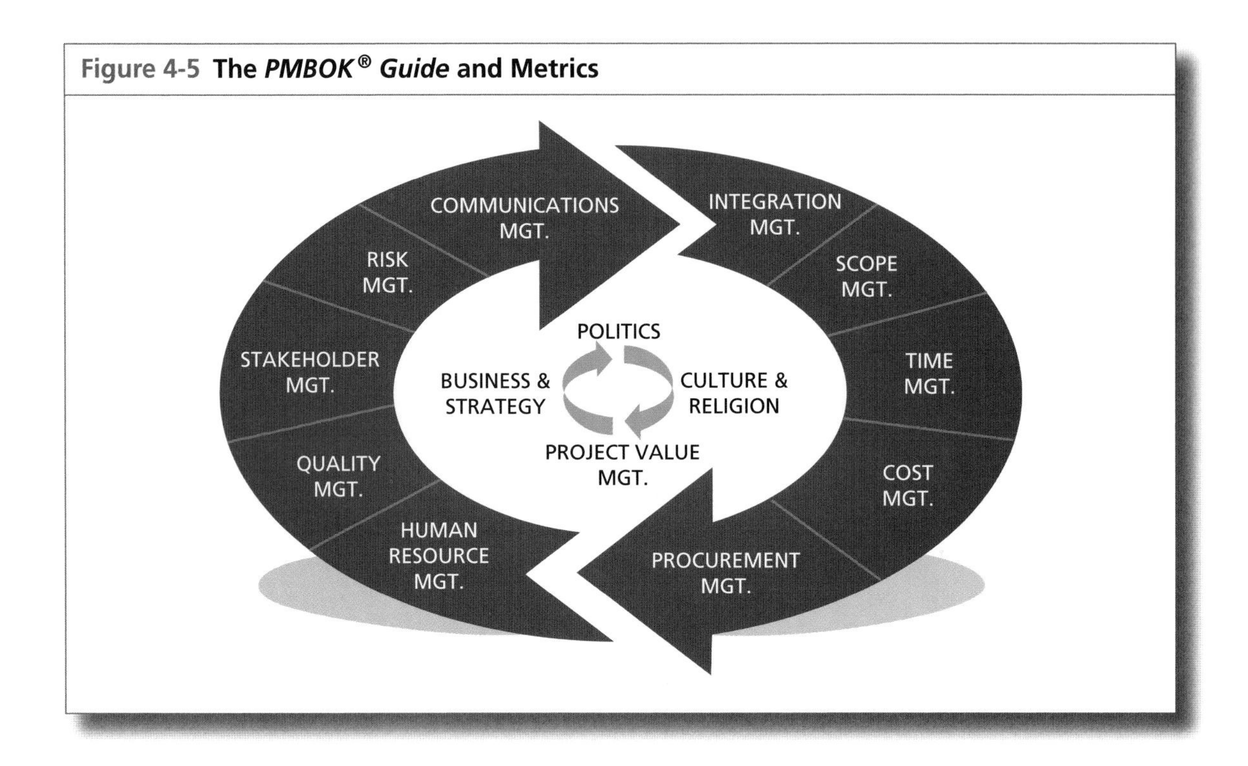

PMBOK® Guide. The reason why the knowledge areas are in the arrows and connected is because the knowledge areas are related and each metric and KPI can probably be related to more than one knowledge area. The subjects in the center of the figure represent additional knowledge that project managers must learn, especially for managing the more complex global projects. These center subjects also introduce addition metrics that are related to the nine knowledge areas.

Another source of intellectual capital can come from benchmarking. As can be seen from Figure 4–6, project management benchmarking activities can accelerate the rate of improvements to the project management processes. It is important that metric benchmarking focus on process effectiveness and process maturity considerations rather than the success of an individual project. KPIs are associated with performance improvement initiatives.

The web site kpilibrary.com has more than 5000 KPIs in their library. They also perform benchmarking surveys on KPIs using the following categories:

- Operational excellence
- Cost leadership
- Product and service differentiation
- Customer intimacy
- Other

Figure 4-6 Project Management Knowledge

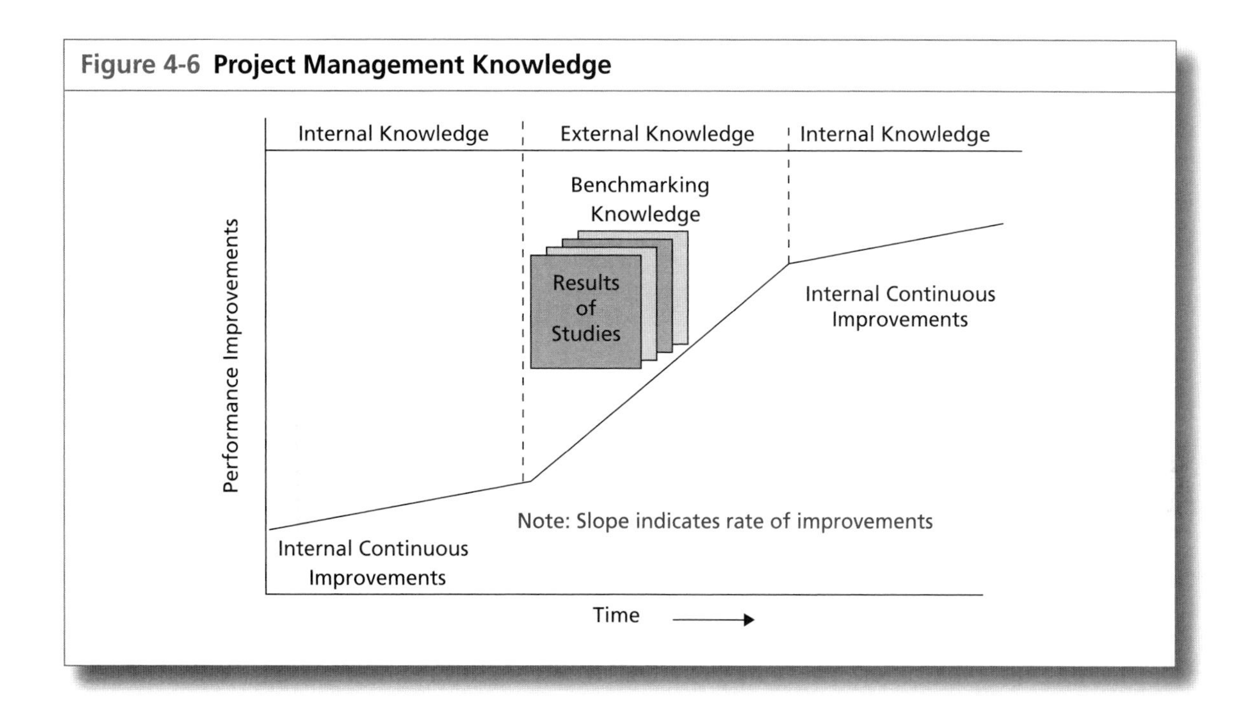

TABLE 4-4 Tracking Metrics for Continuous Improvement

KPI	BENCHMARK STUDIES	CURRENT KPI VALUE	TARGET VALUE
Timing	60 Days	55 Days	50 Days
Cost	$15,000	$14,000	$13,000
Pay Grade	Grade 7	Grade 7	Grade 6
Management Reserve	$100,000	$90,000	$80,000
Manpower	16 People	15 People	12 people
Quality	2 Defects per 3000 Units	2 Defects per 4000 Units	2 Defects per 5000 Units

Benchmarking studies can support continuous improvement efforts as shown in Table 4–4. The target values in Table 4–4 were set up as stretch targets. In Table 4–4, we are comparing the current value of the KPI against the benchmark and the stretch targets. If the current value of the KPI reaches or exceeds the stretch target, then we may be able to assume that continuous improvement has taken place and then raise the bar by inserting new stretch targets.

Figure 4-7 Components of Intellectual Capital

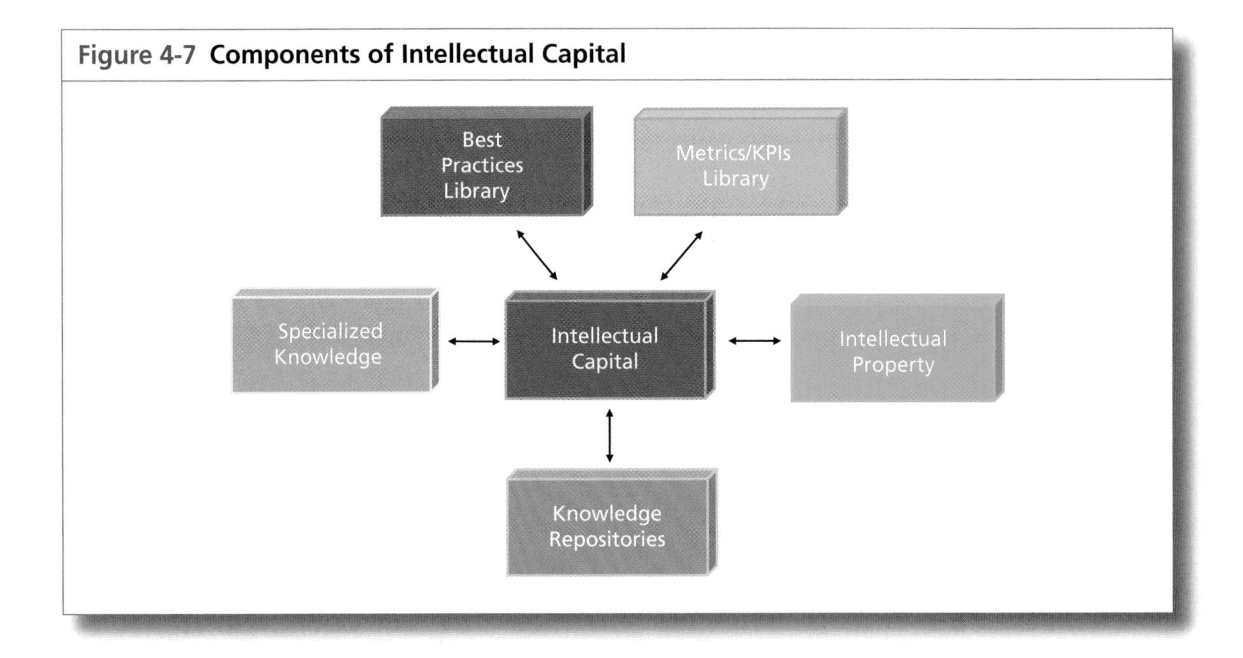

Figure 4–7 shows the components of intellectual capital. As expected, metric/KPI libraries are considered corporate intellectual capital. The PMO is the vehicle by which we convert the metrics and KPIs into intellectual capital and share it appropriately throughout the company.

4.13 KPI BAD HABITS[13]

Stacey Barr is a globally recognized performance measurement expert who has made a career of challenging many of the long-held beliefs and bad habits people have about how performance measures ought to be chosen, created, and used for organizational performance management. Stacey believes that people share similar struggles with performance measurement. They can't find meaningful performance measures, especially for goals that seem immeasurable. They can't get staff engaged in measuring and improving performance. They don't have measures that drive lasting performance improvement. Stacey once struggled with these challenges,

13. This section had been provided by Stacey Barr. ©2012 by Stacey Barr. Reproduced by permission. Additional information on Stacey Barr can be found in the following websites: www.staceybarr.com, www.measureupblog.com, www.performancemeasureblueprintonline.com, www.performancemeasureblueprint.com.

and this became her inspiration for sharing her metric/KPI knowledge with the world. Stacey's research and knowledge of mistakes made in business and strategy performance measurement systems are directly applicable to project management metric management systems.

The KPI Bad Habits Causing Your Performance Measurement Struggles

One of the things that Albert Einstein was famous for defining, over and above $E = MC^2$, was insanity: doing the same thing over and over again and expecting different results.

Clearly, if you want to stop struggling to find meaningful performance measures that align to strategy and engage people in improving performance, then you have to change what you're doing.

A few very fruitful changes you can make are to unlearn some limiting habits that you may not even realise are at the root of most performance measurement struggles.

Bad Habit 1: Using Weasel Words to Articulate Your Goals

The problem with strategy in most organizations and companies is that, in its sanitised and word-smithed published form, it's not measurable.

Look at any strategic plan and the chances are astronomically high that you'll see a glut of words like effective, efficient, productive, responsive, sustainable, engaged, quality, flexible, adaptable, well-being, reliable, key, capability, leverage, robust, and accountable.

They are empty words that sound important and fail to say to anything at all, or at least speak of anything that can be verified in the real world, or measured. It's no wonder people with goals or objectives like the following keep asking "how do you measure that?"

- "Provide efficient, unique, unbiased and responsive, high-quality support"
- "Strengthen student engagement and learning outcomes by enhancing student support and intervention services"

The new habit to replace the weasel-word habit with is to write goals in clear and simple language that evokes in the minds of people an accurate picture of successful achievement of the goal. If you can't see it in your mind, you won't be able to measure it.

Bad Habit 2: Brainstorming to Find KPIs or Performance Measures

For the most part, people are not that conscious or aware of the approach they take to find or choose performance measures. Brainstorming is the most common approach, however. People try to select performance measures for a particular critical success factor or key result area or objective or goal, or for their function or process, by asking a simple question: "So, what measures could we use?" Everyone sits around and randomly suggests potential performance measures for that particular critical success factor or key result area or objective or goal or for their function or process. They might produce a list that looks something like this if they were brainstorming measures for staff engagement:

- Turnover
- Sick days
- Retention Rate
- Introduction of talent management
- Overtime
- Staff Survey
- Engagement Index
- Staff satisfaction with their job
- Leadership development
- Performance management

Brainstorming can generate lots of ideas for measures quite rapidly, it's easy to do, no special knowledge or skill is required, it's familiar so it won't be distracting, and it can be very collaborative and engage people in being part of the measure selection process. But brainstorming rarely produces good measures.

The truth is you're not really finished after the brainstorming is over because you still have to work out how to get a final selection of measures from that long list. And in all honesty, voting or ranking the ideas and skimming the few that rise to the top is not the answer. Ideas for potential measures need to be vetted or tested to weed out the ideas on the list that are not measures at all ("introduction of talent management"), that aren't really relevant to the goal ("overtime"), and that are not feasible to implement.

Instead of habitually brainstorming to find performance measures, think about listing potential measures that are the best observable evidence of the successful achievement of your goal, and then choose the measures that are most relevant balanced with being feasible to implement.

Bad Habit 3: Getting People to Sign Off on Their Acceptance of the Measures

Performance measurement has such a terrible stigma. Many people associate it with the inane drudgery of data collection, with pointing fingers and with big sticks that come beating down on them when things go wrong. They associate it with the embarrassment of being compared with whoever is performing best this month. The emotions people typically feel about performance measurement are frustration, cynicism, defensiveness, anxiety, stress, and fear.

What we really want is for performance measurement to be seen as a natural and essential part of work. We want people to associate it with learning more about what works and what doesn't, with valuable feedback that keeps us on the right track, with continuous improvement of business success. We want people to feel curiosity, pride, confidence, anticipation, and excitement through using performance measures of performance results that matter. This is buy-in, not sign-off.

Buy-in is a natural product of showing people that measurement is about feedback, not judgment; of giving them tools that make measurement easy and fun; of allowing them to decide the measures most useful for their goals.

Of course, you need to stay sensitive to the fact that measurement of performance is an organization-wide system, and each team is only a part of that system. But the trade-off should be biased more toward their buy-in than it is toward sophistication of the measures. You can improve the sophistication of measures on a foundation of buy-in more easily than you can get buy-in to a suite of sophisticated measures.

Bad Habit 4: Assuming Everyone Knows the Right Way to Implement Your Measures

In general, a lot of effort is wasted in bringing performance measures to life. The waste is in the time spent to select measures that are never brought to life, or in the time spent bringing measures to life in the wrong way. Thus the labor of bringing many measures into the world is far more excessive and painful than it needs to be. This, as you no doubt have experienced, breeds cynicism, a feeling of being overwhelmed, and disdain for and disengagement from the process of measuring anything, let alone measuring what matters.

People argue about data or measure validity instead of making decisions about how to improve performance. Measures misinform and mislead decisions because of the wrong calculation or analysis being used. Too many conflicting versions of the same measure result from duplication, and lack of discipline in performance reporting processes and causes confusion and cynicism about the value of measures.

This was the case for Martin, a manager in a freight company, who was receiving 12 different versions of a measure of the cycle time of coal trains from a range of business analysts throughout his department. Because no two of these 12 different measures matched, he had no idea which one was the true and accurate measure of cycle time. Martin had 12 measures and no information.

Defining a performance measure means fleshing out the specifics of its calculation, presentation, interpretation, and ownership. This is the new habit to learn to avoid making assumptions that result in waste and misinformation in performance measurement.

Bad Habit 5: Using Performance Reports

If your performance reports are stacking up in a pile, unread and unused, then they're obviously not "stacking up" well as sources of invaluable insight to guide performance improvement.

It's an emotional thing, performance reporting. Executives give up the precious little time they have for their families and 9 holes of golf to instead paw through piles of strategic reports often more than an inch thick. Or they leave the pile of reports on their desk and make decisions from their gut instead.

Managers earnestly trawl through operational reports to check if anything needs a bit more positive light thrown on it. Supervisors and teams cynically scoff about the volumes of time and effort they waste reporting tables of statistics that track their daily activities to audiences they never see or hear from.

Performance reports need to provide the content that truly matters most, and provide that content so it is fast and easy for managers to digest. But most performance reports are just the opposite:

- They are thrown together in an ad hoc way, making it very hard to navigate to the information most relevant or urgent.
- They are cluttered and cumbersome with too much detail that drowns out the important signals with trivial and inactionable distractions.
- Their information is displayed poorly, using indigestible tables and silly graphs that are designed with entertainment in mind and unwittingly result in dangerous misinterpretation of the information.
- The layout is messy and unprofessional, wasting visual real estate, detracting from the report's importance, and disengaging users before they find the insights they can use.

The habit of using performance reports to justify our existence has to stop. Rather, we ought to get comfortable with making performance reports answer only three simple questions: What's performance doing? Why is it doing that? And what response, if any, should we take?

Bad Habit 6: Comparing This Month's Performance to Last Month's, to the Same Month's Last Year, and to Target

Managers at one of the major saw mills in a timber company thought their performance dashboard was the duck's guts. It tracked a multitude of various performance measures about how the saw mill operations were going, and the data update for many of these measures was almost live, so their dashboard could be updated very regularly.

Traffic lights—red, green, and yellow visual flags that summarize if performance is bad, good, or heading toward unacceptable—were also updated each time the data feed refreshed. The managers and supervisors would react to these traffic lights with unnecessarily large interventions, like changing the settings on timber processing equipment or altering work procedures.

These traffic lights were changing according to data, often from a small sample, like a day. Long-term trends and natural variability in the data were ignored. Rather than looking for signals in their data, managers and supervisors were reacting to the noise, reacting to any variation in the data at all.

Everyone was so busy reacting to data, the key elements that did need changing were left unattended; therefore, the overall performance worsened.

Performance will *always* vary up and down over time—you can safely assume that comparing any two points of performance data will always reveal a difference of *some* magnitude. Drawing a conclusion about whether performance has changed by comparing this month to last month is tantamount to making things up as you go along.

The insightful conclusions—which will lead you to act when you should and not act when you shouldn't—come from the patterns in your performance data. Insights do not come from comparisons between the points of your performance data. Unlearn that bad habit of making point-to-point comparisons and performance will improve.

Bad Habit 7: Treating Performance Measurement Separately from Planning, Reporting, and Strategy Execution

Performance measurement is a process that weaves through your existing management processes. It doesn't stand alone and apart from them.

The steps of selecting performance measures for goals or objectives need to be performed as part of the strategic and operational planning processes, or you end up with goals and objectives that are vague and immeasurable.

Reporting processes, including the business intelligence systems, data warehouses, and information dashboards that support them need to quickly refocus on the data and information for new performance measures and their cause analysis.

Strategy execution needs to be informed by the current performance measures and their targets, and the strategies themselves need to be changed when the measures show they aren't working.

Performance measurement isn't something you do after your strategic plan is cast in stone, and just in time for the annual review. It's not something we do for bureaucratic reasons. We do it because it provides the feedback that about how well we're achieving the endeavours we chose to pursue. If those endeavours are important enough to pursue, they are too important not to measure, and measure well.

4.14 BRIGHTPOINT CONSULTING, INC.—DASHBOARD DESIGN: KEY PERFORMANCE INDICATORS AND METRICS[14]

Introduction

This article will focus on collecting and defining metrics and key performance indicators for executive and operational dashboards. While the techniques discussed here can be used across many different business intelligence requirements gathering efforts, the focus will be collecting and organizing business data into a format for effective dashboard design.

With the explosion of dashboard tools and technologies in the business intelligence market, many people have different understandings of what a dashboard, metric, and key performance indicator (KPI) consist of. In an effort to create a common vocabulary for the scope of this article, we will define a set of terms that will form the basis of our discussion. While the definitions that follow might seem onerous and require a second pass to fully understand, once you have grasped the concepts you will have a powerful set of tools for creating dashboards with effective and meaningful metrics and KPIs.

Metrics and Key Performance Indicators

Metrics and KPIs are the building blocks of many dashboard visualizations; as they are the most effective means of alerting users as to where they are in relationship to their objectives. The definitions that follow form the basic building blocks for dashboard information design, and they build upon themselves, so it is important that you fully understand each definition and the concepts discussed before moving on to the next definition.

Metrics: When we use the term metric we are referring to a direct numerical *measure* that represents a piece of business data in the relationship

14. Material in this section has been taken from BrightPoint Consulting white paper, "Dashboard Design: Key Performance Indicators and Metrics" by Tom Gonzalez, managing director, BrightPoint Consulting, Inc., © 2005 by BrightPoint Consulting, Inc. Reproduced by permission. All rights reserved. Mr. Gonzalez is the founder and Managing Director of BrightPoint Consulting, Inc., serving as a consultant to both Fortune 500 companies and small to medium businesses alike. With over 20 years experience in developing business software applications, Mr. Gonzalez is a recognized expert in the fields of business intelligence and enterprise application integration within the Microsoft technology stack. BrightPoint Consulting, Inc. is a leading technology services firm that delivers corporate dashboard and business intelligence solutions to organizations across the world. BrightPoint Consulting leverages best-of-breed technologies in data visualization, business intelligence and application integration to deliver powerful dashboard and business performance solutions that allow executives and managers to monitor and manage their business with precision and agility. For further company information, visit BrightPoint's Web site at www.brightpointinc.com. To contact Mr. Gonzalez email him at tgonzalez@brightpointinc.com.

of one or more *dimensions*. An example is: "gross sales by week." In this case, the *measure* is dollars (gross sales) and the *dimension* is time (week). For a given measure, you may also want to see the values across different hierarchies within a dimension. For instance, seeing gross sales by day, week, or month would show you the *measure* dollars (gross sales) by different *hierarchies* (day, week, and month) within the time *dimension*. Making the association of a measure to a specific hierarchal level within a dimension refers to the overall *grain* of the metric.

Looking at a measure across more than one dimension, such as gross sales by territory and time, is called multidimensional analysis. Most dashboards will only leverage multidimensional analysis in a limited and static way as opposed to some of the more dynamic "slice-and-dice" tools that exist in the BI market. This is important to note, because if, in your requirements-gathering process, you uncover a significant need for this type of analysis, you may consider supplementing your dashboards with some type of multidimensional analysis tool.

Key Performance Indicators (KPI): A KPI is simply a metric that is tied to a target. Most often a KPI represents how far a metric is above or below a predetermined target. KPIs usually are shown as a ratio of actual to target and are designed to instantly let a business user know if they are on or off their plan without the end user having to consciously focus on the metrics being represented. For instance, we might decide that in order to hit our quarterly sales target we need to be selling $10,000 of widgets per week. The metric would be *widget sales per week*; the target would be $10,000. If we used a percentage gauge visualization to represent this KPI and we had sold $8,000 in widgets by Wednesday, the user would instantly see that they were at 80 percent of their goal. When selecting targets for your KPIs, you need to remember that a target will have to exist for every *grain* you want to view within a metric. Having a dashboard that displays a KPI for gross sales by day, week, and month will require that you have identified targets for each of these associated grains.

Scorecards, Dashboards, and Reports

The difference between a scorecard, dashboard, and report can be one of fine distinctions. Each of these tools can combine elements of the other, but at a high level they all target distinct and separate levels of the business decision-making process.

Scorecards: Starting at the highest, most strategic level of the business decision-making spectrum, we have scorecards. Scorecards are primarily used to help align operational execution with business strategy. The goal of a scorecard is to keep the business focused on a common strategic plan by monitoring real-world execution and mapping the results of that execution back to a specific strategy. The primary measurement used in a scorecard is the key performance indicator. These key performance indicators are often a composite of several metrics or other KPIs that measure the organization's ability to execute a strategic objective. An example of a scorecard KPI is an indicator named "Profitable Sales Growth" that combines several weighted measures such as: new customer acquisition, sales volume, and gross profitability into one final score.

Dashboards: A dashboard falls one level down in the business decision-making process from a scorecard, as it is less focused on a strategic objective and more tied to specific operational goals. An operational goal may directly contribute to one or more higher-level strategic objectives. Within a dashboard, execution of the operational goal itself becomes the focus, not the higher level strategy. The purpose of a dashboard is to provide the user with actionable business information in a format that is both intuitive and insightful. Dashboards leverage operational data primarily in the form of metrics and KPIs.

Reports: Probably the most prevalent BI tool seen in business today is the traditional report. Reports can be very simple and static in nature, such as a list of sales transaction for a given time period, to more sophisticated cross-tab reports with nested grouping, rolling summaries, and dynamic drill-through or linking. Reports are best used when the user needs to look at raw data in an easy to read format. When combined with scorecards and dashboards, reports offer a tremendous way to allow users to analyze the specific data underlying their metrics and key performance indicators.

Gathering KPI and Metric Requirements for a Dashboard

Traditional BI projects will often use a bottom-up approach in determining requirements, where the focus is on the domain of data and the relationships that exist within that data. When collecting metrics and KPIs for your dashboard project, you will want to take a top-down approach. A top-down approach starts with the business decisions that need to be made first and then works its way down into the data needed to support those decisions. In order to take a top-down approach, you MUST involve the actual business users who will be utilizing these dashboards, as these are the only people who can determine the relevancy of specific business data to their decision-making process.

When interviewing business users or stakeholders, the goal is to uncover the metrics and KPIs that lead the user to a specific decision or action. Sometimes users will have a very detailed understanding of what data is important to them, and sometimes they will only have a high-level set of goals. By following the practices outlined below, you will be able to distill the information provided to you by the user into a specific set of KPIs and metrics for your dashboards.

Interviewing Business Users

In our experience working directly with clients and gathering requirements for executive and operational dashboard projects in a variety of industries, we have found that the interview process revolves around two simple questions: "What business questions do you need answers to, and once you have those answers what action would you take or what decision would you make?"

Question 1: "What business questions do you need answers to?"
The purpose here is to help the business user define their requirements in a way that allows us to get to the data behind their question. For instance, a VP of sales might have the question: "Which sales people are my top producers?" or "Are we on target for the month?" In the case of the question "Which sales people are my top producers?" we might then follow up with a couple of questions for the VP and ask her "Would this measure be based on gross sales? Would you like to see this daily, weekly, or monthly?"

We want to identify the specific data components that will make up the KPI or metric. So we need to spend enough time with the user discussing the question until we clearly understand the *measure, dimension, grain,* and *target* (in the case of a KPI) that will be represented.

Question 2: "Based on the answer to Question 1, what other questions would this raise or what action would you take?"
Once we understand the metric or KPI that is needed to answer the user's question, we then need to find out if the user wants to perform further analysis based on that answer, or if he or she would be able to take an action or make a decision. The goal is to have users keep breaking down the question until they have enough information to take action or make a decision. This process of drilling deeper into the question is analogous to peeling back the layers of an onion; we want to keep going deeper until we have gotten to the core, which in this case is the user's ability to make a decision or take action.

As a result of this iterative two-part question process, we are going to quickly filter out the metrics and KPIs that could be considered just interesting from the ones that are truly critical to the user's decision-making process.

Putting It All Together—The KPI Wheel

In order to help with this requirements interview process, BrightPoint Consulting has created a tool called the KPI Wheel (See Figure 4–8). The interview process is very rarely a structured linear conversation, and more often is an organic free-flowing exchange of ideas and questions. The KPI Wheel allows us to have a naturally flowing conversation with the end user, while at the same time keeping us focused on the goal of gathering specific requirements.

Figure 4-8 KPI Wheel *BrightPoint Consulting white paper, "Dashboard Design: Key Performance Indicators and Metrics" by Tom Gonzalez, Managing Director, BrightPoint Consulting, Inc., © 2005 by BrightPoint Consulting, Inc. Reproduced by permission. All rights reserved.*

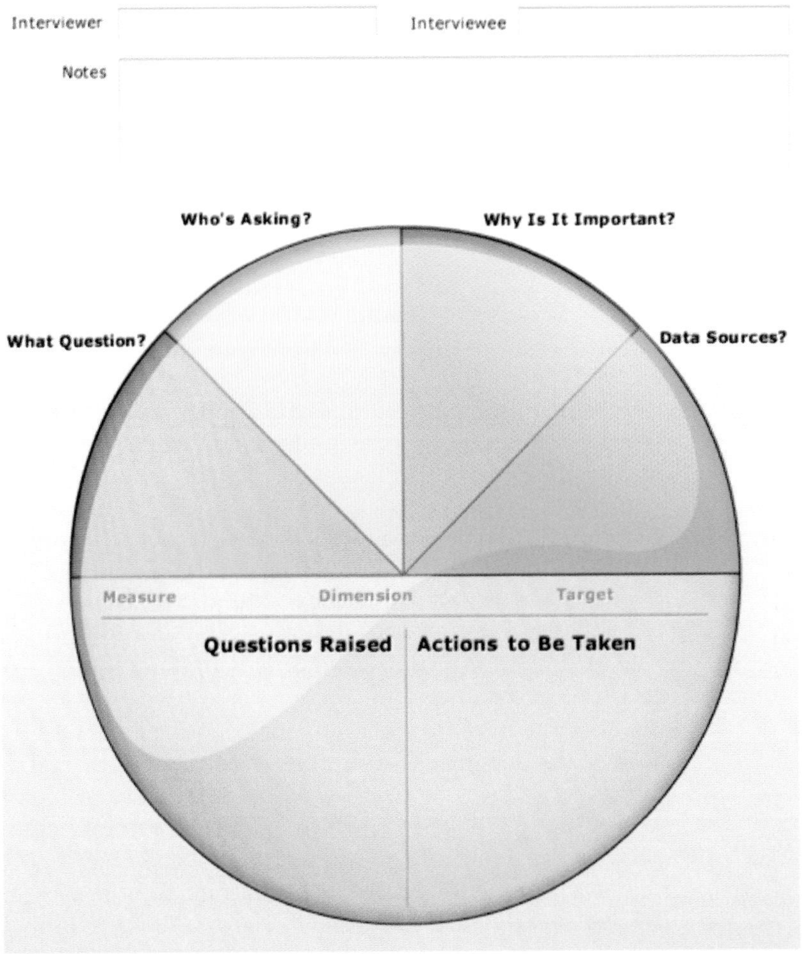

The KPI Wheel is tool that can be used to collect all the specific information that will go into defining and visualizing a metric or KPI. We will use this tool to collect the following information:

1. The business question that we are trying to help the user answer.
2. Which business users this question would apply to.

3. Why the question is important.
4. Where data resides to answer this question.
5. What further questions this metric or KPI could raise.
6. What actions or decisions could be taken with this information.
7. The specific measure, dimension, grain, and target of the metric or KPI.

Start Anywhere, but Go Everywhere

The KPI Wheel is designed as a circle because it embodies the concept that you can start anywhere but go everywhere, thus covering all relevant areas. In the course of an interview session, you will want to refer to the wheel to make sure you are filling in each area, as it is discussed. As your conversation flows, you can simply jot down notes in the appropriate section, and you can make sure to follow up with more questions if some areas remain unfilled. The beauty behind this approach is that a user can start out very high level: "I want to see how sales are doing" or at a very low level, "I need to see product sales broken down by region, time, and gross margins." In either scenario, you are able to start at whatever point the user feels comfortable and then move around the wheel filling in the needed details.

Area 1: What Question?

This area of the wheel refers to the basic "What business question do you need an answer to?" We can often start the interview with this question, or we can circle back to it when the user starts off with a specific metric in mind by asking them "What business question would that metric answer for you?" This segment of the wheel drives the overall context and relevance of the whole metric or KPI.

Area 2: Who's Asking?

For a given metric we want to know who will be using this information to make decisions and take action. It is important to understand the various users within the organization that may be viewing this metric. We can either take note of specific individuals or just refer to a general group of people who would all have similar business needs.

Area 3: Why Is It Important?

Because a truly effective dashboard can become a tool that is used every day we want to validate the importance of each metric and KPI that is displayed. Often times, in going through this requirements-gathering process, we will collect a long list of potential metrics and KPIs, and at some point the user will have to make a choice about what data is truly the most important for them to see on a regular basis. We suggest using a 1–10 scale in conjunction with a description of why the metric is important, so when you begin your dashboard prototyping you will have context as to the importance of this metric.

Area 4: Data Sources

For a given metric or KPI, we also want to identify where the supporting data will come from. Sometimes, in order to calculate a metric along one or more dimensions, we need to aggregate data from several different sources. In the case of the metric "Top Selling products by gross margins," we may need to pull data from both a CRM system and an ERP system. At this stage it is good enough to simply indicate the business system that holds the data; it is unnecessary to dive into actual table/field name descriptions at this point.

Lower Half: Measures, Dimensions, and Targets

We want to make sure that we have captured the three main attributes that create a metric or KPI, and have the user validate the grains of any given dimensions. If we are unable to pin down the measure and dimension for a metric, and/or the target for a KPI, then we will be unable to collect and visualize that data when it comes time to design our dashboard.

Lower Half: Questions Raised

In this section of the KPI wheel, we want to list any other questions that may be raised when we have answered our primary question. This list can serve as the basis for the creation of subsequent KPI Wheels that are used for definition of further metrics and KPIs.

Lower Half: Actions to Be Taken

For any given metric or KPI, we want to understand what types of decisions can be made or what types of actions will be take, depending on the state of the measurement. By filling out this section, we are also able to help validate the importance of the metric and separate the "must-have" KPI from the "nice-to-have" KPI.

Wheels Generate Other Wheels

In filling out a KPI Wheel, the process will often generate the need for several more KPIs and metrics. This is one of the purposes of doing an initial analysis in the first place: to bring all of the user's needs up to the surface. As you work through this requirements-gathering effort, you will find that there is no right path to getting your answers, questions will raise other questions, and you will end up circling back and covering ground already discussed in a new light. It is important to be patient and keep an open mind as this is a process of discovery. The goal is to have a concrete understanding of how you can empower the user through the use of good metrics and KPIs.

As you start to collect a thick stack of KPI Wheels, you will begin to see relationships between the KPIs you have collected. When you feel that you have reached a saturation point and neither you nor the user can think of any more meaningful measurements, you will then want to review all the KPI Wheels in context with each other. It is a good practice to aggregate the KPIs and create logical groupings and hierarchies, so you clearly understand the relationships that exist between various metrics. Once these steps have been accomplished, you will have a solid foundation to start you dashboard visualization and design process upon.

A Word about Gathering Requirements and Business Users

Spending the needed time with a formal requirements-gathering process is often something not well understood by business users, especially senior executives. This process will sometimes be viewed as a lot of unnecessary busy work that interrupts the user's already hectic day. It is important to remember that the decisions you are making now about what data is and is not relevant will have to be done at some point, and the only one who can make this determination is the user himself. The question is whether you spend the time to make those fundamental decisions now, while you are simply moving around ideas or later after you have painstakingly designed the dashboards and built complex data integration services around them.

As with all software development projects, the cost of change grows exponentially as you move through each stage of the development cycle. A great analogy is the one used for home construction. What is the cost to move a wall when it is a line on a drawing versus the cost to move it after you have hung a picture on it?

Wrapping It All Up

While this article touches upon some of the fundamental building blocks that can be used in gathering requirements for a dashboard project, it is by no means a comprehensive methodology. Every business intelligence architect has a set of best practices and design patterns they use when creating a new solution. It is hoped that some of the processes mentioned here can be adapted and used to supplement current best practices for a variety of solutions that leverage dashboard technologies.

5 VALUE-BASED PROJECT MANAGEMENT METRICS[1]

CHAPTER OVERVIEW

For some stakeholders, value is positioned at the top of the priority list. Establishing value metrics is now a necessity. However, there are shortcomings and pitfalls that must be addressed.

CHAPTER OBJECTIVES

- To understand what is meant by value
- To understand the need for measurements of value
- To understand the shortcomings with value measurement
- To understand how value has changed the way we manage projects
- To understand how to create a value-based metric
- To understand the need for creating a value baseline

KEY WORDS

- Boundary box
- Value
- Value baseline
- Value conflicts
- Value measurements
- Value metrics
- Value-driven projects

5.0 INTRODUCTION

For years, the traditional view of project management was that, if you completed the project and adhered to the triple constraints of time, cost, and performance (or scope), the project was successful. Perhaps in the eyes of the project manager the project appeared to be a success. In the eyes of the customer or the stakeholders, however, the project might be regarded as a failure.

As stated in Chapter 1, project managers are now becoming more business oriented. Projects are being viewed as part of a business for the purpose

1. Portions of the material in this chapter has been adapter from H. Kerzner and F. Saladis, *Value-Driven Project Management*, Hoboken, NJ: John Wiley & Sons and the International Institute for Learning, 2009.

of providing value to both the ultimate customer and the parent corporation. Project managers are expected to understand business operations more so today than in the past. As the project managers become more business oriented, our definition of success on a project now includes both a business and value component. The business component may be directly related to value.

SITUATION: The IT group of a large public utility would always service all IT requests without question. All requests were added to the queue and would eventually get done. The utility implemented a PMO that was assigned to develop a template for establishing a business case for the request, clearly indicating the value to the company if the project were completed. In the first year of using the business case template, one-third of all of the projects in the queue were tossed out.

Projects must provide some degree of value when completed as well as meeting the competing constraints. Perhaps the project manager's belief is that meeting the competing constraints provides value, but that's not always the case. Why should a company work on projects that provide no near-term or long-term value? Too many companies either are working on the wrong projects or simply have a poor project portfolio selection process, and no real value appears at the completion of the projects even though the competing constraints have been met.

Assigning resources that have critical skills that are in demand on other projects to projects that provide no appreciable value is an example of truly inept management and poor decision making. Yet selecting projects that will guarantee value or an acceptable ROI is very challenging because some of today's projects do not provide the targeted value until years into the future. This is particularly true for R&D and new product development, where as many as 50 or more ideas must be explored to generate one commercially successful product. Predicting the value at the start and tracking the value during execution is difficult. In the pharmaceutical industry, the cost of developing a new drug could run about $850 million, take 3000 days to go from exploration to commercialization, and provide no meaningful return on investment. In the pharmaceutical industry, less than 3 percent of the R&D projects are ever viewed as a commercial success and generate more that $400 million per year in revenue.

There are multiple views of the definition of value. For the most part, value is like beauty; it is in the eyes of the beholder. In other words, value may be viewed as a perception at project selection and initiation based on data available at the time. At project completion, however, the actual value becomes a reality that may not meet the expectations that had initially been perceived.

Another problem is that the achieved value of a project may not satisfy all of the stakeholders, since each stakeholder may have had a different perception of value as it relates to his/her business function. Because of the money invested in some projects, establishing value-based metrics is essential. The definition of value, along with the metrics, can be industry-specific, company-specific, or even dependent on the size, nature, and business base of the firm. Some stakeholders may view value as job security or profitability. Others might view value as image, reputation, or the creation of intellectual property. Satisfying all stakeholders is a formidable task often difficult to achieve and, in some cases, may simply be impossible. In any event, value-based metrics must be established along with the traditional metrics.

TIP Do not make promises to stakeholders about final value unless you have metrics that will confirm that their expectations can or will be met.

5.1 VALUE OVER THE YEARS

Before discussing value-based metrics, it is important to understand how the necessity for value identification has evolved. Surprisingly enough, numerous research on value has taken place over the past 15 years. Some of the items covered in the research include:

- Value Dynamics
- Value Gap Analysis
- Intellectual Capital Valuation
- Human Capital Valuation
- Economic Value-Based Analysis
- Intangible Value Streams
- Customer Value Management/Mapping
- Competitive Value Matrix
- Value Chain Analysis
- Valuation of IT Projects
- Balanced Scorecard

Following are some of the models that have occurred over the past 15 years as a result of the research:

- Intellectual Capital Valuation
- Intellectual Property Scoring
- Balanced Scorecard
- Future Value Management™
- Intellectual Capital Rating™
- Intangible Value Stream Modeling
- Inclusive Value Measurement™
- Value Performance Framework
- Value Measurement Methodology (VMM)

The reason why these models have become so popular in recent years is because we have developed techniques for the measurement and determination of value. This is essential in order to have value metrics on projects.

TABLE 5-1 Application of VPR to Project Management

VPF ELEMENT	PROJECT MANAGEMENT APPLICATION
Understand key principles of valuation	Working with the project's stakeholders to define value
Identification of key value drivers for the company	Identification of key value drivers for the project
Assessing performance on critical business processes and measures through evaluation and external benchmarking	Assessing performance of the enterprise project management methodology and continuous improvement using the PMO
Creating a link between shareholder value and critical business processes and employee activities	Creating a link between project values, stakeholder values and team member values
Aligning employee and corporate goals	Aligning employee, project and corporate goals
Identification of key "pressure points" (high leverage improvement opportunities) and estimating potential impact on value	Capturing lessons learned and best practices that can be used for continuous improvement activities
Implementation of a performance management system to improve visibility and accountability in critical activities	Establish and implement a series or project-based dashboards for customer and stakeholder visibility of key performance indicators
Development of performance dashboards with high level visual impact	Development of performance dashboards for stakeholder, team, and senior management visibility

There is some commonality among many of these models such that they can be applied to project management. For example, Jack Alexander created a model entitle Value Performance Framework (VPF).[2] The model focuses on building shareholder value and is heavily biased toward financial key performance indicators. However, the key elements of VPF can be applied to project management, as shown in Table 5-1. The first column contains the key elements of VPF from Jack Alexander's book and the second column illustrates the application to project management.[3]

5.2 VALUES AND LEADERSHIP

The importance of value can have a significant impact on the leadership style of project managers even though we do not always create value leadership metrics. Historically, project management leadership was perceived as the inevitable conflict between individual values and organizational values. Today, companies are looking for ways to get employees to align their personal values with the organization's values.

2. Jack Alexander, *Performance Dashboards and Analysis for Value Creation*, Hoboken, NJ: John Wiley & Sons, 2007, p. 5.
3. Ibid., p. 6.

TABLE 5-2 Changing Values

MOVING AWAY FROM: INEFFECTIVE VALUES	MOVING TOWARD: EFFECTIVE VALUES
Mistrust	Trust
Job Descriptions	Competency Models
Power and Authority	Teamwork
Internal Focus	Stakeholder Focus
Security	Taking Risks
Conformity	Innovation
Predictability	Flexibility
Internal Competition	Internal Collaboration
Reactive Management	Proactive Management
Bureaucracy	Boundaryless
Traditional Education	Lifelong Education
Hierarchical Leadership	Multidirectional Leadership
Tactical Thinking	Strategic Thinking
Compliance	Commitment
Meeting Standards	Continuous Improvements

Several books have been written on this subject, and the best one, in this author's opinion, is *Balancing Individual and Organizational Values* by Ken Hultman and Bill Gellerman.[4] Table 5–2, adapted from Hultman and Gellerman, shows how our concept of value has changed over the years.[5] If you look closely at the items in Table 5–2, you can see that the changing values affect more than just individual versus organizational values. Instead, it is more likely to be a conflict among four groups, as shown in Figure 5–1. The needs of each group might be:

- Project Manager:
  - Accomplishment of objectives
  - Demonstration of creativity
  - Demonstration of innovation

4. Ken Hultman and Bill Gellerman, *Balancing Individual and Organizational Values*, Jossey-Bass/Pfeiffer/John Wiley and Sons, 2002.
5. Ibid., pp. 105–106.

Figure 5-1 Project Management Value Conflicts

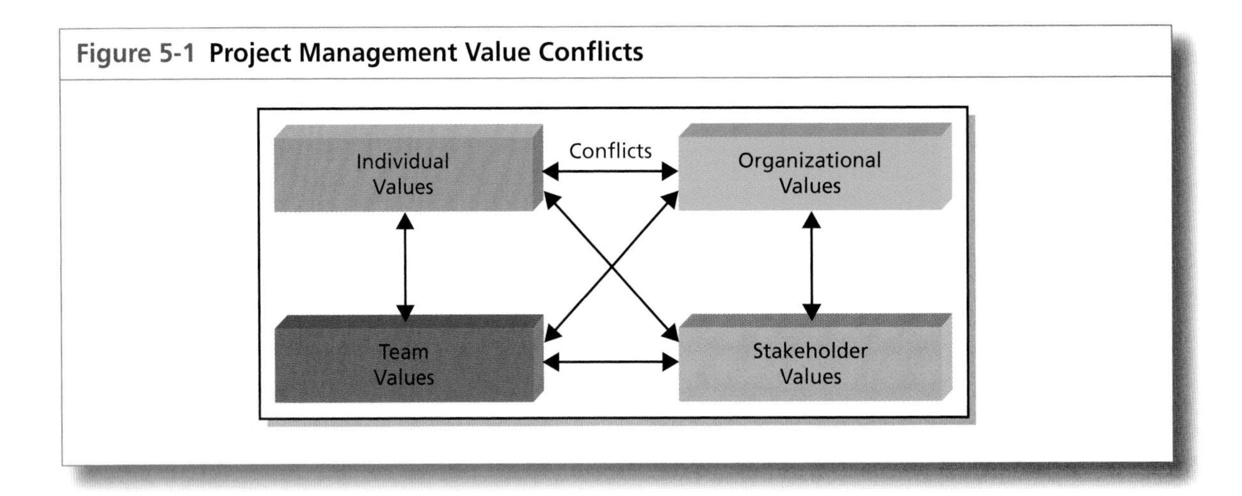

- Team Members:
  - Achievement
  - Advancement
  - Ambition
  - Credentials
  - Recognition
- Organization:
  - Continuous improvement
  - Learning
  - Quality
  - Strategic focus
  - Morality and ethics
  - Profitability
  - Recognition and Image
- Stakeholders:
  - Organizational stakeholders: Job security.
  - Product/market stakeholders: Quality performance and product usefulness.
  - Capital markets: Financial growth.

There are several reasons why the role of the project manager and the accompanying leadership style have changed. Some reasons include:

- We are now managing our business as though it is a series of projects.
- Project management is now viewed as a full-time profession.
- Project manager are now viewed as both business managers and project managers, and are expected to make decisions in both areas.

- The value of a project is measured more so in business terms rather than solely technical terms.
- Project management is now being applied to parts of the business that traditionally haven't used project management.

5.3 COMBINING SUCCESS AND VALUE

Based upon some of the value models discussed previously, such as the Balanced Scorecard Model, we can identify a classification system for projects. The types of projects, combined with a heavy focus on business alignment and value, can be classified as:

- **Enhancement or internal projects:** These are projects designed to update processes, improve efficiency and effectiveness, and possibly improve morale.
- **Financial projects:** Companies require some form of cash flow for survival. These are projects for clients external to the firm and have an assigned profit margin.
- **Future related projects:** These are long-term projects to produce a future stream of products or services capable of generating a future cash flow. These projects may be an enormous cash drain for years with no guarantee of success.
- **Customer-related projects:** Some projects may be performed, even at a financial loss, to maintain or build a customer relationship. However, performing too many of these projects can lead to financial disaster.

Today, these projects focus more on value than on the competing constraints. With the value-driven constraints, we emphasize stakeholder satisfaction and decisions, and the value that is expected on the project. In others words, success is when the value is obtained, hopefully within the triple or competing constraints. As a result, we can define the four cornerstones of success using Figure 5–2.

Very few projects are completed without some tradeoffs. Metrics provide some of the necessary information needed for decisions on tradeoffs. This holds true for both the traditional projects and those that are based upon value components and metrics. Traditional tradeoffs result in an elongation of the schedule and an increase in the budget. The same holds true for the value-driven projects, but the major difference is with performance. With traditional tradeoffs, we tend to reduce performance to satisfy others requirements. With value-driven projects, we tend to increase performance in hopes of providing added value, and this tends to cause much larger cost overruns and schedule slippages than with traditional tradeoffs. The amount of additional time and funding that the stakeholders will allow is dependent on the tracking of the metrics.

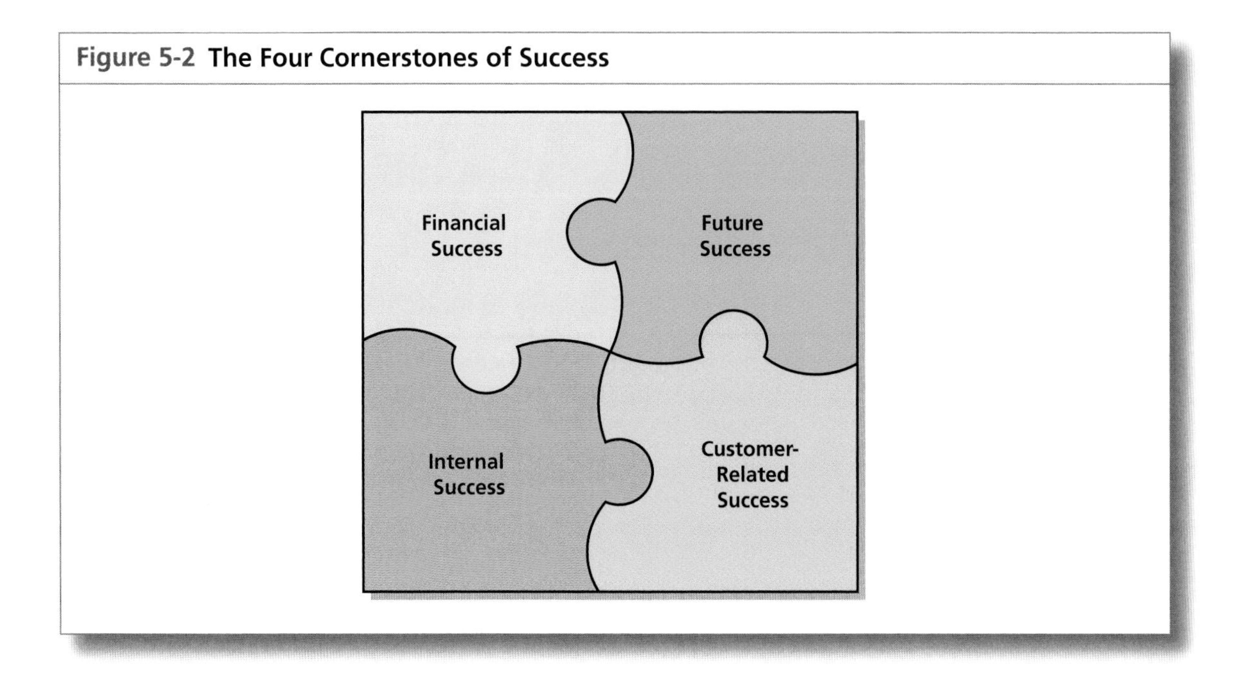

Figure 5-2 The Four Cornerstones of Success

Projects managers generally do not have the sole authority for scope or performance increases or decreases. For traditional tradeoffs, the project manager and the project sponsor, working together, may have the authority to make tradeoff decisions.

However, for value-driven projects, all or most of the stakeholders may need to be involved. This can create additional issues such as:

- It may not be possible to get all of the stakeholders to agree on a value target during project initiation.
- It may not be possible to get all of the stakeholders to agree on the metrics or key performance indicators.
- Getting agreement on scope changes, extra costs, and schedule elongations is significantly more difficult the further along you are in the project.
- Stakeholders must be informed of this at project initiation and continuously briefed as the project progresses; that is, no surprises!

Conflicts among the stakeholders may occur. As an example:

- During project initiation, conflicts among stakeholder are usually resolved in favor of the largest financial contributors.
- During execution, conflicts over future value are more complex, especially if major contributors threaten to pull out of the project.

For projects that have a large number of stakeholders, project sponsorship may not be effective with a single sponsor. Therefore, committee sponsorship may be necessary. Membership in the committee may include:

- Perhaps a representative from all stakeholder groups
- Influential executives
- Critical strategic partners and contractors
- Others based upon the type of value

Responsibilities for the sponsorship committee may include:

- Taking a lead role in the definition of the targeted value
- Taking a lead role in the acceptance of the actual value
- Ability to provide additional funding
- Ability to assess changes in the enterprise environment factors
- Ability to validate and revalidate the assumptions

Sponsorship committees may have significantly more expertise than the project manager in defining and evaluating the value in a project.

Each of the quadrants in Figure 5–2 can have its own unique set of critical success factors and likewise its own unique metrics and key performance indicators. Following are typical CSFs for each quadrant:

Internal Success:
- Adherence to schedule, budget, and quality/scope (triple constraint)
- Mutually agreed upon scope change control process
- Without disturbing the main flow of work
- Clear understanding of the objectives (end-user involvement)
- Maintaining the timing of sign-offs
- Execution without disturbing the corporate culture
- Building lasting internal working relationships
- Consistently respecting each other's opinions
- Searching for value-added opportunities

Financial Success:
- Integrating program and project success into one definition
- Maintaining ethical conduct
- Adherence to regulatory agency requirements
- Adherence to health, safety, and environmental laws
- Maintaining or increasing market share
- Maintaining or improving ROI, NPV, IRR, payback period, etc.
- Maintaining or improving net operating margins

Future Success:

- Improving the processes needed for commercialization
- Emphasizing follow-on opportunities
- Maintaining technical superiority
- Protecting the company image and reputation
- Maintaining a knowledge repository
- Retaining presale and postsale knowledge
- Aligning projects with long-term strategic objectives
- Informing the teams about the strategic plans
- Team members willing to work with this project manager again

Customer-Related Success:

- Keeping promises made to the customers over and over again
- Maintaining customer contact and interfacing continuously
- Focusing upon customer satisfaction from start to finish
- Improving customer satisfaction ratings on a continuous basis
- Using every customer's name as a reference
- Measuring variances against customer-promised best practices
- Maintaining or improving on customer delivery requirements
- Building long-term relationships between organizations

In Chapter 3, we identified the different type of metrics. We can now identify which type of metric is most suitable for each success quadrant. This is shown in Figure 5–3.

Figure 5-3 Categories of Success Metrics

Financial Success	Future Success
• Quantitative • Directional • Financial	• Quantitative • Financial
Internal Success	**Customer Related Success**
• Quantitative • Practical • Directional • Actionable • Milestone	• Directional • End Result

5.4 RECOGNIZING THE NEED FOR VALUE METRICS

The importance of the value component in the definition of success cannot be overstated. Consider the following eight postulates:

- **Postulate #1:** Completing a project on time and within budget does not guarantee success if you were working on the wrong project.
- **Postulate #2:** Completing a project on time and within budget is not necessarily success.
- **Postulate #3:** Completing a project within the triple constraints does not guarantee that the necessary business value will be there at project completion.
- **Postulate #4:** Having the greatest enterprise project management methodology in the world cannot guarantee that value will be there at the end of the project.
- **Postulate #5:** Price is what you pay. Value is what you get (Warren Buffett).
- **Postulate #6:** Business value is what your customer perceives as worth paying for.
- **Postulate #7:** Success is when business value is achieved.
- **Postulate #8:** Following a project plan to conclusion is not always success if business-related changes were necessary but never implemented.

These eight postulates lead us to believe that perhaps value may become the dominating factor in the selection of a project portfolio. Project requestors must now clearly articulate the value component in the project's business case or run the risk that the project will not be considered. If the project is approved, then value metrics must be established and tracked. However, it is important to understand that value may be looked at differently during the portfolio selection of projects because the tradeoffs that take place are among projects rather than the value attributes of a single project.

In Postulate #1, we can see what happens when management makes poor decisions during project selection, establishment of a project portfolio, and when managing project portfolios. We end up working on the wrong project or projects. What is unfortunate about this scenario is that we can produce the deliverable that was requested but:

- There's no market for the product.
- The product cannot be manufactured as engineered.
- The assumptions may have changed.
- The marketplace may have changed.
- Valuable resources were wasted on the wrong project.
- Stakeholders may be displeased with management's performance.
- The project selection and portfolio management process is flawed and needs to be improved.
- Organizational morale has diminished.

Postulate #2 is the corollary to Postulate #1. Completing a project on time and on budget:

- Does not guarantee a satisfied client/customer
- Does not guarantee that the customer will accept the product/service
- Does not guarantee that performance expectations will be met
- Does not guarantee that value exists in the deliverable
- Does not guarantee marketplace acceptance
- Does not guarantee follow-on work
- Does not guarantee success

SITUATION: During the initiation of the project, the project manager and the stakeholders defined project success and established metrics for each of the competing constraints. When it became obvious that all of the constraints could not be met, the project manager concluded that the best alternative was a tradeoff on value. The stakeholders became irate upon hearing the news and decided to prioritize the competing constraints themselves. This took time and delayed the project.

Postulate #3 focuses on value. Simply because the deliverable is provided according to a set of constraints is no guarantee that the client will perceive value in the deliverable. The ultimate objective of all projects should be to produce a deliverable that meets expectations and achieves the desired value. While we always seem to emphasize the importance of the competing constraints when defining the project, we spend very little time in defining the value characteristics and resulting metrics that we expect in the final deliverable. The value component or definition must be agreed upon jointly by the customer and the contractor (buyer/seller) during the initiation stage of the project.

Most companies today have some type of project management methodology in place. Unfortunately all too often there is a mistaken belief that the methodology will guarantee project success. Methodologies:

- Cannot guarantee success
- Cannot guarantee value in the deliverable
- Cannot guarantee that the time constraint will be adhered to
- Cannot guarantee that the quality constraint will be met
- Cannot guarantee any level of performance
- Are not a substitute for effective planning
- Are not the ultimate panacea to cure all project ills
- Are not a replacement for effective human behavior

TIP The definition of value must be aligned with the strategic objectives of both the customer and the contractor.

Methodologies can improve the chances for success but cannot guarantee success. Methodologies are tools and, as such, do not manage projects. Projects are managed by people and, likewise, tools are managed by people. Methodologies do not replace the people component in project management. They are designed to enhance the performance of people.

In Postulate #5, we have a quote from Warren Buffet that emphasizes the difference between price and perceived value. Most people believe that customers pay for deliverables. This is not necessarily true. Customers pay for the value they expect to receive from the deliverable. If the deliverable has not achieved value or has limited value, the result is a dissatisfied customer.

Some people believe that a customer's greatest interest is quality. In other words, "Quality comes first! While that may seem to be true on the surface, the customer generally does not expect to pay an extraordinary amount of money just for high quality. Quality is just one component in the value equation. Value is significantly more than just quality.

When customers agree to a contract with a contractor/supplier for a deliverable, the customer is actually looking for the value in the deliverable. The customer's definition of success is "value achieved."

Unfortunately, unpleasant things can happen when the project manager's definition of success is the achievement of the deliverable (and possibly the triple constraint) and the customer's definition of success is value. This is particularly true when customers want value and you, as the contractor, focus on the profit margins of your projects.

Postulate #7 is a summation of Postulates #1 through #6. Perhaps the standard definition of success using just the triple constraints should be modified to include a business component such as value, or even be replaced by a more specific definition of value.

Sometimes the value of a project can change over time, and the project manager may not recognize that these changes have occurred. Failure to establish value expectations or lack of value in a deliverable can result from:

- Market unpredictability
- Market demand that has changed, thus changing constraints and assumptions
- Technology advances or inability to achieve functionality
- Critical resources were that were not available or resources who lacked the necessary skills

Establishing value metrics early on can identify if a project should be canceled. The earlier the project is canceled, the quicker we can assign the resources to those projects that have a higher perceived value and probability of success. Unfortunately early warning signs are not always present to indicate that the value will not be achieved. The most difficult metrics to establish are value-driven metrics.

TIP Degradation in value metrics is a clear indication that the project is in trouble and that it may be canceled.

5.5 THE NEED FOR EFFECTIVE MEASUREMENT TECHNIQUES

Selecting metrics and KPIs are not that difficult provided they can be measured. This is the major obstacle with the value-driven metrics. On the surface, they look easy to measure, but there are complexities. However, even though value appears in the present and in the future, it does not mean that the value outcome cannot be quantified. Table 5–3 illustrates some of the metrics that are often treated as value-driven KPIs.

Traditionally, business plans identified the benefits expected from the project, and the benefits were the criteria for project selection. Today, portfolio management techniques require identification of the value as well as the benefits. However, conversion from benefits to value is not easy.[6]

There are shortcomings in the conversion process that can make the conversion difficult. Figure 5–4 illustrates several common shortcomings.

There are other shortcomings with the measurement of KPIs. KPI are metrics for assessing value. With traditional project management, metrics are established by the enterprise project management methodology and fixed for the duration of the project's life cycle. With value-driven project management, however, metrics, can change from project to project, during a life cycle phase and over time because of:

- The way the customer and contractor jointly define success and value at project initiation

TABLE 5-3 Measuring Value

VALUE METRIC	MEASUREMENT
Profitability	Easy
Customer Satisfaction	Hard
Goodwill	Hard
Penetrate New Markets	Easy
Develop New Technology	Moderate
Technology Transfer	Moderate
Reputation	Hard
Stabilize Work Force	Easy
Utilize Unused Capacity	Easy

6. For additional information on the complexities of conversion, see Jack J. Phillips, Timothy W. Bothell, and G. Lynne Snead, *The Project Management Scorecard*, Oxford, UK: Butterworth Heinemann, An Imprint of Elsevier, 2002, Chapter 13.

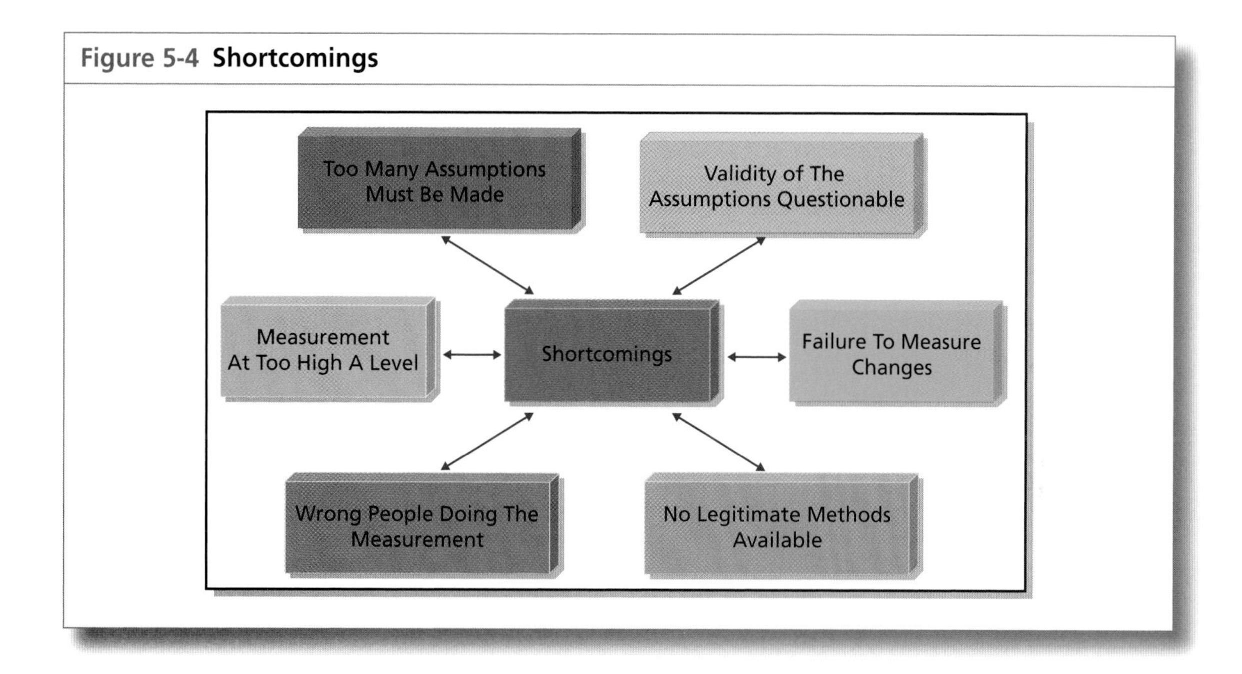

Figure 5-4 **Shortcomings**

- The way the customer and contractor come to an agreement at project initiation as to what metrics should be used on a given project
- The way the company defines value
- New or updated versions of tracking software
- Improvements to the enterprise project management methodology and the accompanying project management information system
- Changes in the enterprise environmental factors

Even with the best possible metrics, measuring value can be difficult. Benefits less costs indicate the value and determine if the project should be done. The challenge is that not all costs are quantifiable. Some values are easy to measure, while others are more difficult. The easy value metrics to measure are often called soft or are established by the enterprise project management methodology and fixed for the duration of the project's life cycle. With value-driven project management, however, metrics can change from project to project, during a life cycle phase and over time because of: tangible value metrics, whereas the hard values are often considered as intangible value metrics. Table 5–4 illustrates some of the easy and hard value metrics to measure. Table 5–5 shows some of the problems associated with measuring both hard and soft value metrics.

The intangible elements are now considered by some to be more important than tangible elements. This appears to be happening on IT projects where executives are giving significantly more attention to intangible

TABLE 5-4 Typical Financial Value Metrics

EASY (SOFT/TANGIBLE) VALUE METRICS	HARD (INTANGIBLE) VALUE METRICS
ROI Calculators	Stockholder Satisfaction
Net Present Value (NPV)	Stakeholder Satisfaction
Internal Rate of return (IRR)	Customer Satisfaction
Cash Flow	Employee Retention
Payback Period	Brand Loyalty
Profitability	Time-to-Market
Market Share	Business Relationships
	Safety
	Reliability
	Goodwill
	Image

TABLE 5-5 Problems with Measuring Value Metrics

EASY (SOFT/TANGIBLE) VALUE METRICS	HARD (INTANGIBLE) VALUE METRICS
Assumptions are often not fully disclosed and can affect decision making.	Value is almost always based upon subjective attributes of the person doing the measurement.
Measurement is very generic.	It is more of an art than a science.
Measurement never meaningfully captures the correct data.	Limited models are available to perform measurement.

values. The critical issue with intangible values is not necessarily in the end result, but in the way that the intangibles were calculated.[7]

Tangible values are usually expressed quantitatively where as intangible values may be expressed through a qualitative assessment. There are three schools of thought for value measurement:

- **School #1:** The only thing that is important is ROI.
- **School #2:** ROI can never be calculated effectively; only the intangibles are important.
- **School #3:** If you cannot measure it, then it does not matter.

The three schools of thought appear to be an all-or-nothing approach where value is either 100 percent quantitative or 100 percent qualitative. The best approach is most likely a compromise between a quantitative and qualitative assessment of value. It may be necessary to establish an effective

7. For additional information on the complexities of measuring intangibles, see Jack J. Phillips, Timothy W. Bothell, and G. Lynne Snead, *The Project Management Scorecard*, Oxford, UK: Butterworth Heinemann, An Imprint of Elsevier, 2002, Chapter 10. The authors emphasize that the true impact on a business must be measured in business units.

Figure 5-5 Quantitative versus Qualitative Assessment

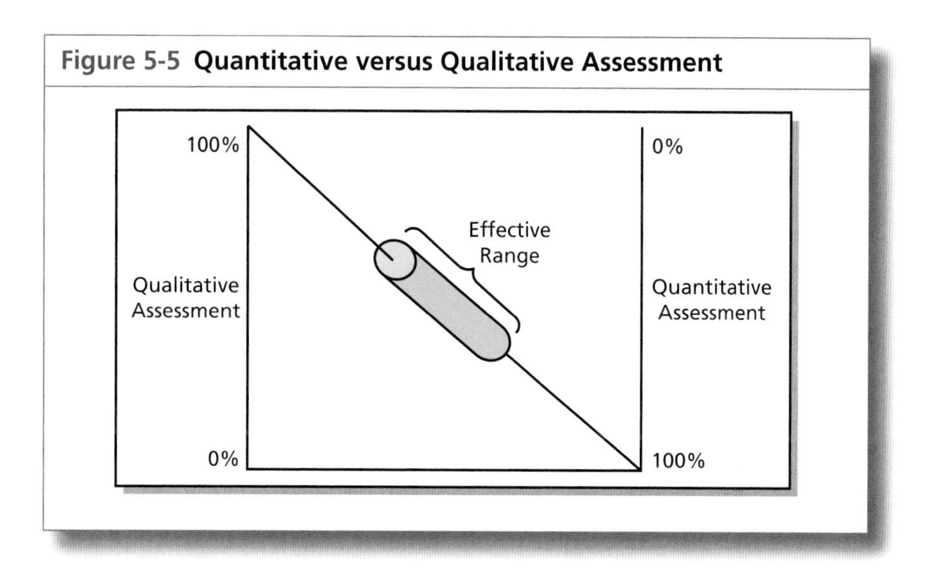

range, as show in Figure 5–5, which is a compromise among the three schools of thought. The effective range can expand or contract.

The timing of value measurement is absolutely critical. During the life cycle of a project, it may be necessary to switch back and forth from qualitative to quantitative assessment of the metric and, as stated previously, the actual metrics or KPIs may then be subject to change. Certain critical questions must be addressed:

- When or how far along the project life cycle can we establish concrete metrics, assuming this can be done at all?
- Can value be simply perceived and, therefore, no value metrics are required?
- Even if we have value metrics, are they concrete enough to reasonably predict actual value?
- Will we be forced to use value-driven project management metrics on all projects or are there some projects where this approach is not necessary?
  - Well defined versus ill defined
  - Strategic versus tactical
  - Internal versus external
- Can we develop a criterion for when to use value-driven project management, or should we use it on all projects but at a lower intensity level?

For some projects, using metrics to assess value at project closure may be difficult. We must establish a time frame for how long we are willing to wait to measure the final or real value or benefits from a project. This is particularly important if the actual value cannot be identified until some time after the project has been completed. Therefore, it may not be possible to appraise the success of a project at closure if the true economic values cannot be realized until some time in the future.

Some practitioners of value measurement question whether value measurement is better using boundary boxes instead of life cycle phases. For value-driven projects, the potential problems with life cycle phase metrics include:

- Metrics can change between phases and even during a phase.
- Inability to account for changes in the enterprise environmental factors.
- Focus may be on the value at the end of the phase rather than the value at the end of the project.
- Team members may get frustrated not being able to quantitatively calculate value.

Boundary boxes, as shown in Figure 5–6, have some degree of similarity to statistic process control charts and can assist in metric measurements. Upper and lower strategic targets for the value of the metrics are established. As long as the KPIs indicate that the project is still within the upper and lower value targets, the project's objectives and deliverables may not undergo any scope changes or tradeoffs.

Value-driven projects must undergo value health checks to confirm that the project will make a contribution of value to the company. Value metrics, such as KPIs, indicate the current value. What is also needed is an extrapolation from the present into the future. Using traditional project management combined with the traditional enterprise project management methodology, we can calculate the time at completion and the cost at completion. These are common terms that are part of earned value measurement systems. However, as stated previously, being on time and within budget is no guarantee that the perceived value will be there at project completion.

Therefore, instead of using an enterprise project management methodology, which focuses on earned value measurement, we may need to

Figure 5-6 The Boundary Box

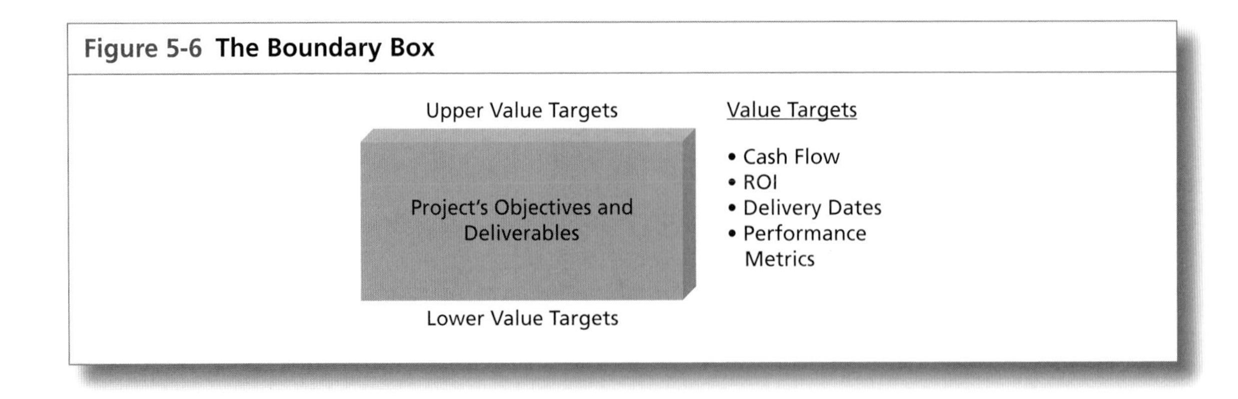

create a value management methodology (VMM), which stresses the value variables. With VMM, time to complete and cost to complete are still used, but we introduce a new term entitled value (or benefits) at completion. Determination of value at completion must be done periodically throughout the project. However, periodic reevaluation of benefits and value at completion may be difficult because:

- There may be no reexamination process.
- Management is not committed and believes that the reexamination process is unreal.
- Management is overoptimistic and complacent with existing performance.
- Management is blinded by unusually high profits on other projects (misinterpretation).
- Management believes that the past is an indication of the future.

5.6 CUSTOMER/STAKEHOLDER IMPACT ON VALUE METRICS

For years, customers and contractors have been working toward different definitions of project success. The project manager's definition of success was profitability and tracked through financial metrics. The customer's definition of success was usually the quality of the deliverables. Unfortunately, quality was measured at the closure of the project because it was difficult to track throughout the project. Yet quality was often considered the only measurement of success.

Today, clients and stakeholders appear to be more interested in the value they will receive at the end of the project. If you were to ask 10 people, including project personnel, the meaning of value, you would probably get 10 different answers. Likewise, if you were to ask which critical success factor has the greatest impact on value, you would get different answers. Each answer would be related to the individual's work environment and industry. Today, companies seem to have more of an interest in value than in quality. This does not mean that we are giving up on quality. Quality is part of value. Some people believe that value is simply quality divided by the cost of obtaining that quality. In other words, the less you pay for obtaining the customer's desired level of quality, the greater the value to the customer.

The problem with this argument is that we assume that quality is the only attribute of value that is important to the client and, therefore, we need to determine better ways of measuring and predicting just quality.[8] Unfortunately, there are other attributes of value and many of these other

8. Throughout this chapter, when we refer to a "client," the client could be internal to your company, external to your company, or the customers of your external client. You might also consider stakeholders as your clients.

attributes are equally as difficult to measure and predict. Customers can have many attributes that they consider to represent value, but not all of the value attributes are equal in importance.

Unlike the use of quality as the solitary parameter, value allows a company to better measure the degree to which the project will satisfy its objectives. Quality can be regarded as an attribute of value along with other attributes. Today, every company has quality and produces quality in some form. This is necessary for survival. What differentiates one company from another, however, are the other attributes, components, or factors used to define value. Some of these attributes might include price, timing, image, reputation, customer service, and sustainability.

In today's world, customers make decisions to hire a contractor based upon the value they expect to receive and the price they must pay to receive this value. Actually, it is more of a "perceived" value that may be based upon tradeoffs on the attributes of the client's definition of value. *The client may perceive the value of your project to be used internally in their company or pass it on to their customers through their customer value management program.* If your organization does not or cannot offer recognized value to your clients and stakeholders, then you will not be able to extract value (i.e., loyalty) from them in return. Over time, they will defect to other contractors.

The importance of value is clear. According to a study by the American Productivity and Quality Center (APQC):

> Although customer satisfaction is still measured and used in decision making, the majority of partner organizations [used in this study] have shifted their focus from customer satisfaction to customer value.[9]

Project managers in the future must consider themselves as the creators of value. The definition of a project that I use in my courses is "a set of values scheduled for sustainable realization." As a project manager, you must, therefore, establish the correct metrics so that the client and stakeholders can track the value that you will be creating. Measuring and reporting customer value throughout the project is now a competitive necessity. If it is done correctly, it will build emotional bonds with your clients.

5.7 CUSTOMER VALUE MANAGEMENT (CVM)

For decades, many companies believed that they had an endless supply of potential customers. Companies called this the "door knob" approach, whereby they expected to find a potential customer behind every door.

9. "Customer Value Measurement: Gaining Strategic Advantage," The American Productivity and Quality Center (APQC), 1999, p. 8.

Under this approach, customer loyalty was nice to have but not a necessity. Customers were plentiful, often with little regard for the quality of your deliverables. Those days may be gone.

As the quality movement began to take hold during the 1980s, so did the need for effective customer relations management (CRM), as shown in Figure 5–7. The focus of most CRM programs was to: (1) find the right customers, (2) develop the right relationships with these customers, and (3) retain the customers. This included stakeholder relations management and seeking out ways to maintain customer loyalty.

Historically, sales and marketing were responsible for CRM activities. Today, project managers are doing more than simply managing a project; they are managing part of a business. Therefore, they are expected to make business decisions as well as project decisions, and this includes managing activities related to CRM. Project managers soon found themselves managing projects that now required effective stakeholder relations management, as well as customer relations management. Satisfying the needs of both the client and various stakeholders was difficult.

As CRM began to evolve, companies soon found that there were different perceptions among their client base as to the meaning of quality and value. In order to resolve these issues, companies created customer value management (CVM) programs. Customer value management programs address the critical question, *Why should customers purchase from you rather*

Figure 5-7 Growth in the Importance of Value

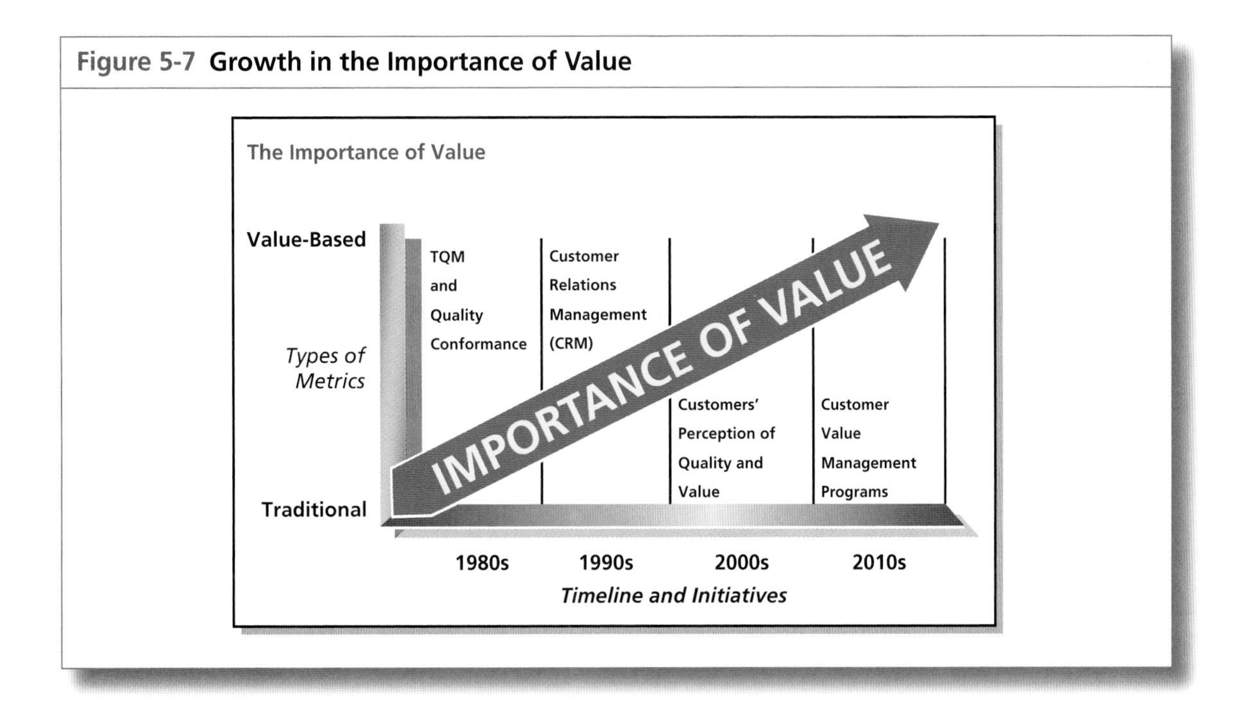

than from the competition? The answer was most often the value that you provided through your products and deliverables. Loyal customers appeared to be more value sensitive than price sensitive. Loyal customers are today a scarce resource and also a source of value for project managers and their organizations. Value breeds loyalty.

There are other items, such as trust and intangibles, that customers may see as a form of value. As stated by a technical consultant:

> The business between the vendor and the customer is critical. It's situational but in technical consulting for instance the customer may really value only the technical prowess of a vendor's team; project management is expected to be competent in this case. If project management itself is adding value then isn't that really a matter of the customer's view of the project manager providing services above and beyond the normal view of functional responsibility? That would come down to the relationship with the customer. Ask any customer what they truly value in a vendor and they will tell you it is trust because there is a reliance on the customer's business strategy succeeding based on how well the vendor executes.
>
> For example, in answer to the question, Why do you value vendor X?, you could imagine the following answers from customers: "So and so always delivers for me, and I can count on them to deliver a quality (defined however) product on time and at the agreed price," or "Vendor X really helped me be successful with my management by pulling in the schedule by x weeks,." or "I really appreciated a recent project done by vendor Y because they handled our unexpected design changes with professionalism and competence."
>
> Now most project management is within the sphere of operations in a vendor's organization. The customer facing business relationship is handled by some company representative in most cases. The project manager would be brought in once the work is underway and typically the direct reporting is to someone underneath the person who authorized the work; so in this case the project manager has the opportunity to build value with the underling but not the executive.
>
> We discount the personality of the project managers as though this isn't an issue. It is a major issue and people need to realize that it is. Understanding one's own personality and the personality of the customer is vital to getting a label of value added from a customer. If the project manager isn't flexible in this area then creating value with the customer becomes more difficult.
>
> Anyway, there are as many variations to this theme as there are projects since personality and other interpersonal relationship nuances are involved. However, so much of successful project management is all about these intangibles.

CVM today focuses on maximizing customer value, regardless of the form. In some cases, CVM must measure and increase the lifetime value of the deliverables of the project for each customer and stakeholder. By doing this, the project manager is helping the customer manage their profitability as well.

CVM performed correctly can and will lead to profitability, but being profitable does not mean that you are performing CVM correctly. There are benefits to implementing CVM effectively as shown in Table 5-6. CVM is the leveraging of customer and stakeholder business relationships throughout the project. Because each project will have different customers and different stakeholders, CVM must be custom fit to each organization and possibly each project.

If CVM is to be effective, then presenting the right information to the client and stakeholders becomes critical. CVM introduces a value mind-set into decision making. Many CVM programs fail because of poor metrics and not measuring the right things. Just focusing on the end result does not tell you if what you are doing is right or wrong. Having the correct value-based metrics is essential. Value is in the eyes of the beholder, which is why there can be different value metrics.

CVM relies heavily upon customer value assessment. Traditional CVM models are light on data and heavy on assumptions. For CVM to work effectively, it must be heavy on data and light on assumptions. Most successful CVM programs perform "data mining," where the correct attributes of value are found along with data that supports the use of those attributes. However, we should not waste valuable resources calculating value metrics unless the client perceives the value of using the metric. Project management success in the future will be measured by how well the project manager provides superior customer value. To do this, you must know what motivates the customers as well as the customers' customers.

TABLE 5-6 Before and after CVM Implementation

FACTOR	BEFORE CVM	AFTER CVM (WITH METRICS)
Stakeholder communications	Loosely structured	Structured using a network of metrics
Decision-making	Based upon partial information	Value-based informed decision making
Priorities	Partial agreements	Common agreements using metrics
Tradeoffs	Less structured	Structured around value contributions
Resource allocation	Less structured	Structured around value contributions
Business objectives	Projects poorly aligned to business	Better alignment to business strategy
Competitiveness	Market underperformer	Market outperformer

Executives generally have a better understanding of the customers' needs than do the workers at the bottom of the organizational chart where the work takes place. Therefore, the workers may not see, understand or appreciate the customer's need for value. Without the use of value metrics, we focus on the results of the process rather than the process itself and miss opportunities to add value. Only when a crisis occurs do we put the process under a microscope. Value metrics provide the workers with a better understanding of the customer's definition of value.

Understanding the customer's perception of value means looking at the disconnects or activities that are not value-added work. Value is created when we can eliminate non-value-added work rather than looking at ways to streamline project management processes. As an example, consider the following situation:

> **SITUATION:** A company had a project management methodology that mandated that a risk management plan be developed on each and every project regardless of the magnitude of the risk. The risk management plan was clearly defined as a line item in the work breakdown structure and the clients eventually paid the cost of this. On one project, the project manager created a value metric and concluded that the risks on the project were so low that risk should not be included as one of the attributes of the value metric. The client agreed with this, and the time and money needed to develop a risk management plan was eliminated from the project plan. On this project, the risk management plan was seen as a disconnect and eliminated so that added value could be provided to the client. The company recognized that the risk management plans might be a disconnect on some projects and made the risk management plan optional at the discretion of the project manager and the clients.

This situation is a clear example that value is created when outputs are produced using fewer inputs or more outputs are created with the same number of inputs. However, care must be taken to make sure that the right disconnects are targeted for elimination.

Project managers must work closely with the customers for CVM to be effective. This includes:

- Understanding the customer's definition of satisfaction and effective performance
- Knowing how the customer perceives your price/value relationship (some clients still believe that value is simply quality divided by price)
- Making sure that the customers understand that value can be expressed in both nonfinancial and financial terms
- Seeing if the client can tell you what your distinctive competencies are and determining if they are appropriate candidates for value attributes

- Being prepared to debrief the customer and stakeholders on a regular basis for potential improvements and best practices
- Validating that the client is currently using or is willing to use the value metrics for their own informed decision making
- Understanding which value attributes are most important to your customers
- Building a customer project management value model or framework possibly unique for each customer
- Making sure that your model or framework fits the customer's internal business model
- Designing metrics that interface with the customer's business model
- Recognizing that CVM can maximize your lifetime profitability with each customer
- Changing from product-centric or service-centric marketing to project management-centric marketing
- Maximizing the economics of customer loyalty

5.8 THE RELATIONSHIP BETWEEN PROJECT MANAGEMENT AND VALUE

Companies today are trying to link quality, value, and loyalty. These initiatives, which many call customer value management initiatives, first appeared as business initiatives performed by marketing and sales personnel rather than project management initiatives. Today, however, project managers are slowly becoming more involved in business decisions, and value has become extremely important.

Quality and customer value initiatives are part of CVM activities and are a necessity if a company wishes to obtain a competitive advantage. Competitive advantages in project management do not come just from being on time and within budget at the completion of each project. Offering something that your competitors do not offer may help. However, true competitive advantage is found when your efforts are directly linked to the customer's value initiatives, and whatever means by which you can show this will give you a step up on the competition. Projects managers must develop value-creating strategies. Project managers must also know how to create a project value baseline as will be discussed in Section 5.21.

Customers today have become more demanding and are requiring the contractor to accept the customer's definition of value according to the attributes selected by the customer. Each customer can, therefore, have a different definition of value. Contractors may wish to establish their own approach for obtaining this value based upon their company's organizational process assets rather than having the customer dictate it. If you establish your own approach to obtaining the desired value, do not assume that your customer will understand the approach. They may need to be educated. Customers who recognize and understand the value that you are providing are more likely to want a long-term relationship with your firm.

To understand the complexities with introducing value to project management activities, let's assume that companies develop and commercialize products according to the phases in Figure 5–8. Once the project management phase is completed, the deliverables are turned over to someone in marketing and sales responsible for program management and ultimately commercialization of the product. Program management and commercialization may be done for products developed internally for your own company or they may be done for the client, or even for the client's own customers. In any event, it is during or after commercialization when companies survey their customers as to feedback on customer satisfaction and the value of the product. If the customers are unhappy with the final value, it is an expensive process to go back to the project stage and repeat the project in order to try to improve the value for the end-of-line customers.

The failures from Figure 5–8 can be attributed to:

- Project managers are allowed to make only project decisions rather than both project and business decisions.
- Project managers are not informed as to the client's business plans as they relate to what the project manager is developing.
- Customers do not clearly articulate to the project manager, either verbally or through documented requirements, the exact value that they expect.
- Customers fund projects without fully understanding the value needed at the completion of the project.
- Project managers interface with the wrong people on the project.
- No value-based metrics are established in the project management phase whereby informed decision making can take place to improve the final value.

Figure 5-8 Simplified Product Stages of Development

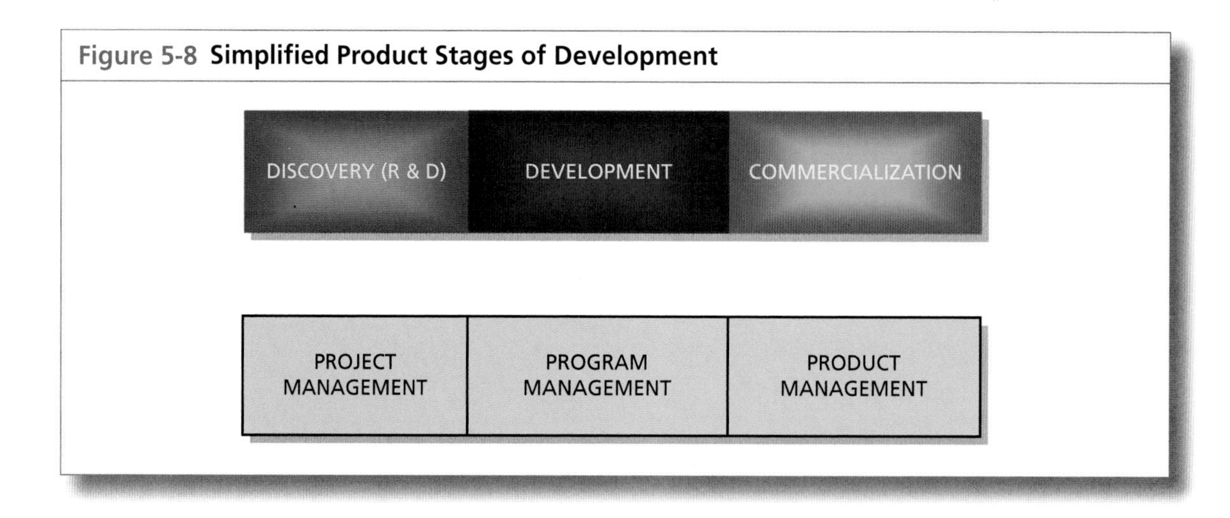

- Tradeoffs are made without considering the impact on the final value.
- Quality and value are considered as synonymous; quality is considered as the only value attribute.

Returning to Figure 5–7, we can see that as we approach CVM activities, traditional metrics are replaced with value-based metrics. The value metric should not be used to replace other metrics that people may be comfortable using for tracking project performance. Instead, it should be used to support other metrics being used. As stated by a healthcare IT consulting company:

> [The need for a value metric] "hits directly on the point" that so many of us consultants encounter on a regular basis surrounding the fact that often the metrics that are utilized to ensure success of projects and programs are too complex, don't really demonstrate the key factors of what the sponsors, stakeholders, investors, client, etc. "need" in order to demonstrate value and success in a project and do not always "circle back" to the mission, vision, and goals of the project at the initial kick-off. What we have found is that the majority of our clients are constantly in search of something simple, streamlined, and transparent (truly just as simple as that). In addition, they generally request a "pretty picture" format; something that is clear, concise and easy and quick to glance at in order to determine areas that require immediate attention. They crave for the red, yellow, green. They yearn for a crisp measure to state what the "current state" truly is [and whether the desired value is being achieved]. They are tired and weary of the "circles," they are exhausted of the mounds of paperwork, and they want to take it back to the straight facts.

We are now in the infancy stage of determining how to define and measure value. For internal projects, we are struggling in determining the right value metrics to assist us in project portfolio management with the selection of one project over another. For external projects, the picture is more complex. Unlike traditional metrics used in the past, value-based metrics are different for each client and each stakeholder. In Figure 5–9, you can see the three dimensions of values: your parent company's values, your client's values, your client's customers' values, and we could even add in a fourth dimension, namely stakeholder values. It should be understood that value that the completion of the project brings to your organization may not be as important as the total value that the project brings to the client's organization and the client's customer base.

For some companies, the use of value metrics will create additional challenges. As stated by a global IT consulting company:

> This will be a cultural change for us and for the customer. Both sides will need to have staff competent in identifying the right metrics to use and weightings; and then be able to explain in layman's language what the value metric is about.

Figure 5-9 Dimensions of Value

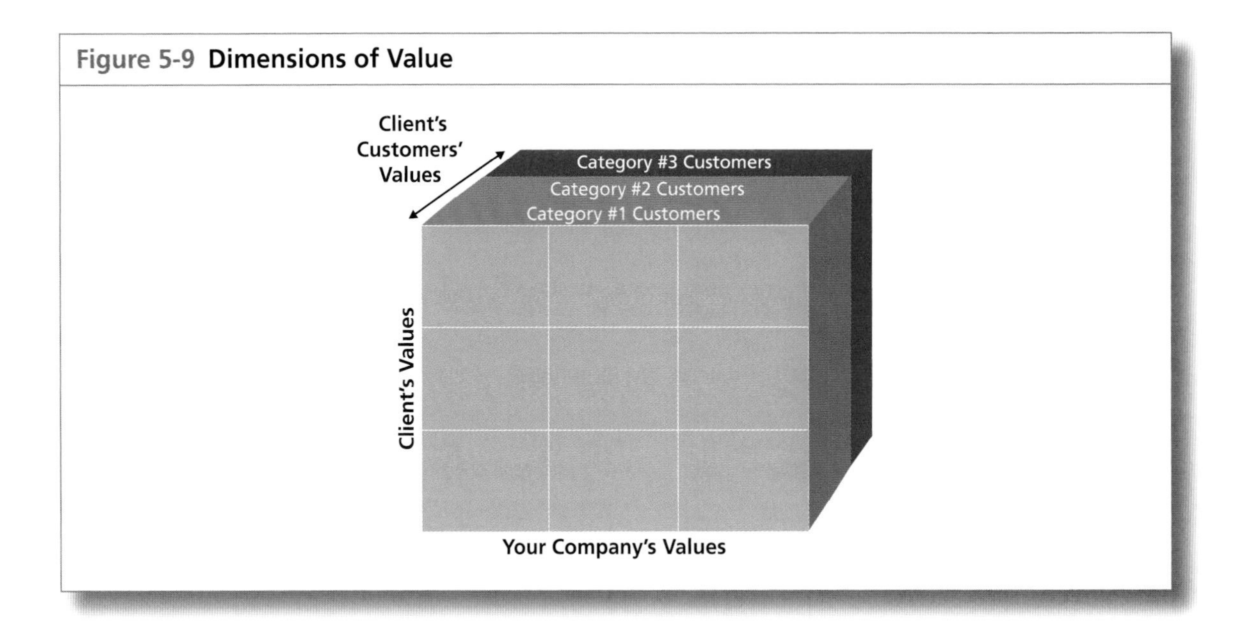

The necessity for such value initiatives is clear. As stated by a senior manager:

> I fully agree on the need for such value initiatives and also the need to make the importance clear to senior management. If we do not work in that direction, it will be very difficult for the companies to clearly state if they working efficiently enough, providing value to the customer and the stockholders, and in consequence being sufficiently predictive with regard to its future in the market.

The window into the future now seems to be getting clearer. As a guess as to what might happen, consider the following:

- Your clients will perform CVM activities with their clients to discover what value attributes are considered as important. Your client's success is achieved by providing superior value to their customers.
- These attributes will be presented to you at the initiation of the project such that you can create value-based metrics on your project using these attributes if possible. You must interact with the client to understand their value dimensions.
- You must then create value metrics. Be prepared to educate your client on the use of the metrics. It is a mistake to believe that your client will fully understand your value metrics approach.

Figure 5-10 Core Components of Project Management Value

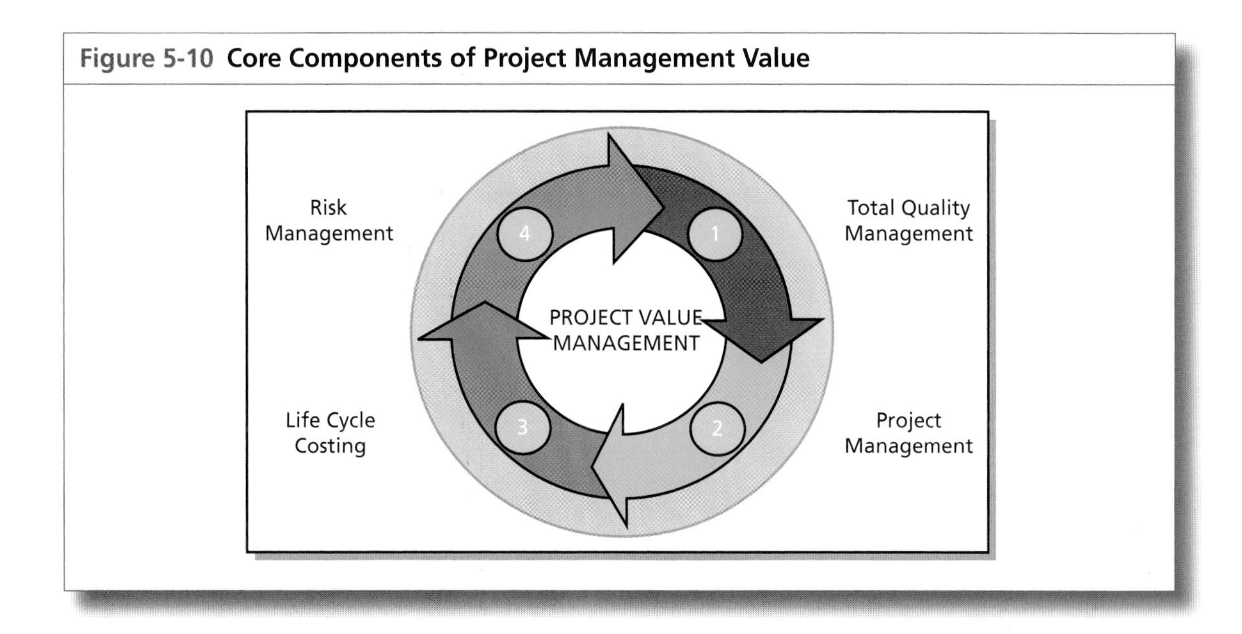

- Interact closely with your client to make sure you are fully aware of any changes in the value attributes they are finding in their customer value management efforts.
- Since value creation is a series of key and informed decisions, be prepared for value attribute tradeoffs and changes to your value metrics.

Value is now being introduced into project management practices. Value management practices have been with us for several decades and have been hidden under the radar screen in many companies. Some companies performed these practices in value engineering departments. Today the primary processes of value management, as shown in Figure 5–10, are becoming readily apparent.[10]

Although only four processes appear in Figure 5–10, it could be argued that all of the areas of knowledge in the *PMBOK® Guide* are part of project value management.

10. For an excellent book discussing the evolution of value management, see Michel Thiry, *Value Management Practice*, Newtown Square, PA: The Project Management Institute, 1997. Another good reference, which discusses how value can be used in making project decisions is Thomas G. Lechler and John C. Byrne, *The Mindset for Creating Project Value*, Newtown Square, PA: The Project Management Institute, 2010. Mel Schnapper and Steven Rollins created an excellent book, *Value-Based Metrics for Improving Results; An Enterprise Project Management Toolkit*, Ft. Lauderdale, FL: J. Ross Publishers, 2006. The authors discuss how this technique was applied at 3M Corporation. Chapters 17–25 directly relate value-based metrics to the areas of knowledge in the *PMBOK® Guide*.

5.9 BACKGROUND TO METRICS

You have been managing a project for the past several months. During that time, the client appeared quite happy with your performance, especially because all of the status reports indicated that your performance was within time and cost. You patted yourself on the back for doing a good job, you received accolades from your team and management, and then reality set in as the end of the project neared; the client was unhappy with the end result and didn't believe that they were receiving the value that they expected when the project was initiated. The client even commented that they probably should have canceled the project before wasting all of this money. What went wrong?

The problem can be addressed in one word, *metrics*, or perhaps we should say using the wrong metrics or the lack of metrics that could have projected or demonstrated value throughout the project. Having some metrics is certainly better than having no metrics at all. However, having the right metrics, especially the inclusion of metrics that can in some manner describe and communicate the value of the project, is best. Metrics must fully communicate what was needed by the customer.

When projects fail, we conduct debriefing sessions that focus more on blame-laying and finger-pointing than identifying the root cause of the failure. Sometimes, we go through meticulous pain to identify every possible reason for a failure without identifying the most critical and real causes of the failure. As an example, in some industries such as IT, surveys disclose a multitude of causes of failure. Well-known IT surveys include:

- The Standish Group Chaos Reports (1995–2010)
- The OASIG Study (1995)
- The KPMG Canada Survey (1997)
- The Bull Survey (1998)

Yet in each of these surveys, very little effort is expended on the fact that the wrong metrics may have been used. Also, many of the causes for failure could have been prevented if metrics had been established to track the potential causes of failure. Then we wonder why these surveys show that less than 30 percent of the IT projects are completed successfully.

Every year we publish surveys that repeatedly show basically the same reasons for the failures of IT projects. We have to wonder why the same causes appear year after year, and yet we refuse to identify new metrics to track these causes of failure. Just imagine how many billions of dollars could have been saved on IT projects if we started with the right metrics.

Measuring performance on a project is more than looking at just time and cost. For more than half a century, we taught project managers that the "holy grail" of performance metrics was getting the job done within time and cost. Unfortunately, achieving time and cost does not guarantee that business value will be there at the end of the project nor does it guarantee that you have been working on the right project. Instead, it simply means that you spent what you thought that you would spend, and it took you as long as you thought. It does not mean that the project was or will be a success.

Redefining Success

Our definition of project success for decades was meeting the proverbial triple constraints, as shown in Figure 5–11. Time and cost were two of the three sides to the triangle, and the third side was scope, technology, performance, or quality, depending on who was defining success. Whenever other constraints appeared, such as risk, business value, image reputation, safety and sustainability, they were inserted into the center of the triangle with the belief that they either elongated or compressed the boundary triple constraints.

Today, project management practitioners agree that there are more than three constraints on most projects, and we refer to the constraints as competing constraints. The old adage of working with just the triple constraints has gone by the wayside. We also prioritize the competing constraints. As an example of prioritization, the design and development of new attractions at Disneyworld and Disneyland were characterized by six constraints; time, cost, scope, safety, aesthetic value, and quality. During my consulting with Disney, it was apparent the three priority constraints were safety, aesthetic value, and quality. If and when tradeoffs were required, the only options were tradeoffs on time, cost, and scope. Safety, aesthetic value, and quality constraints were considered untouchable.

Figure 5-11 The Traditional Triple Constraints

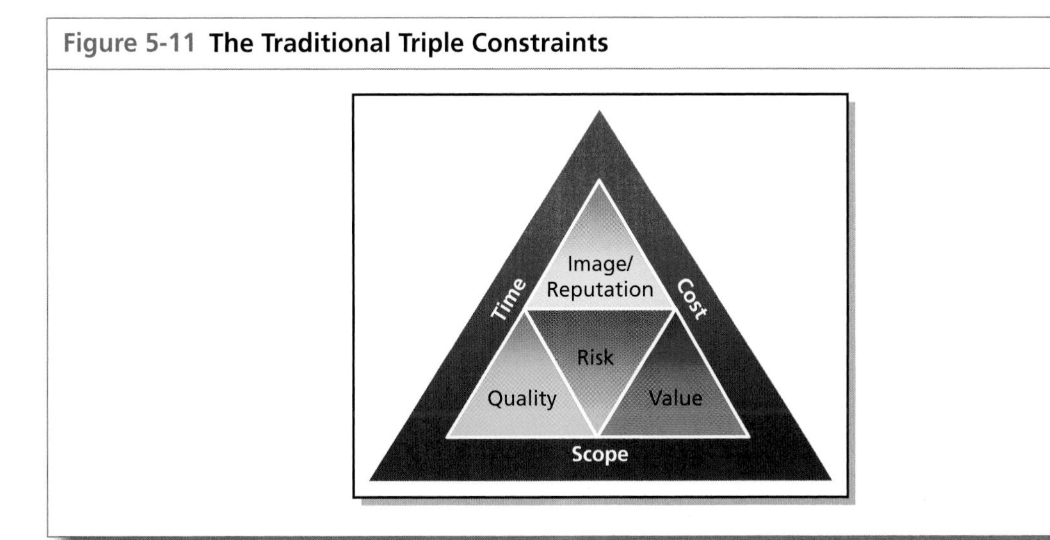

One of the first steps on a project is the agreement between the contractor and the client/stakeholders on the requirements of the project and the various limitations or constraints. For years, contractors, clients, and stakeholders each had their own definitions of success, creating havoc with performance reporting. Every project can have different constraints and a different definition of success, but the final definition of project success must be agreed to jointly by the contractor and the client/stakeholders. Otherwise, confusion will reign as to what is or is not important.

The need today for more than just three constraints to define success is quite apparent. Unfortunately, as with most changes to the way we do business, new headaches appear. For example, new constraints require new and more sophisticated metrics. For years we abided by the rule of inversion, which stated that time and cost were the only two metrics that needed to be tracked because they were the easiest to measure. The more difficult metrics, which could provide an indication of the project's true value, were omitted because they could not be measured. All project management software packages track time and cost, and many of the packages track only time and cost as the core metrics. However, as stated previously, time and cost alone cannot accurately determine performance and the success or value of a project. They tell only part of the story.

As we become more experienced in the use of project management and undertake more complex projects, we realize the necessity for using additional constraints to validate performance and success. This is a necessity to create a project value baseline, and will be discussed later in Section 5.21. For every competing constraint that appears on projects today, we must have one or more metrics to track that constraint. This will require major enhancements to many of the project management software packages currently in the marketplace.

The Growth in the Use of Metrics

For more than two decades, we reluctantly recognized and understood the cost of paperwork on projects and yet did nothing about it. Some companies estimated that 8 to 10 hours were needed for each page of a report or handout we gave to the customer. This included organizing the report, typing, proofing, editing, retyping, graphic arts, approvals, reproduction, distribution, classification, storage, and destruction. Based upon the cost per billable hour, it was not uncommon for a company to spend $1200 to $2000 per page.

The solution to this problem was quite simple; let's go to paperless project management, but in order to minimize our cost of paperwork, we must go to other forms of conveying information such as dashboards. The purpose of a dashboard is to convert raw data into meaningful information that can be easily understood and used for informed decision making.

Thus, the need existed for effective metrics that could appear in dashboard reporting systems. The value of dashboards was now quite clear:

- Reduction or consolidation of reports
- Less time wasted in preparing and reading reports
- Reduction in time needed for project monitoring and control
- Informed decision-making based upon current or real-time data
- More time available for important project management work

Unfortunately, we tend today to add in more artwork than we need to the dashboards, a trend that has resulted in a new term, "infographics." Some problems with the growth of infographics include:

- There is a heavy focus on designs, colors, images, and text rather than the quality of the information being presented.
- A decline in the quality of the information makes it difficult for stakeholders to use the data properly.
- There are too many pretty graphics that can be misleading and hard to understand.
- The dashboard has been converted from a project management performance tool to a marketing/sales tool.
- Some graphic artists do not understand or utilize information visualization best practices.
- Allowing political factors to dictate the design.
- Rushing into often complex technology before understanding the needs of the user group.
- Allowing metric scope creep because of to a lack of a formalized metric selection process.
- Failing to understand which metrics are:
  - Must have—fundamental to the requirements
  - Nice to have—but not a necessity (partial value)
  - Can wait—does not add value
  - Not needed—forget it (no value at all)

We must have better and clearer representation of the metrics we select.

Companies have been using metrics for business applications and developing business strategies for some time. However, the application of business metrics to project management is difficult to achieve because of the differences as shown in Table 5–7. Business and project metrics both have important use, but in different contexts.

For project management applications, companies used the rule of inversion (i.e., selecting only metrics that were easy to measure and track) and created a list of six core metrics as shown in Table 5–8. Some companies use additional or supporting metrics such as:

TABLE 5-7 Business versus Project Metrics/ KPIs

VARIABLE	BUSINESS/FINANCIAL	PROJECT
Focus	Financial Measurement	Project Performance
Intent	Meeting Strategic Goals	Meeting Project Objectives, Milestones, and Deliverables
Reporting	Monthly or Quarterly	Real-Time Data
Items to be looked at	Profitability, market share, repeat business, number of new customers, etc.	Adherence to competing constraints, validation, and verification of performance
Length of use	Decades of even longer	Life of the project
Use of the data	Information flow and changes to the strategy	Corrective action to maintain baselines
Target audience	Executive management	Stakeholders and working levels

TABLE 5-8 The Core Metrics

MEASURE	INDICATOR
Time	Schedule Performance Index
Cost	Cost Performance Index
Resources	Quality and Number of Actual versus Planned Staff
Scope	Number of Change Requests
Quality	Number of Defects against User Acceptance Criteria
Action Items	Number of Action Items behind Schedule

- **Deliverables (schedule):** Late versus on time
- **Deliverables (quality):** Accepted versus rejected
- **Management reserve:** Amount available versus amount used
- **Risks:** Number of risks in each core metric category
- **Action items:** Action items in each core category
- **Action items aging:** How many action item not completed are over 1 month, 2 months, 3 or more months

The core metrics are often represented by a dashboard labeled "Health Metrics," as shown in Figure 5–12. Most upper-management and customer organizations look for a single, simple metric that will indicate to them if a given project is on the road to success. If a single value-based metric does

Figure 5-12 Core Project Health Metrics

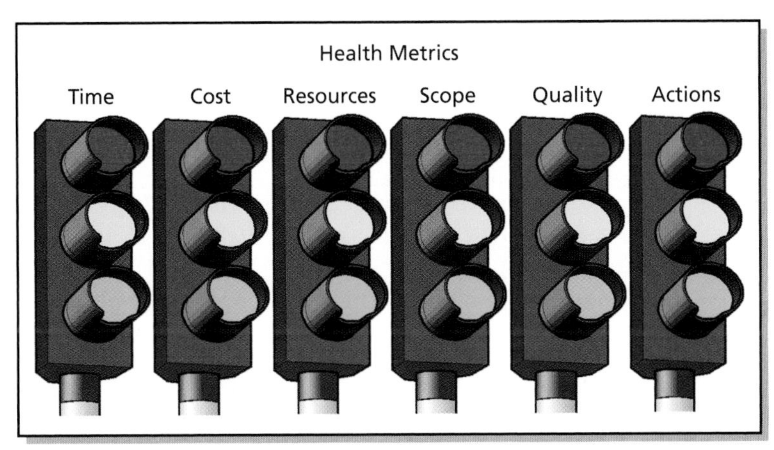

not exist, then they usually accept some or all of the core metrics shown in Figure 5-12. A simple signal light is all they may be looking for:

- **Green:** On the road to success (No action required)
- **Yellow:** Problems have surfaced that can derail the progress of the project. (Upper management and the customer will want to know what is being done to solve these problems)
- **Red:** Progress has stopped because the project is derailed. (Upper management and the customer want to know what resources they should apply to get the project healthy or is it beyond repair so that it has to be abandoned. *Is it no longer of value?*)

This is the management/customer value metric in its simplest form. According to an aerospace firm:

> The challenge is identifying what the intrinsic value of the project is. This can only be defined by a dedicated process to obtain consensus from management and the customer. Once this is defined and agreed to, then the project manager can select the minimum set of metrics that will allow him to gauge the success of his efforts. The set of metrics must be at a minimum so that the tracking of metrics does not in itself become a burden to the successful progress of the project.

Unfortunately, there are problems with using just the core metrics to explain the health of a project:

- The core metrics are usually interdependent and must be considered together to get an accurate picture of status
- Additional metrics may have to be added specific to the project at hand
- Explaining the core metrics by the colors of red, yellow, and green can be confusing
- Core metrics are similar to vital signs taken when visiting a doctor's office. Doctors always take the same core metrics; height, weight, temperature, and blood pressure. From these core metrics alone, the doctor usually cannot diagnose the problem or prescribe a corrective course of action. Additional metrics such as blood work and X-rays may be required.

We have known for some time about the importance of using metrics other than just time and cost, or even the core metrics. Knowing it and doing something about it are two different things, however. In the past, we avoided using more metrics because we did not know how to measure them. Today, books in the marketplace promote the concept that "anything can be measured."

Techniques have been developed by which we can measure image, reputation, goodwill, and customer satisfaction, just to name a few. Some of the measurement techniques include:[11]

- Observations
- Ordinal (e.g., four or five stars) and nominal (e.g., male or female) data tables
- Ranges, sets of value, number, headcount, percentages
- Simulation
- Statistical measurement
- Calibration estimates and confidence limits
- Decision models (EV, EVPI, etc.)
- Sampling techniques
- Decomposition techniques
- Direct versus indirect measurement
- Human judgment

5.10 SELECTING THE RIGHT METRICS

Because of these measurement techniques, companies are now tracking a dozen or more metrics on projects. While this sounds good, it has created the additional problem of potential information overload. Having too

11. For a description of several of these techniques, see Douglas W. Hubbard, *How to Measure Anything*, Hoboken, NJ: John Wiley & Sons, 2007.

many performance metrics may provide the viewers with more information than they actually need and; therefore, they may not be able to discern the true status or what information is really important. It may be hard to ascertain what is important and what is not, especially if decisions must be made. Providing too few metrics can make it difficult for the viewers to make informed decisions. There is also a cost associated with metric measurement, and we must determine if the benefits of using this many metrics outweigh the costs of measurement. Cost is important because we tend to select more metrics than we actually need.

There are three categories of metrics:

- **Traditional metrics:** These metrics are used for measuring the performance of the applied project management discipline more so than the results of the project and how well we are managing according to the predetermined baselines. (e.g., cost variance and schedule variance)
- **Key performance indicators (KPIs):** These are the few selected metrics that can be used to track and predict whether the project will be a success. These KPIs are used to validate that the critical success factors (CSFs) defined at the initiation of the project are being met. (e.g., time-at-completion, cost-at-completion, and customer satisfaction surveys)
- **Value (or value reflective) metrics:** These are special metrics that are used to indicate whether the stakeholders' expectations of project value are or will be met. Value metrics can be a combination of traditional metrics and KPIs. (value-at-completion and time to achieve full value)

Each type of metric has a primary audience, as shown in Table 5–9.

There can be three information systems on a project:

- One for the project manager
- One for the project manager's superior or parent company
- One for the stakeholders and the client

There can be a different set of metrics and KPIs for each of these information systems.

TABLE 5-9 Audiences for Various Metrics

TYPE OF METRIC	AUDIENCE
Traditional metrics	Primarily the project manager and the team, but may include the internal sponsor(s) as well
Key performance indicators	Some internal usage but mainly used for status reporting for the client and the stakeholders
Value metrics	Can be useful for everyone but primarily for the client

Traditional metrics, such baselines as the cost, scope, and schedule, track and can provide information on how well we are performing according to the processes in each Knowledge Area or Domain Area in the *PMBOK® Guide*. Project manager must be careful not to micromanage their project and establish 40 to 50 metrics.

Typical metrics may include:

- Number of assigned versus planned resources
- Quality of assigned versus planned resources
- Project complexity factor
- Customer satisfaction rating
- Number of critical constraints
- Number of cost revisions
- Number of critical assumptions
- Number of unstaffed hours
- Percent of total labor hours on overtime
- Cost variance
- Schedule performance index
- Cost performance index

This is obviously not an all-inclusive list. These metrics may have some importance for the project manager but not necessarily the same degree of importance for the client and the stakeholders.

Clients and stakeholders are interested in critical metrics or KPIs. These chosen few metrics are reported to the client and stakeholders and provide an indication of whether or not success is possible; however, they do not necessarily identify if the desired value will be achieved. The number of KPIs is usually determined by the amount of real estate on a computer screen. Most dashboards can display between six and ten icons or images where the information can be readily seen with reasonable ease.

To understand what a KPI means requires a dissection of each of the terms:

- **Key:** A major contributor to success or failure
- **Performance:** Measurable, quantifiable, adjustable, and controllable elements
- **Indicator:** Reasonable representation of present and future performance

Obviously, not all metrics are KPIs. There are six attributes of a KPI, and these attributes are important when identifying and selecting the KPIs.

- **Predictive:** Able to predict the future of this trend
- **Measurable:** Can be expressed quantitatively
- **Actionable:** Triggers changes that may be necessary
- **Relevant:** The KPI is directly related to the success or failure of the project

- **Automated:** Reporting minimizes the chance of human error
- **Few in number:** Only what is necessary

Applying these six attributes to traditional metrics is highly subjective and will be based upon the agreed -upon definition of success, the critical success factors (CSFs) that were selected, and possibly the whims of the stakeholders. There can be a different set of KPIs for each stakeholder based upon each stakeholder's definition of project success and final project value. This could significantly increase the costs of measurement and reporting, especially if each stakeholder requires a different dashboard with different metrics.

Previously we identified 12 possible metrics that could be used on projects, but how many of those 12 are actually regarded as a KPI? If we apply the first five of the six KPI attributes identified above, we could end up with the representation as shown in Table 5–10.

Only six of the twelve metrics (1, 3, 4, 8, 11, and 12) may be regarded as a KPI and, once again, this is often a highly subjective selection process. In this example, these would be the critical metrics that would be shown on the project dashboard and could be a necessity for informed decision making. The other metrics can still be used, but the reader might need to "drill down" on the screens to get access to the traditional metrics.

TABLE 5-10 Selecting the KPIs

	PREDICTIVE	MEASURABLE	ACTIONABLE	RELEVANT	AUTOMATED
1. # of assigned vs. planned res.	Yes	Yes	Yes	Yes	Yes
2. Quality of assigned vs. planned res.		Yes		Yes	Yes
3. Project complexity factor	Yes	Yes	Yes	Yes	Yes
4. Customer satisfaction rating	Yes	Yes	Yes	Yes	Yes
5. # of critical constraints		Yes	Yes	Yes	
6. # of cost revisions		Yes	Yes	Yes	Yes
7. # of critical assumptions		Yes	Yes	Yes	
8. # of unstaffed hours	Yes	Yes	Yes	Yes	Yes
9. % of overtime labor hours		Yes	Yes		Yes
10. Cost variance		Yes			Yes
11. Schedule performance index	Yes	Yes	Yes	Yes	Yes
12. Cost performance index	Yes	Yes	Yes	Yes	Yes

5.11 THE FAILURE OF TRADITIONAL METRICS AND KPIs

While some people swear by metric and KPIs, there are probably more failures than success stories. Typical causes of metric failure include:

- Performance is expressed in traditional or financial terms only
- The use of measurement inversion; using the wrong metrics
- No link of performance metrics to requirements, objectives, and success criteria
- No link to whether or not the customer was satisfied
- Lack of understanding as to which metrics indicate project value

Metrics used for business purposes tend to express all information in financial terms. Project management metrics cannot always be expressed in financial terms. Also, in project management we often identify metrics that cannot effectively predict project success and/or failure and are not linked to the customer's requirements.

Perhaps the biggest issue today is in which part of the value chain metrics are used. Michael Porter, in his book *Competitive Advantage* (Free Press, 1985), used the term value chain to illustrate how companies interact with upstream suppliers, the internal infrastructure, downstream distributors and end-of-the-line customers. While metrics can be established for all aspects of the value chain, most companies do not establish metrics for how the end-of-the-line customer perceives the value of the deliverable. Those companies that have developed metrics for this part of the value chain are more likely doing better than those that have not. These are identified as customer-related value metrics.

5.12 THE NEED FOR VALUE METRICS

In project management, it is now essential to create metrics that focus not only on business (internal) performance but also on performance toward customer satisfaction. If the customer cannot see the value in the project, then the project may be canceled and repeat business will not be forthcoming. Good value metrics can also result in less customer and stakeholder interference and meddling in the project.

The performance metric process for project management is shown in Figure 5–13.

The need for an effective metrics management program that focuses on value-based metrics is clear:

- There must be a customer/contractor/stakeholders agreement on how a set of metrics will be used to define success or failure; otherwise, you have just best guesses. Value metrics will allow for a better agreement on the definitions of success and failure.

Figure 5-13 **Typical Steps in the Performance Metrics Process**

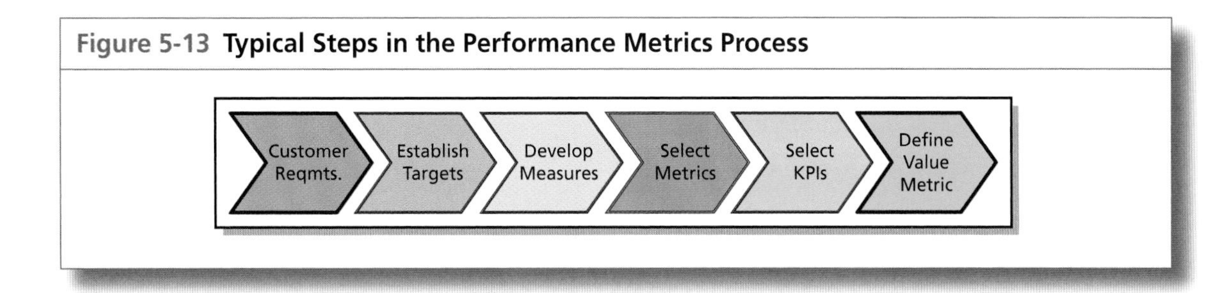

- Metric selection must cover the reality of the entire project; this can be accomplished with a set of core metrics supported by a value metric.
- A failure in effective metrics management, especially value metrics, can lead to stakeholder challenges and a loss of credibility.

We need to develop value-based metrics that can forecast stakeholder value, possibly shareholder value, and most certainly project value. Most models for creating this metric are highly subjective and are based upon assumptions that must be agreed upon upfront by all parties. Traditional value-based models that are used as part of a business intelligence application are derivatives of the QCD Model (Quality, Cost, and Delivery).

5.13 CREATING A VALUE METRIC

The ideal situation would be the creation of a single value metric that the stakeholders can use to make sure that the project is meeting or exceeding the stakeholder's expectation of value. The value metric can be a combination of traditional metrics and KPIs. Discussing the meaning of a single value metric may be more meaningful than discussing the individual components; the whole is often greater than the sum of the parts.

There must be support for the concept of creating a value metric. According to a global IT consulting company:

> There has to be buy-in from both sides on the importance and substance of a value metric; it can't be the latest fad—it has to be understood as a way of tracking the value of the project.

Typical criteria for a value metric may be:

- Every project will have at least one value metric or value KPI. In some industries, it may not be possible to use just one value metric.
- There may be a limit, such as five, for the number of value attributes that are part of the value metric. As we mature in the use of value metrics, the

number of attributes can grow or be reduced. Not all attributes that we would like to have will be appropriate or practical.
- There will be weighting factors assigned to each component.
- A project value baseline (discussed in Section 5.21) will be tracked using the value metrics
- The weighting factors and the component measurement techniques will be established by the project manager and the stakeholders at the onset of the project. There may be company policies on assigning the weighting factors.
- The target boundary boxes for the metrics will be established by the project manager and possibly the project management office (PMO). If a PMO does not exist, then there may be a project management committee taking responsibility for accomplishing this, or it may be established by the funding organization.

To illustrate how this might work, let's assume that, for the IT projects you perform for your stakeholders, the attributes of the value metric will be:

- Quality (of the final software package)
- Cost (of development)
- Safety protocols (for security of information)
- Features (functionality)
- Schedule or timing (for delivery and implementation)

These attributes are agreed to by you, the client, and the stakeholders at the onset of the project. The attributes may come from your metric/KPI library or may be new attributes. Care must be taken to make sure that your organizational process assets can track, measure, and report on each attribute. Otherwise, additional costs may be incurred and these costs must be addressed up front so that they can be included in the contract price.

Time and cost are generally attributes of every value metric. However, there may be special situations where neither time, nor cost, nor both are value metric attributes:

- The project must be completed by law, such as environmental projects, where failure to perform could result in stiff penalties.
- The project is in trouble, but necessary, and we must salvage. whatever value we can.
- We must introduce a new product to keep up with the competition regardless of the cost.
- Safety, aesthetic value, and quality are more important than time, cost, or scope.

Other attributes are almost always included in the value metric to support time and cost.

The next step is to set up targets with thresholds for each attribute or component. This is shown in Figure 5–14. If the attribute is cost, then we might say that performing within ±10 percent of the cost baseline is normal performance. Performing at greater than 20 percent over budget could be disastrous, whereas performing at more than 20 percent below budget is superior performance. However, there are cases where a +20 percent variance could be good and a −20 percent variance could be bad.

The exact definition or range of the performance characteristics could be established by the PMO if company standardization is necessary or through an agreement with the client and the stakeholders. In any event, targets and thresholds must be established.

The next step is to assign value points for each of the cells in Figure 5–14, as shown in Figure 5–15. In this case, two value points were assigned to the cell labeled "Performance Target." The standard approach is to then assign points in a linear manner above and below the target cell. Nonlinear applications are also possible, especially when thresholds are exceeded.

In Table 5–11, weighting factors are assigned to each of the attributes of the value metric. As before, the weighting percentages could be established by the PMO or through an agreement with the client (i.e., funding organization) and the stakeholders. The use of the PMO might be for company standardization on the weighting factors. However, it sets a dangerous precedence when the weighting factors are allowed to change indiscriminately.

Figure 5-14 The Value Metric/KPI Boundary Box

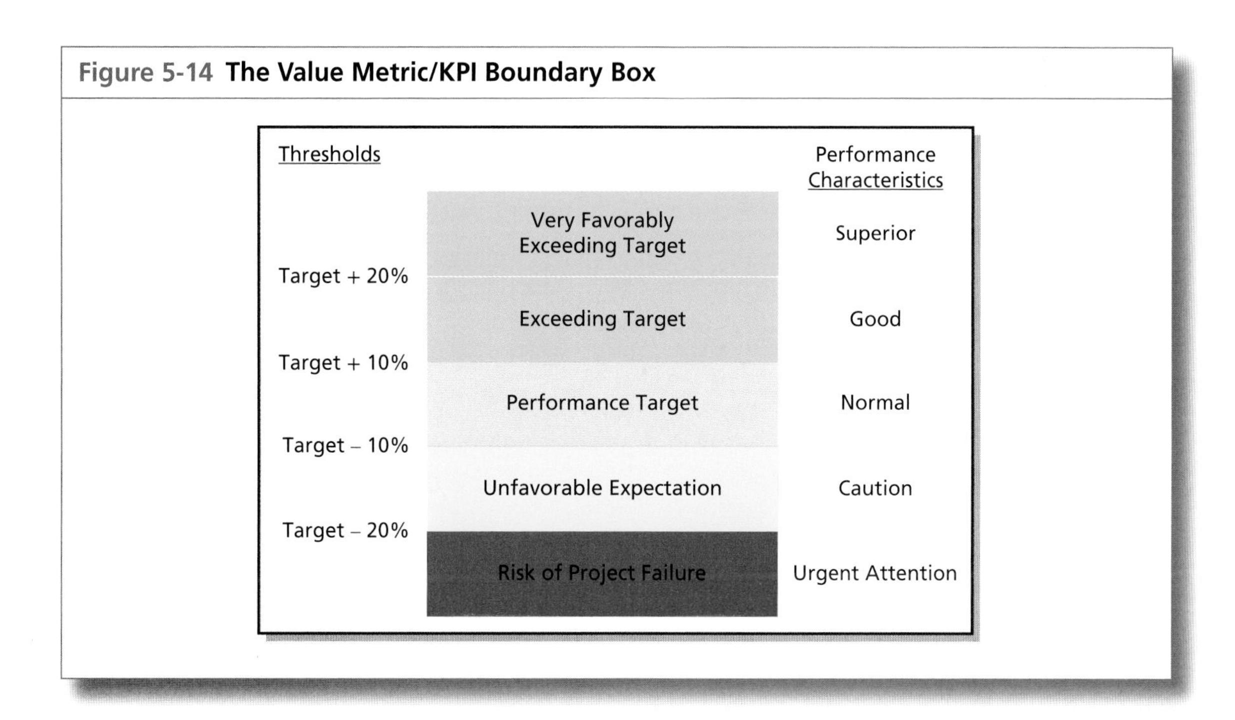

Figure 5-15 Value Points for a Boundary Box

	Performance Characteristics	Value Points
Very Favorably Exceeding Target	Superior	4
Exceeding Target	Good	3
Performance Target	Normal	2
Unfavorable Expectation	Caution	1
Risk of Project Failure	Urgent Attention	0

TABLE 5-11 Value Metric Measurement

VALUE COMPONENT	WEIGHTING FACTOR	VALUE MEASUREMENT	VALUE CONTRIBUTION
Quality	10%	3	0.3
Cost	20%	2	0.4
Safety	20%	4	0.8
Features	30%	2	0.6
Schedule	20%	3	0.6
			TOTAL = 2.7

Now, we can multiply the weighting factors by the value points and sum them up to get the total value contribution. If all of the value measurements indicated that we were meeting our performance targets, then 2.0 would be the worth of the value metric. However, in this case, we are exceeding performance with regard to quality, safety and schedule, and therefore the final worth of the value metric is 2.7. This implies that the stakeholders are receiving additional value that is most likely meeting or exceeding expectations.

There are still several issues that must be considered when using this technique:

- We must clearly define what is meant by normal performance. The users must understand what this means. Is this level actually our target level or is it the minimal acceptable level for the client? If it is our target level, then having a value below 2.0 might still be acceptable to the client if our target were greater than what the requirements asked for.
- The users must understand the real meaning of the value metric. When the metric goes from 2.0 to 2.1 how significant is that? Statistically, this is a 5 percent increase. Does it mean that that the value increased 5 percent? How can we explain to a layman the significance of such an increase and the impact on value?

Value metrics generally focus on the present and/or future value of the project and may not provide sufficient information as to other factors that may affect the health of the project. As an example, let's assume that the value metric is quantitatively assessed at a value of 2.7. From the customer's perspective, they are receiving more value than they anticipated. But other metrics may indicate that the project should be considered for termination. For example,

- The value metric is 2.7 but the remaining cost of development is so high that the product may be over-priced for the market.
- The value metric is 2.7 but the time-to-market will be too late.
- The value metric is 2.7 but a large portion of the remaining work packages have a very high critical risk designation.
- The value metric is 2.7 but significantly more critical assumptions are being introduced.
- The value metric is 2.7 but the project no longer satisfies the client's needs.
- The value metric is 2.7 but your competitors have introduced a product with a higher value and quality.

In Table 5–12, we reduced the number of features in the deliverable, which allowed us to improve quality and safety as well as accelerate the schedule. Since the worth of the value metric is 2.4, we are still providing additional value to the stakeholders.

In Table 5–13, we have added additional features as well as improving quality and safety. However, to do this, we have incurred a schedule slippage and a cost overrun. The worth of the value metric is now 2.7, which implies that the stakeholders are still receiving added value. The stakeholders may be willing to incur the added cost and schedule slippage because of the added value.

TABLE 5-12 A Value Metric with a Reduction in Features

VALUE COMPONENT	WEIGHTING FACTOR	VALUE MEASUREMENT	VALUE CONTRIBUTION
Quality	10%	3	0.3
Cost	20%	2	0.4
Safety	20%	4	0.8
Features	30%	1	0.3
Schedule	20%	3	0.6
			TOTAL = 2.4

TABLE 5-13 A Value Metric with Improved Quality, Features, and Safety

VALUE COMPONENT	WEIGHTING FACTOR	VALUE MEASUREMENT	VALUE CONTRIBUTION
Quality	10%	3	0.3
Cost	20%	1	0.2
Safety	20%	4	0.8
Features	30%	4	1.2
Schedule	20%	1	0.2
			TOTAL = 2.7

TABLE 5-14 Changing the Weighting Factors

VALUE COMPONENT	NORMAL WEIGHTING FACTOR	WEIGHTING FACTORS IF WE HAVE A SIGNIFICANT SCHEDULE SLIPPAGE	WEIGHTING FACTORS IF WE HAVE A SIGNIFICANT COST OVERRUN
Quality	10%	10%	10%
Cost	20%	20%	40%
Safety	20%	10%	10%
Features	30%	20%	20%
Schedule	20%	40%	20%

Whenever it appears that we may be over budget or behind schedule, we can change the weighting factors and overweigh those components that are in trouble. As an example, Table 5–14 shows how the weighting factors can be adjusted. Now, if the overall worth of the value metric exceeds 2.0 with the adjusted weighting factors, the stakeholders may still consider the

TABLE 5-15 Weighting Factor Ranges

VALUE COMPONENT	MINIMAL WEIGHTING VALUE	MAXIMUM WEIGHTING VALUE	NOMINAL WEIGHTING VALUE
Quality	10%	40%	20%
Cost	10%	50%	20%
Safety	10%	40%	20%
Features	20%	40%	30%
Schedule	10%	50%	20%

continuation of the project. Sometimes, companies identify minimum and maximum weights for each component, as shown in Table 5–15. However, there is a risk that management may not be able to adjust to and accept weighting factors that can change from project to project, or even during a project. Also, standardization and repeatability of the solution may disappear with changing weighting factors.

Companies are generally reluctant to allow project managers to change weighting factors once the project is under way and may establish policies to prevent unwanted changes from occurring. The fear is that the project manager may change the weighting factors just to make the project look good. However, there are situations where a change may be necessary:

- Customers and stakeholders are demanding a change in weighting factors possibly to justify the continuation of project funding
- The risks of project have changed in downstream life cycle phases and a change in weighting factors is necessary
- As the project progresses, new value attributes are added into the value metric
- As the project progresses, some value attributes no longer apply and must be removed from the value metric
- The enterprise environment factors have changed requiring a change in the weighting factors
- The assumptions have changed over time
- The number of critical constraints have changed over time

We must remember that project management metrics and KPIs can change over the life of a project and, therefore, the weighting factors for the value metric may likewise be susceptible to changes.

Sometimes, because of the subjectivity of this approach, when the information is presented to the client, we should include identification of which measurement technique was used for each target. This is shown in

TABLE 5-16 Weighting Factors and Measurement Techniques

VALUE COMPONENT	WEIGHTING FACTOR	MEASUREMENT TECHNIQUE	VALUE MEASUREMENT	VALUE CONTRIBUTION
Quality	10%	Sampling Techniques	3	0.3
Cost	20%	Direct Measurement	2	0.4
Safety	20%	Simulation	4	0.8
Features	30%	Observation	2	0.6
Schedule	20%	Direct Measurement	3	0.6

Table 5–16. The measurement techniques may be subject to negotiations at the beginning of the project.

The use of metrics and KPIs has been with us for decades, but the use of a value metric is relatively new. Therefore, failures in the use of this technique are still common and may include:

- Is not forward looking; the value metric focuses on the present rather than the future
- Does not go beyond financial metrics and, thus, fails to consider the value in knowledge gained, organizational capability, customer satisfaction, and political impacts
- Believing that value metrics (and the results) that other companies use will be the same for your company
- Not considering how the client and stakeholders define value
- Allowing the weighting factors to change too often, to make the project's results look better

As with any new technique, additional issues always arise. Typical questions that we are now trying to answer in regard to the use of a value metric include:

- What if only three of the five components can be measured, for example, in the early life cycle phases of a project?
- In such a case where only some components can be measured, should the weighting factors be changed or normalized to 100 percent, or left alone?
- Should the project be a certain percent complete before the value metric has any real meaning?
- Who will make decisions as to changes in the weighting factors as the project progresses through its life cycle phases?
- Can the measurement technique for a given component change over each life cycle phase or must it be the same throughout the project?
- Can we reduce the subjectivity of the process?

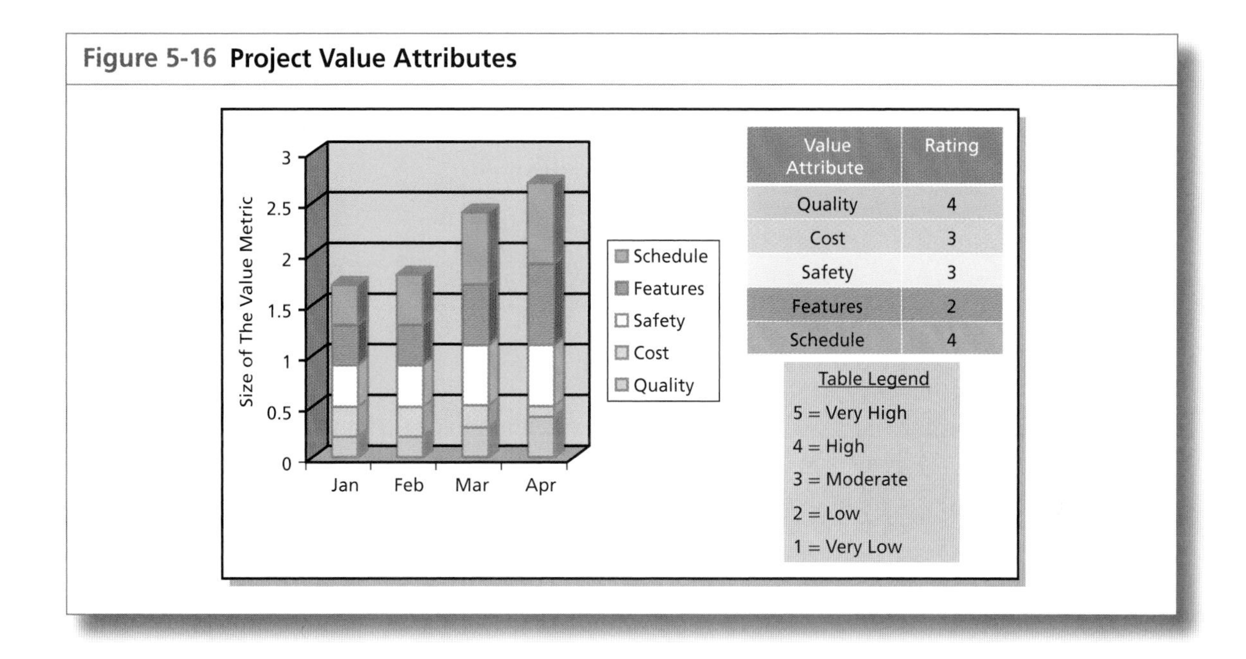

Figure 5-16 Project Value Attributes

5.14 PRESENTING THE VALUE METRIC IN A DASHBOARD

Figure 5–16 illustrates how the value metric may appear on a dashboard. The value attributes and ratings in the table in the upper-right corner reflect the values in the month of April. In January, the magnitude of the value metric was about 1.7. In April, the magnitude is 2.7.

The stakeholders can easily see the growth in value over the past four months. They can also see that four of the five attributes have increased their value over this time period, whereas the cost attribute appears to have diminished in value.

Eventually, as we become more knowledgeable in the use of value metrics, we may end up with a single value metric that can be obtained through an objective and automated process. In the near term, however, we can expect the value metric process to be more qualitative than quantitative and highly subjective based upon the value attributes that were selected.

5.15 INDUSTRY EXAMPLES OF VALUE METRICS

This section provides examples of how various companies use value metrics. The number of companies using value metrics is still quite small. Some of the companies surveyed could not provide accurate weighting factors because the factors can change for each project. Other companies differentiated between project success and product success and stated that the value attributes and weighting factors were different for each as well. (Note: The names of the companies were withheld at the request of the companies.)

In several of the examples, there are descriptions of the attributes. In most cases, the attributes of the final value metric are a composite of various KPIs as discussed previously in this chapter.

Aerospace and Defense: (Company 1)

- Schedule: 25% (This would result from objective data via the project's earned value measurement system.)
- Cost: 25% (This would result from objective data via the project's earned value measurement system.)
- Technical factor: 30% (This would be based upon technical performance measures established at the beginning of the project.)
- Quality factor: 10% (This results from an ongoing audit of the adherence to the established quality standards and procedures.)
- Risk factor: 10% (This is based upon how well risk mitigation plans are implemented and followed.)

These percentages are for a generic project. Depending upon the nature of the project and the constraints imposed by the client, the percentages would be adjusted. The actual percentages would be coordinated with the client so that they would know and agree to the weighting to be used. The projects that this would apply to would result in a product or system, or both. Most of the clients in aerospace and defense are extremely sensitive to cost and schedule, which is why the percentages for those are as high as they are.

Aerospace and Defense: (Company 2)

- Quality: 35%
- Delivery: 25%
- Cost: 20%
- Technology: 5%
- Responsiveness: 10%
- General Management: 5%

Capital Projects:

- Revenue growth/generation: 30% (Our primary focus has been on growing revenue through leveraging alternative options.)
- Cost efficiencies: 30% (This has a direct bottom-line impact and seen almost equal to generating revenue.)
- Handle/market share growth: 20% (There's a revenue impact here as well.)
- Project schedule: 10% (We have many time-constrained projects because of our core business operating model. We have the natural tradeoff decisions; however, the first three factors are weighted more heavily.)
- Project cost: 10% (Generally tracked to ensure increasing costs do not overwhelm expected benefits)

IT Consulting [External Clients]: (Company 1) No percentages provided

- Risks
- Scope
- Resources
- Quality
- Schedule
- Overall status

IT Consulting [External Clients]: (Company 2)

- Quality: 40% (as determined by the feedback from internal engagement leaders review of project deliverables before they are submitted to the client, the number of iterations before the deliverables are client-ready, and eventually, by the client's satisfaction with the provided deliverables, and the manner in which client interactions and expectations were managed throughout the engagement, that is, did we make and keep our promises, did we offer the client a "no-surprise" experience, and did we demonstrate professionalism through effective interaction and communications?)
- Talent: 20% (as determined by feedback from Firm and engagement leaders as well as satisfaction of project team members in terms of how the project was planned and delivered, that is, was there a positive working environment that was created; were the views and opinions of individual team members valued and considered; were the roles and contributions of individual team members well defined, properly communicated, and understood by all involved; did the project provide an opportunity for personal and professional growth and development, etc.?)
- Marketplace: 10% (as determined by the evaluation of the Firm and engagement leadership as to the extent to which the given project demonstrated understanding of the client and the client's industry, as well as the extent to which the project contributed to the establishing or supporting the firm's preeminence within the given service line or industry)
- Financial: 30% (as determined by on-time and on-budget delivery of the project, the profitability of the engagement, the achieved recovery rate or the level of applied discount to standard rates, and the ability to submit invoices and collect payment within a reasonable amount of time)

IT Consulting [External Clients]: (Company 3)

- Customer Satisfaction: 30% (their perception on how well the project is going and satisfaction with the team and solution)
- Budget: 20%
- Schedule: 10%
- Solution Deployed: 20% (solution is being used by the customer in production and providing value)
- Support Issues: 10%
- Opportunity Generation: 10% (a successful project for us would also generate additional opportunities with the customer—difficult to measure until well after project completion)

These factors would apply to all industries and solutions. The only differences are that the Solution Deployed is not applicable in certain project types (such as a Health Check), and Opportunity Generation is only applicable when looking back in time after a project is completed.

IT Consulting [External Clients]: (Company 4)

- Customer Satisfaction/Conditions of Satisfaction: 30%
- Manage Expectations/Communication: 20%
- Usability/Performance: 20%
- Quality: 20%
- Cost: 10%

It is less about time and cost now and more about can/will the product/service be used and is the client happy with the delivery of the final product/service?

IT Consulting [External Clients]: (Company 5)

These value metric attributes are primarily for consulting in healthcare.

- Quality: 25%
- Cost: 20%
- Durations/Timeliness: 15%
- Resource Utilization: 10%
- Incorporation of Processes (clinical, technical, business focused): 30%

IT Consulting [External Clients]: (Company 6)

Similar to Company 5, this company also provided IT consulting services to healthcare but did not provide a breakdown of the value metric. They believe that the value metric (whether specifically outlined or not on a project) is the true reason why these projects continue to be funded beyond their schedule and budget.

IT Consulting [Internal]:

- Scope: 25%
- Project Client Satisfaction: 22.5%
- Schedule: 17.5%
- Budget: 17.5%
- Quality: 17.5%

Software Development [Internal]: No percentages provided

- Code: number of lines of code
- Language understandability: language and/or code is easy to understand and read
- Movability/immovability: The ease by which information can be moved
- Complexity: loops, conditional statements, etc . . .

- Math complexity: time and money to execute algorithms
- Input/output understandability: How difficult is it to understand the program?

Telecommunications: (Company 1)
- Financial: 35%
- Quality/Customer Satisfaction: 35%
- Process Adherence: 15%
- Teamwork: 15%

Telecommunications: (Company 2) No percentages provided
- Customer satisfaction
- Employee satisfaction
- Quality
- Financial
- Cost

New Product Development:
- Features/Functions: 35% (This is where the company believes it can differentiate itself from its competitors.)
- Time to market: 25%
- Quality: 25%
- Cost: 15%

Automotive Suppliers:
- Quality: 100%
- Cost: 100%
- Safety: 100%
- Timing: 100%

It is interesting to note that in this company there were four value metrics rather than just one, and the company believed that the four value metrics could not be combined into a single value metric. This is why 100 percent is assigned to each. In the auto industry, being less than 100 percent on each value metric could delay the launch of a product and create financial problems for the client and all of the suppliers.

Global Consulting: (Company 1) [Not industry specific and no weights]

This company's observations are that soft skills and personal attributes are some of the key factors affecting the outcome of projects and very little association has so far been made between that and project failure. This is why some of the soft skills are part of their value metric. The challenge is in quantifying these soft skills factors for customer value management. This will vary for each project, which is why percentages are not provided.

- **Management:** Consideration given to quick resolution of issues, minimal time wasted, minimal recycling of ideas, timely escalations.
- **Communications/relationship building:** Verbal and written status reports, weekly meetings, and the like. Agreed to at the start of the project.
- **Competency:** Impact and influence as well as performance and knowledge of project management.
- **Flexibility and commitment:** Balancing clients' requests and requirements with that of suppliers and third-party contractors—ability to monitor and control to bring optimum results.
- **Quality:** The quality definition would vary based on project deliverables. This would be influenced by the industry.
- **Usability:** How much of what is implemented can give immediate added value—this is on the premise that few organizations progress beyond the implemented subset in under two years by which time the original users have moved on and their successors would not have had the vision to move the systems to the next level. On the contrary, the reverse happens, as less functionality is utilized by the inheritors of the systems.
- **Delivery strategy:** This speaks directly to the strategy or approach for implementation of the solution offered to the client—the solution on paper could have seemed to meet the client's need, but if it is not executed optimally, it can cause added pain rather than gains. This speaks directly to governance approaches, production of specified deliverables, performing tasks on schedule, and the like.
- **Customer focus:** To what extent does the PM team seek the interest of the customer/client.
- **HSE:** The satisfaction level with the project team's compliance with health, safety, and environmental policies while performing duties at the customer's site.

Global Consulting: (Company 2) [Not industry-specific and no weights]

- Profitability
- Schedule
- Impact/Result
- Customer Satisfaction
- Safety
- Erosion

Erosion is the value of work performed in excess of what was originally estimated in terms of effort or duration, and not recovered through project change management procedures. In simple terms, erosion is the difference between billable work (that which has been estimated, proposed and accepted by the customer) and nonbillable work (that, which has not been estimated, proposed, and accepted by the customer).

5.16 USE OF CRISIS DASHBOARDS FOR OUT-OF-RANGE VALUE ATTRIBUTES

Most companies today that use dashboards to communicate with clients and stakeholders include drill-down capability on the dashboards. The top dashboard contains the KPIs and the value metric. If additional information is required, the drill-down capability allows more detailed information to appear on the screen.

Some companies have taken this concept a step further. Rather than requiring the viewer to drill down to find the cause of a potential problem, companies are creating a crisis dashboard that is similar to an exception report. All KPIs and value metric attributes that exceed the minimum threshold limits unfavorably will appear in the crisis dashboard. According to one of the contributing companies:

> What we have developed is a kind of algorithm that summarizes the status of the project. It is based upon the following:
>
> There are indicators or semaphores associated to several aspects of the project that we believe are contributors to value (cost, schedule, margin variation, risk, issues, pending invoicing, pending payments, milestones, etc.)
>
> The indicators get a green, yellow, or red colors, depending upon thresholds and variation percentages that are characteristic of each business unit.
>
> Colors have an allocated value (0,1,4).
>
> Then, a new overall indicator or metric is created called "**CLOA, calculated level of attention**" that goes from "very low" to "very high" or even to "requiring intensive care". The value of CLOA is assigned through a formula that takes into account the colors of the indicators and some of the absolute figures. The intent is not to create unnecessary alarms if the amounts involved are not significant.
>
> This single indicator is mainly used (until now) by the finance controllers, the local PMOs, the QA departments, portfolio managers, etc. to claim their attention on specific projects. The goal is of course to provide an **early detection** mechanism of failing projects, and to establish the corresponding corrective actions.
>
> The algorithm is not perfect, but it is becoming more and more useful as we refine it. In any case, the "owners" of this indicator are the finance controllers of the business unit, not the project managers. Only the finance controllers are able to modifying its value, and to report the reasons for modification. The report is then stored in the system for historical backup.

Rather than using a crisis dashboard, some companies simply use alerts. Alerts or attention-getters are indications that the KPI's threshold boundary condition or integrity level has been reached. In general, alerts serve as early warning signs that some action must be taken before the situation gets worse. The situation might actually be worse than it appears and stakeholders may assume the worst possible scenario. Alerts are similar to exception reports.

Not all KPIs will trigger alerts. Also, not all alerts indicate unfavorable trends. One company established an alert trigger if the company became too efficient and produced more units than the warehouse could handle. Malik describes several types of alerts that may be appropriate for enterprise dashboards:[12]

- Critical Alert
- Important Alert
- Informational Alert
- Public Alert
- Private Alert
- Unread Alert

In a project management environment, there may be three types of alerts:

- Project Team Alert
- Management Alert
- Stakeholder Alert

5.17 ESTABLISHING A METRICS MANAGEMENT PROGRAM

The future of project management must include metrics management. We can now identify certain facts about metrics management:

- You cannot effectively promise deliverables to a stakeholder unless you can also identify measurable metrics.
- Good metrics allow you to catch mistakes before they lead to other mistakes.
- Unless you identify a metrics program that can be understood and used, you are destined to fail.
- Metrics programs may require change, and people tend to dislike change.
- Good metrics are rallying points for the project management team and the stakeholders.

There are also significant challenges facing organizations in the establishment of value-based metrics:

- Project risks and uncertainties may make it difficult for the project team to identify the right attributes and perform effective measurement of the value attributes.

12. Shadan Malik, *Enterprise Dashboards*, Hoboken, NJ: John Wiley & Sons, Hoboken, 2005, p.66.

- The more complex the project, the greater the difficulty is in establishing a single value metric.
- Competition and conflicting priorities among projects can lead to havoc in creating a value metrics program.
- Added pressure by management and the stakeholders to reduce the budget and compress the schedule may have a serious impact on the value metrics.

Metric management programs must be cultivated. Some facts to consider in establishing such a program include:

- There must be an institutional belief in the value of a metrics management program.
- The belief must be visibly supported by senior management.
- The metrics must be used for informed decision making.
- The metrics must be aligned with corporate objectives as well as project objectives.
- People must be open and receptive to change.
- The organization must be open to using metrics to identify areas of and for performance improvement.
- The organization must be willing to support the identification, collection, measurement, and reporting of metrics.

There are best practices and benefits that can be identified as a result of using metrics management correctly and effectively. Some of the best practices include:

- Confidence in metrics management can be built using success stories.
- Displaying a "wall" of metrics for employees to see is a motivational force.
- Senior management support is essential.
- People must not overreact if the wrong metrics are occasionally chosen.
- Specialized metrics generally provide more meaningful results than generic or core metrics.
- The minimization of the bias in metrics measurement is essential.
- Companies must be able to differentiate between long-term, short-term, and lifetime value.

The benefits of metrics include:

- Companies that support metrics management generally outperform those that do not.
- Companies that establish value-based metrics are able to link the value metrics to employee satisfaction and better business performance.

5.18 USING VALUE METRICS FOR FORECASTING

Performance reporting is essential for effective decision making to take place. In general, there are three types of performance reports:

- **Progress reports:** These reports describe the work accomplished to date. This includes:
  - The planned amount work up to the timeline of the report
  - The actual amount of work accomplished up to the timeline
  - The actual cost accumulated up to the timeline
- **Status reports:** These reports indicate the status by comparing the progress to the baselines and determining the variances. This includes:
  - The schedule variance up to the timeline of the report
  - The cost variance up to the timeline.
- **Forecast reports:** The progress reports and status reports are snapshots of where we are today. The forecast reports, which are usually of significant importance to the stakeholders, indicate where we will end up. This includes:
  - The expected cost at the completion of the project
  - The expected time duration or date at completion of the project

There are other items that can be included in these reports. However, our main concern is with the Forecast Reports and the metrics used to make forecasts.

Traditional forecast reports provide information on the time and cost expected at the completion of the project. This data can be calculated from extrapolation of trends or formulas, or projections of the metrics and KPIs. Unfortunately, this data may not be sufficient to provide management with the necessary information to make effective business decisions and to decide whether or not to continue on with the project or consider termination.

Two additional pieces of information may be necessary; the expected benefits at completion and the expected value at completion. Most earned value measurements systems in use today do not report these two additional pieces of information, probably because there are no standard formulas for them. Also, the value and benefits at completion may not be known until months after the project is completed.

The benefits and value at completion must be calculated periodically. However, depending upon which life cycle phase you are in, there may be insufficient data to perform the calculation quantitatively. In such cases, a qualitative assessment of benefits and value at completion may be necessary, assuming of course that information exists to support the assessment. Expected benefits and value are more appropriate for business decision making and usually provide a strong basis for continuation or cancellation of the project.

Using value metrics, an assessment of value at completion can tell us if value tradeoffs are necessary. Reasons for value tradeoffs include:

- Changes in the enterprise environmental factors
- Changes in the assumptions
- Better approaches have been found, possibly with less risk
- Availability of highly skilled labor
- A breakthrough in technology

As stated previously, most value tradeoffs are accompanied by an elongation of the schedule. Two critical factors that must be considered before schedule elongation takes place are:

- Elongating a project for the desired or added value may incur risks.
- Elongating a project consumes resources that may have already been committed to other projects in the portfolio.

Traditional tools and techniques may not work well on value-driven projects. The creation of a value measurement methodology (VMM) may be necessary to achieve the desired results. A VMM can include the features of earned value measurement systems (EVMS) and enterprise project management systems (EPM), as shown in Table 5–17. However, additional variables must be included for the capturing, measurement, and reporting of value and possibly value metrics.

TABLE 5-17 A Comparison of EVMS, EPM, and VMM

VARIABLE	EVMS	EPM	VMM
Time	Yes	Yes	Yes
Cost	Yes	Yes	Yes
Quality		Yes	Yes
Scope		Yes	Yes
Risks		Yes	Yes
Tangibles			Yes
Intangibles			Yes
Benefits			Yes
Value			Yes
Tradeoffs			Yes

TABLE 5-18 Placing Metrics Knowledge into Job Descriptions

GRADE LEVEL	COMPETENCY
1	Understand project metrics and key performance indicators
2	Be able to identify and create project-specific metrics
3	Be able to track and report metrics on a project
4	Be able to measure and evaluate metrics
5	Be able to extract best practices from metrics

5.19 METRICS AND JOB DESCRIPTIONS

Because project managers are now expected to be knowledgeable in metrics, job descriptions for project managers are being revised to include a level of knowledge in metrics management. Table 5–18 illustrates some of the expectations for a company that has five levels of job descriptions for project managers.

The growth of measurement techniques has accelerated the importance metrics, and this includes both tangible and intangible forms of value metrics. Project management is slowly becoming metrics-driven project management. The traditional metrics that we used for decades no longer satisfy the needs of the clients and stakeholders. Value-based metrics will become critical in stakeholder relations management. In addition, metrics management will lead us into a better understanding of the necessity for the development of more sophisticated knowledge management techniques.

5.20 GRAPHICAL REPRESENTATION OF METRICS

The old adage "a picture is worth one thousand words" certainly holds true when graphically displaying metrics. In Figure 5–17, we show the resources that are assigned versus the planned resources. For Work Package #1, five people were scheduled to be assigned, but only four people are currently working on this work package. The metric can also be used to show if excess labor is assigned to a work package.

Figure 5–17 shows a shortage of resources, but that may not be bad if the workers are assigned with higher skills than originally planned for. This is shown in Figure 5–18. From this figure, the assigned resources are pay grades 6, 7, and 8. If we planned on using only pay grades 5 and 6, then this may be good. However, if we anticipated some pay grade 9 workers, then we may have a problem.

In Figure 5–19, we are looking at regular time, overtime and unstaffed hours. In January, people worked 600 hours on regular shift and 50 hours

Figure 5-17 Planned versus Assigned Labor

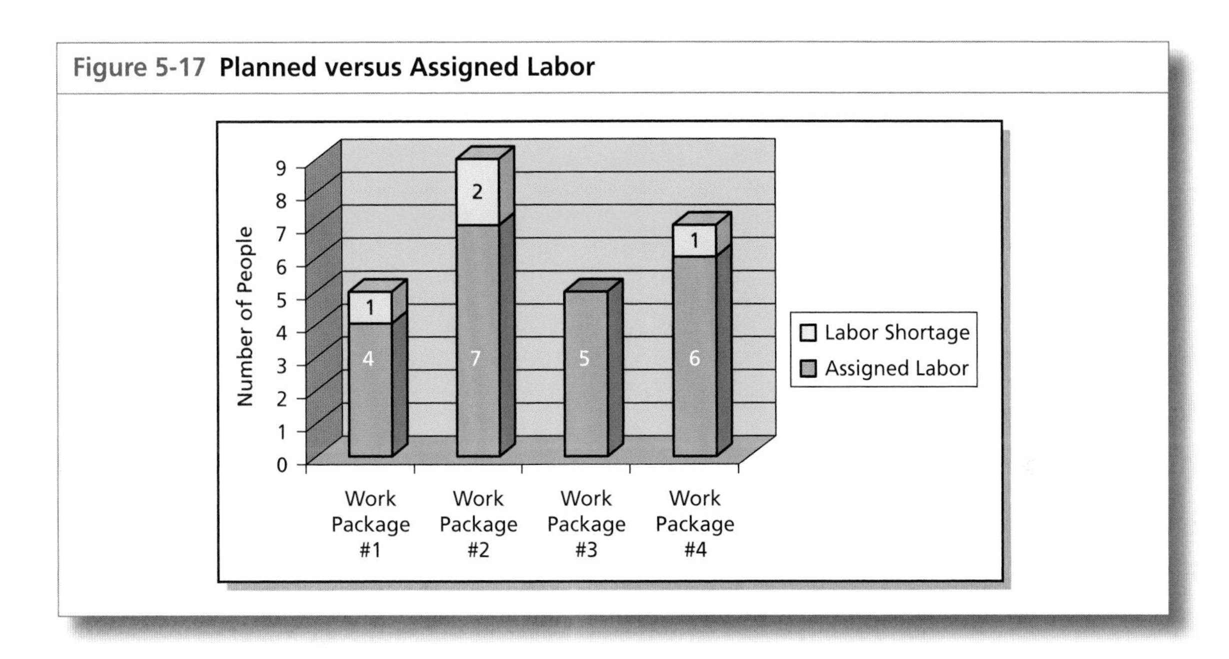

Figure 5-18 Pay Grade of the Assigned Resources

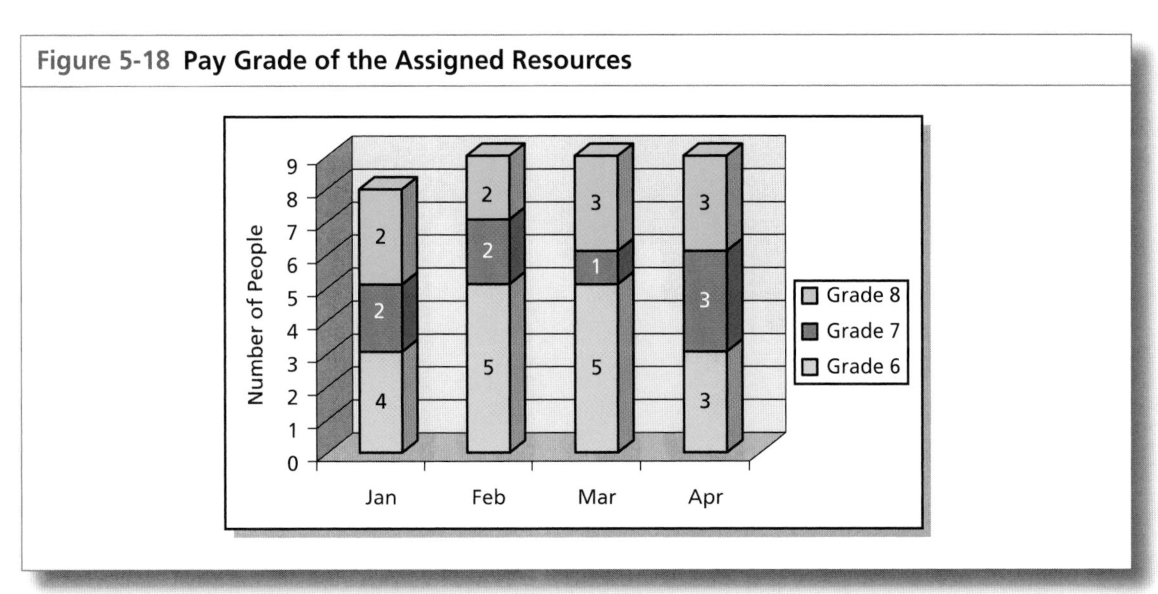

on overtime. We were still short some 100 hours of work that needed to be done. This could lead to a schedule slippage.

Stakeholders often want to know what percentage of the work packages scheduled have been completed and how many are still open, or possibly late. This is shown in Figure 5–20. It is most likely a good sign that the work packages late are shrinking each month. This representation could also be

Figure 5-19 Hours Worked on Regular Time, Overtime, and Unstaffed Hours

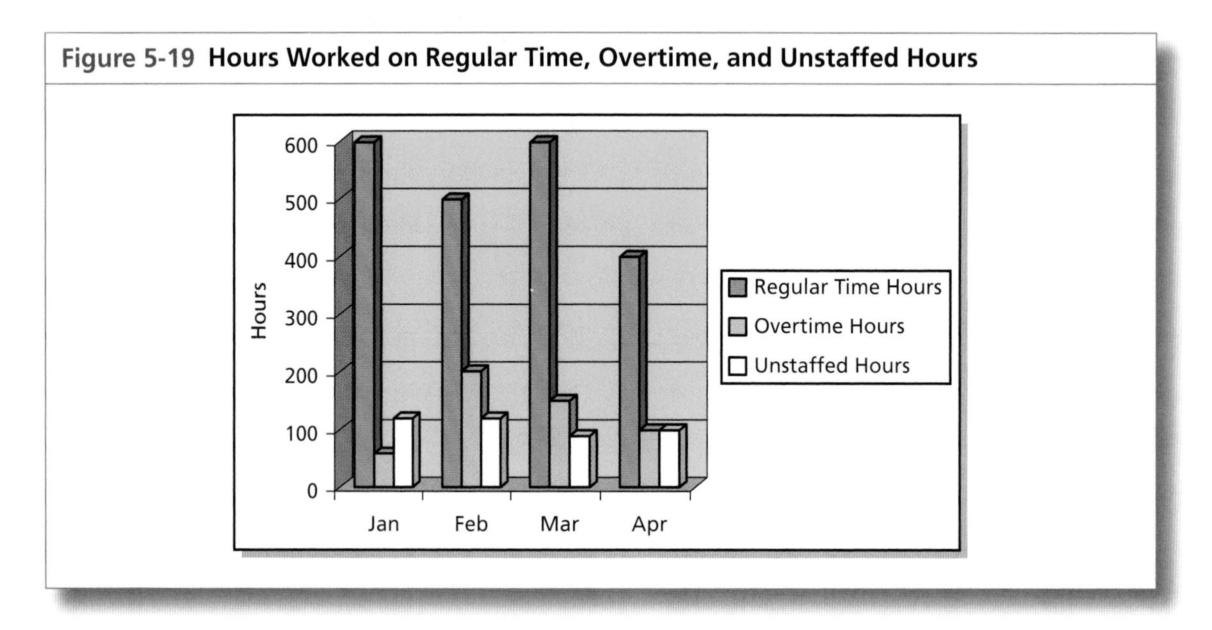

Figure 5-20 Work Packages Scheduled for Completion, Including Those Completed and Those Still Open

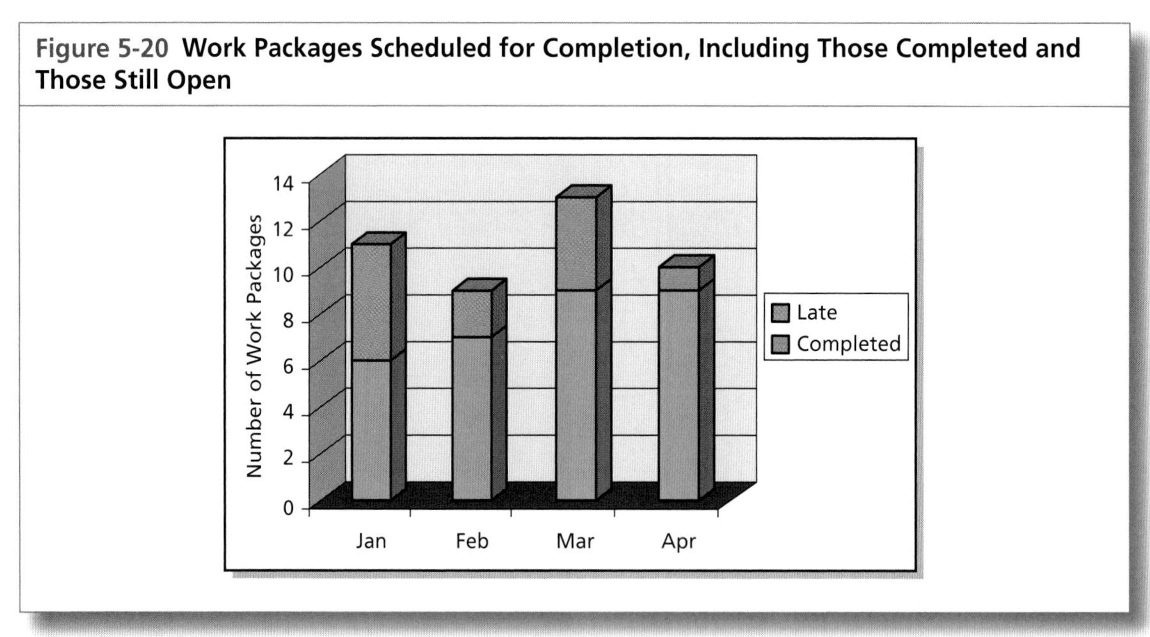

used to illustrate a percentage of all of the work packages, regardless of whether or not they have already started.

Some companies provide critical risk designations to each work package. In Figure 5-21, the number of work packages with critical risk designations is diminishing over time. This is a good sign.

Figure 5-21 **Work Packages with a Critical Risk Designation**

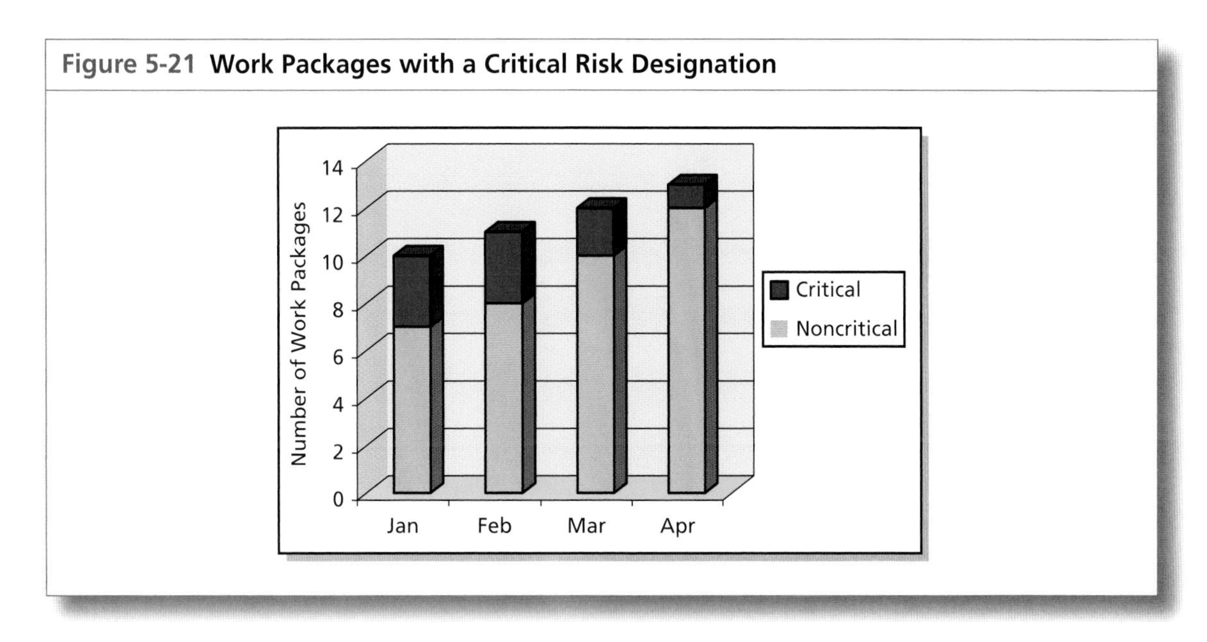

Figure 5-22 **Work Packages Adhering to the Budget**

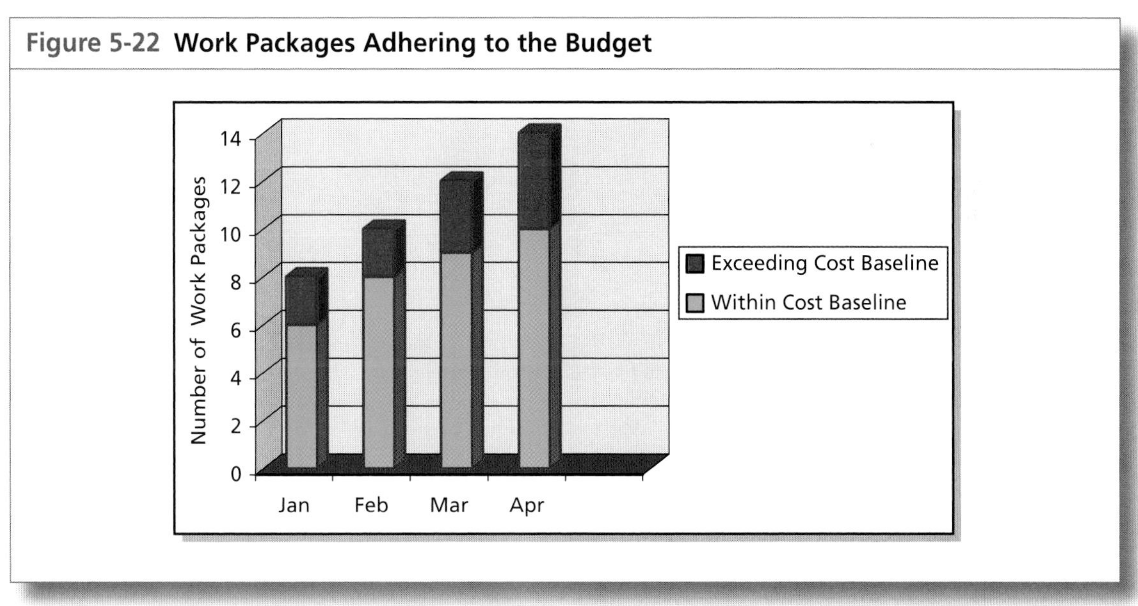

In Figure 5–22, we see the number of work packages adhering to the cost baseline. Since the number is increasing over time, costs seem under control. When costs, scope, and schedules need to be revised, we have baseline changes. The number of baseline changes is usually an indication of the quality of the upfront planning process and/or the company's estimating capability. This is seen in Figure 5–23. This type of metric is important because it illustrates the

Figure 5-23 Number of Baseline Revisions

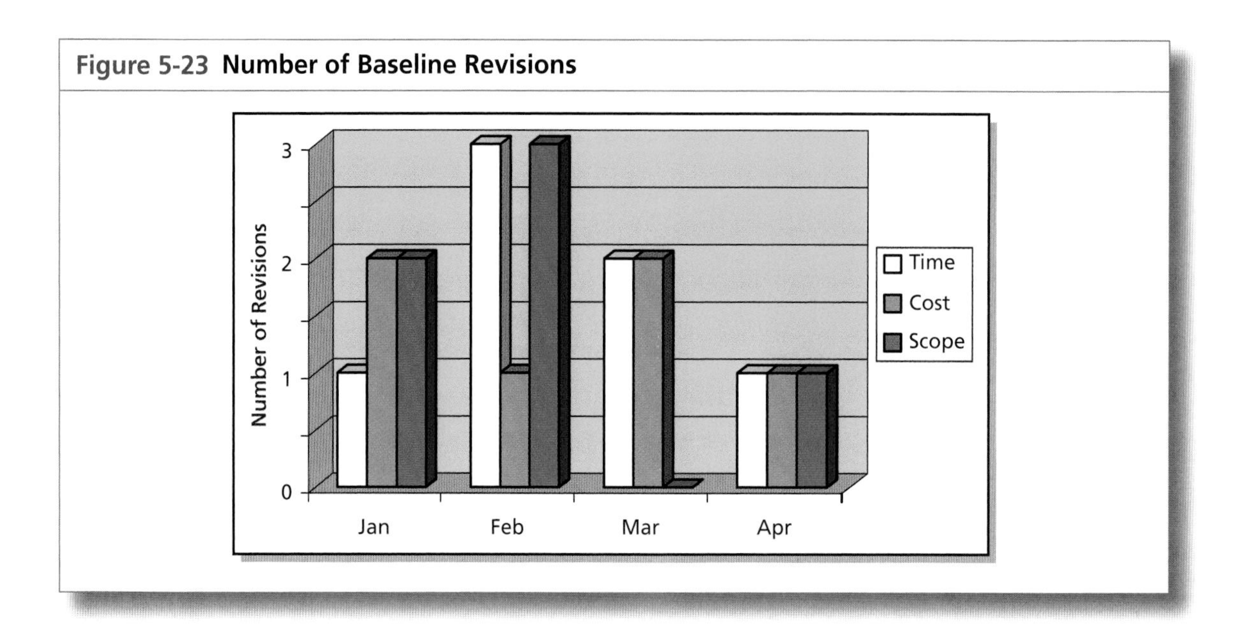

Figure 5-24 Number of Scope Changes Pending, Approved, and Denied

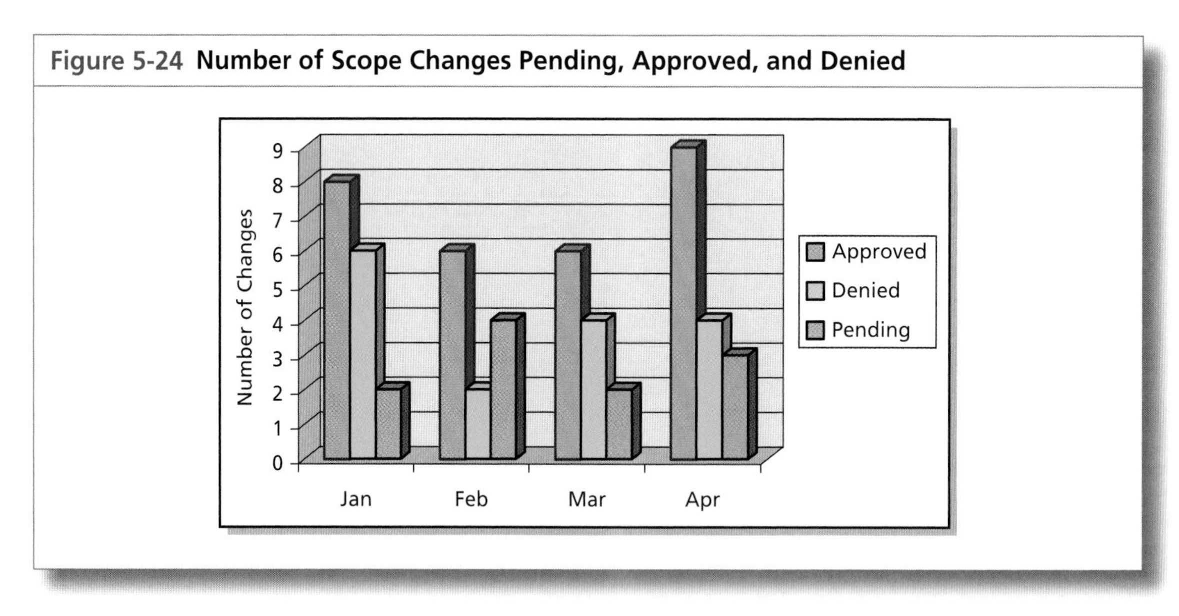

rate of change in the requirements over time. The number of baseline revisions can be the result of scope changes being made throughout the project. This is shown with the metric displayed in Figure 5–24.

Cost and schedule slippages can be the result of poor governance on a project and the inability to close out action items in a timely manner. This is shown in Figure 5–25. Action items that remain open for two or three months can have a serious impact on the final deliverables of the project as

Figure 5-25 **Number of Action Items Open Each Month and How Long They Remained Open**

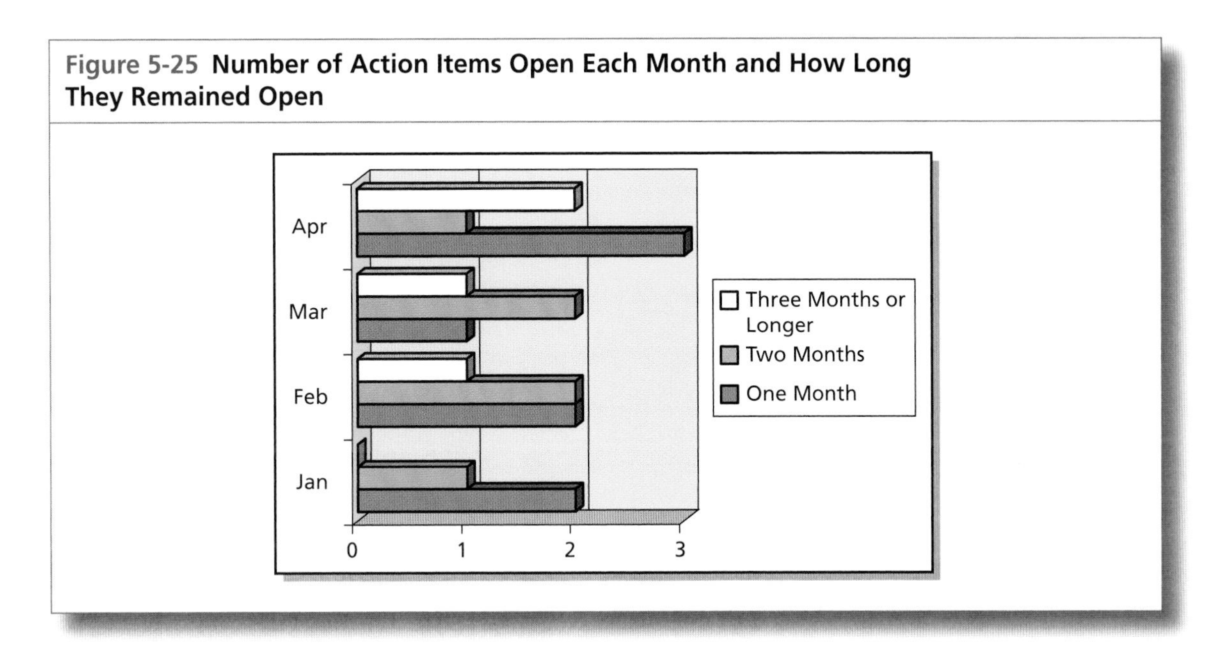

Figure 5-26 **Number of Critical Constraints Each Month**

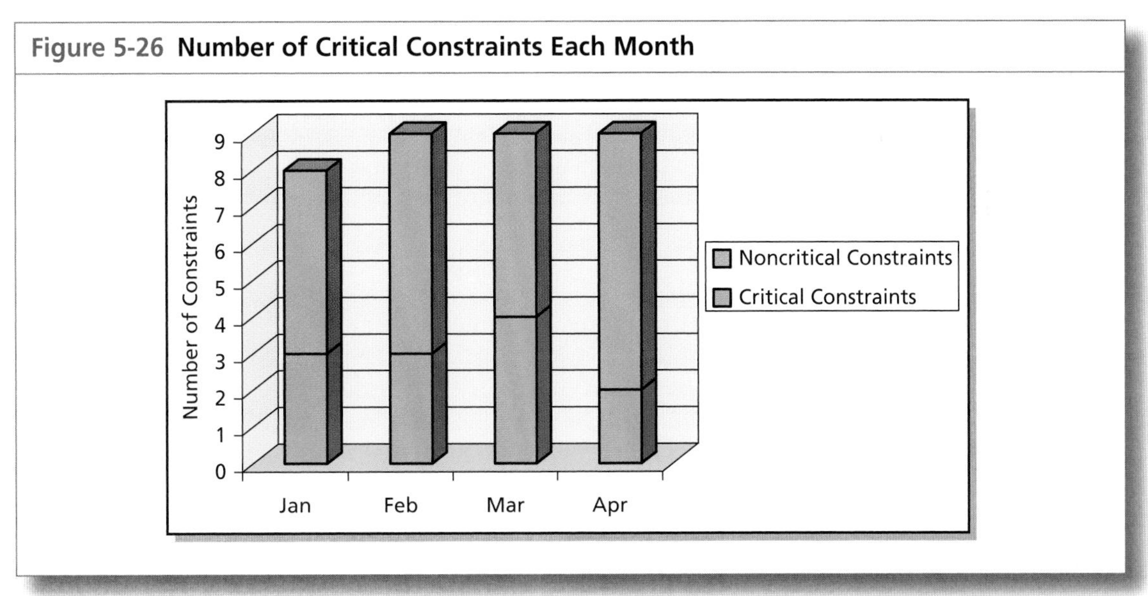

well as stakeholder satisfaction. However, there are situations where action items may need to be open for an extended period of time. Two such examples include taking advantage of a business opportunity and looking at risk mitigation techniques.

Not all constraints are equal. Some companies designate their constraints as critical and noncritical. Both the critical and noncritical constraints, as shown in Figure 5–26, must be tracked closely throughout the project.

Figure 5-27 Number of Critical Assumptions That Are New or Have Been Changed

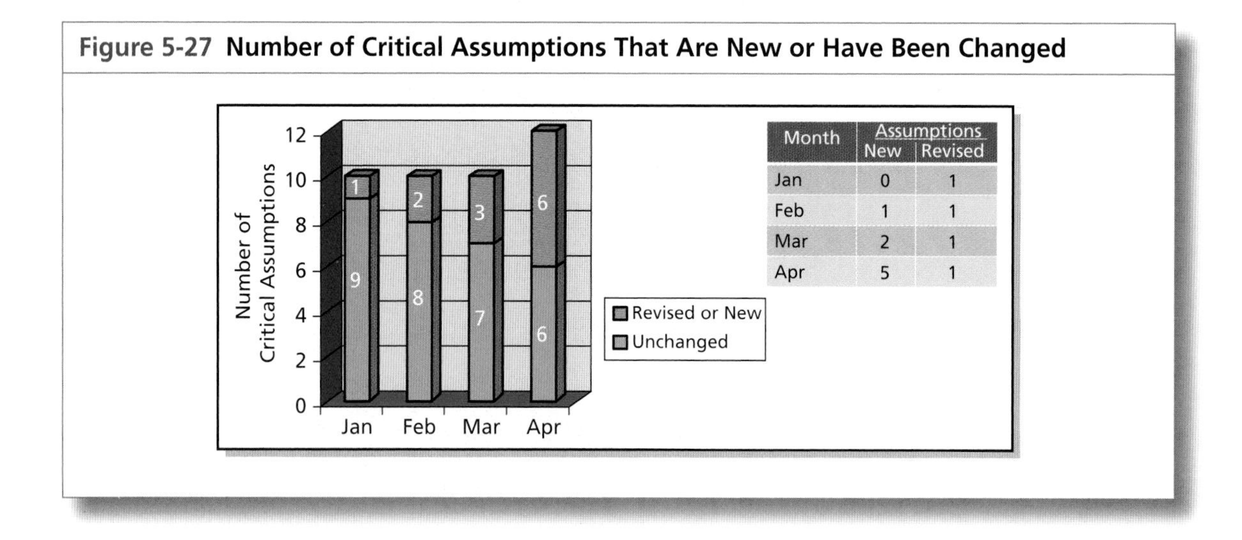

Month	Assumptions New	Assumptions Revised
Jan	0	1
Feb	1	1
Mar	2	1
Apr	5	1

Changes in the importance of the constraints can have a serious impact on the final value the stakeholders expect to receive.

Assumptions can also change during a project. The longer the project, the greater the likelihood that the number of assumptions will change. The assumptions must be tracked closely. Significant changes in the assumptions can cause a project to be canceled. This is shown in Figure 5–27. In February, eight of the assumptions were the same as in January. Two of the assumptions changed. From the table in the right of Figure 5–27, one of the two assumptions changed was new and the second assumption was modified.

Most companies today maintain a best practices library. When bidding on a contract, companies often make promises to the client that all of the best practices in the library that relate to this project will be used. In order to validate that these promises are being kept, the project team can track the number of best practices that are being used as compared to what was promised. This is shown in Figure 5–28. In this figure, 10 best practices were promised to the customer. Seven of the best practices have already been used, two will be used in the future, and one best practice will not be used.

Some companies assign a project complexity factor to each project based upon the aggregate risks. This is shown in Figure 5–29. In this example, a complexity factor of 15 would indicate a very serious risk. Risks are normally the greatest at the beginning of a project where the least amount of information is known. Since the complexity factor appears to be diminishing over time, things appear to be proceeding well.

Sometimes, companies that maintain a metrics library identify the artwork along with a description of the metric. This is shown in Figure 5–30.

Figure 5-28 Actual versus Promised Best Practices Used

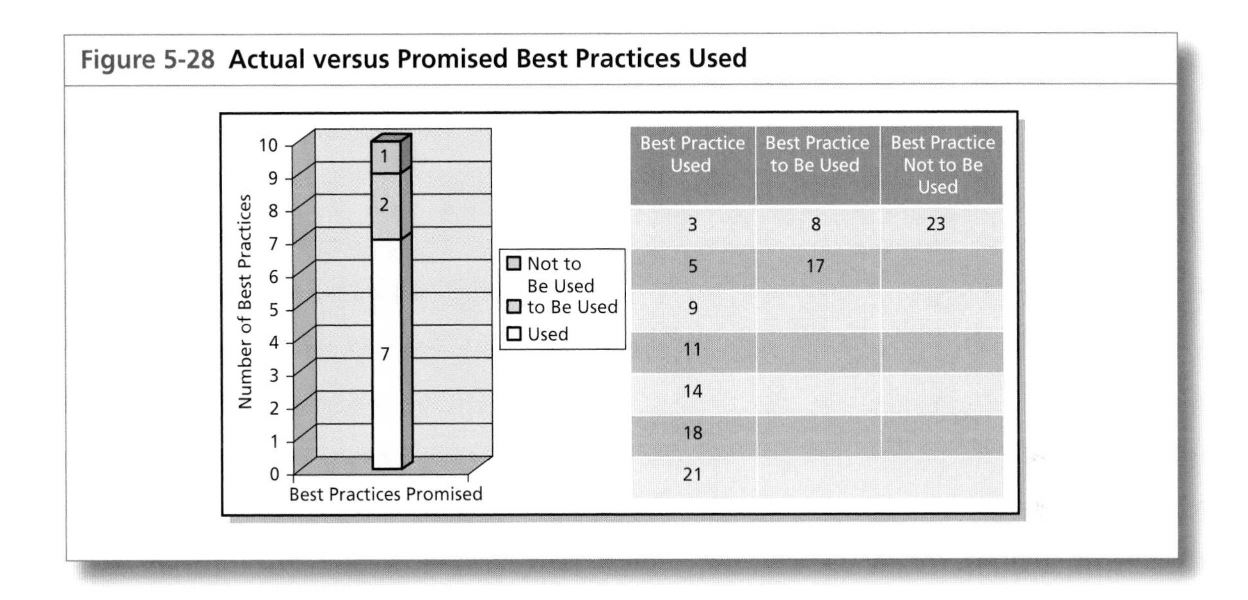

Best Practice Used	Best Practice to Be Used	Best Practice Not to Be Used
3	8	23
5	17	
9		
11		
14		
18		
21		

Figure 5-29 Project Complexity Factor

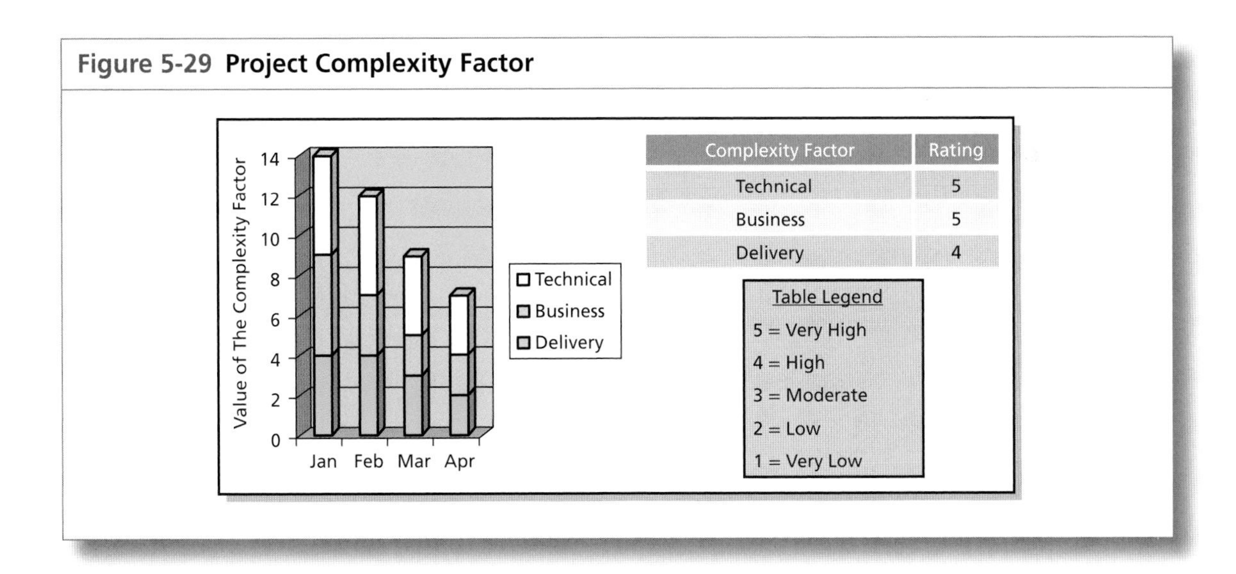

Complexity Factor	Rating
Technical	5
Business	5
Delivery	4

Table Legend
5 = Very High
4 = High
3 = Moderate
2 = Low
1 = Very Low

Figure 5–30 is how Figure 5–29, the project complexity factor, would appear in the metric library. The column on the right in Figure 5–30 shows some of the information that is used to describe the metric.

On large projects, stakeholders are often interested in the total manpower that is assigned to the project. Figure 5–31 shows this in the form of a metric. Another important metric is the rate at which the management

Figure 5-30 **Project Complexity Factor Appearing in the Metric Library**

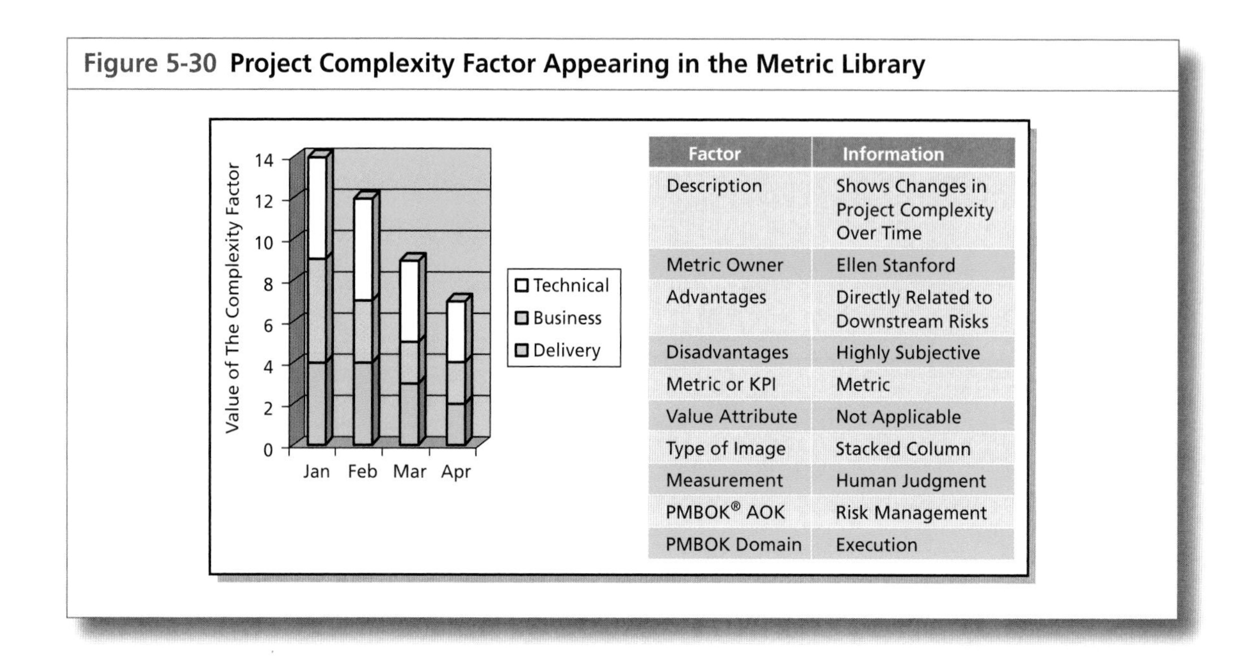

Factor	Information
Description	Shows Changes in Project Complexity Over Time
Metric Owner	Ellen Stanford
Advantages	Directly Related to Downstream Risks
Disadvantages	Highly Subjective
Metric or KPI	Metric
Value Attribute	Not Applicable
Type of Image	Stacked Column
Measurement	Human Judgment
PMBOK® AOK	Risk Management
PMBOK Domain	Execution

Figure 5-31 **Total Project Manpower**

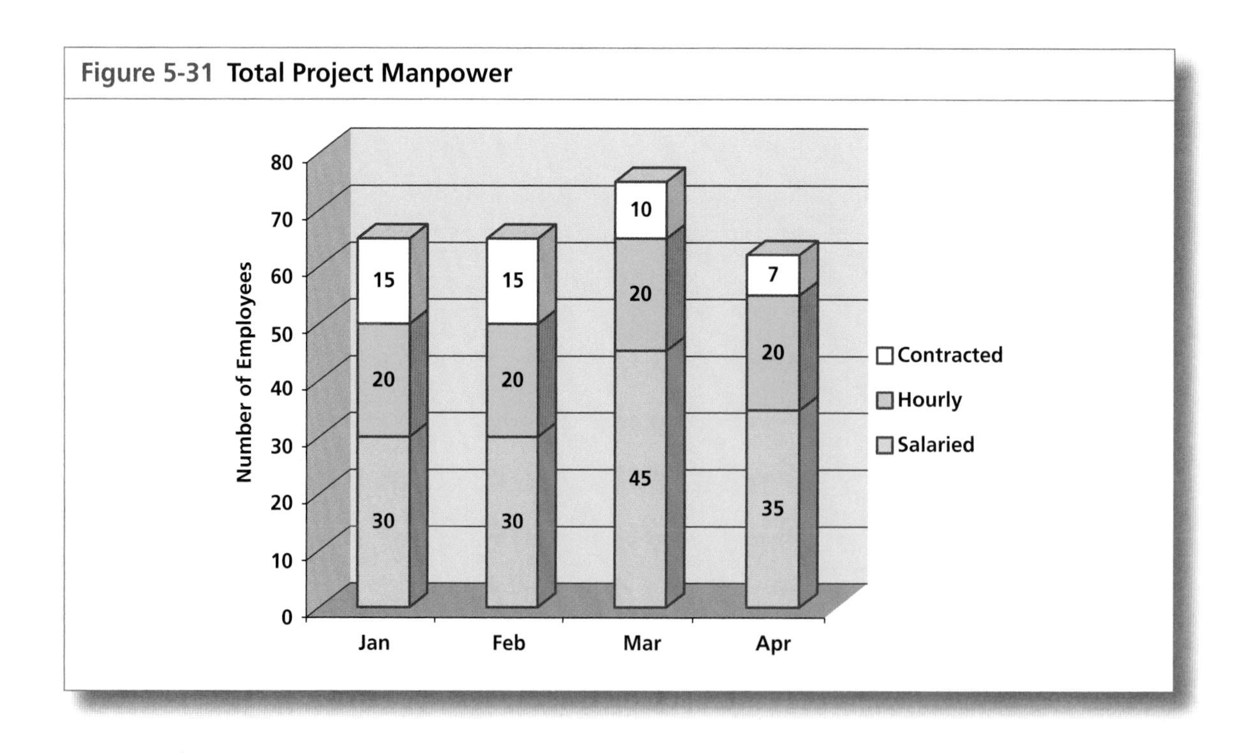

reserve is being used up. This is shown in Figure 5–32. In January, we established a management reserve of $100,000. By the end of March, $60,000 had been used and $40,000 remained. In April, we added in another $30,000 to the reserve fund and a total of $80,000 had been used.

Figure 5–33 shows the number of deliverables that were delivered on time or late each month. This metric is usually accompanied by another

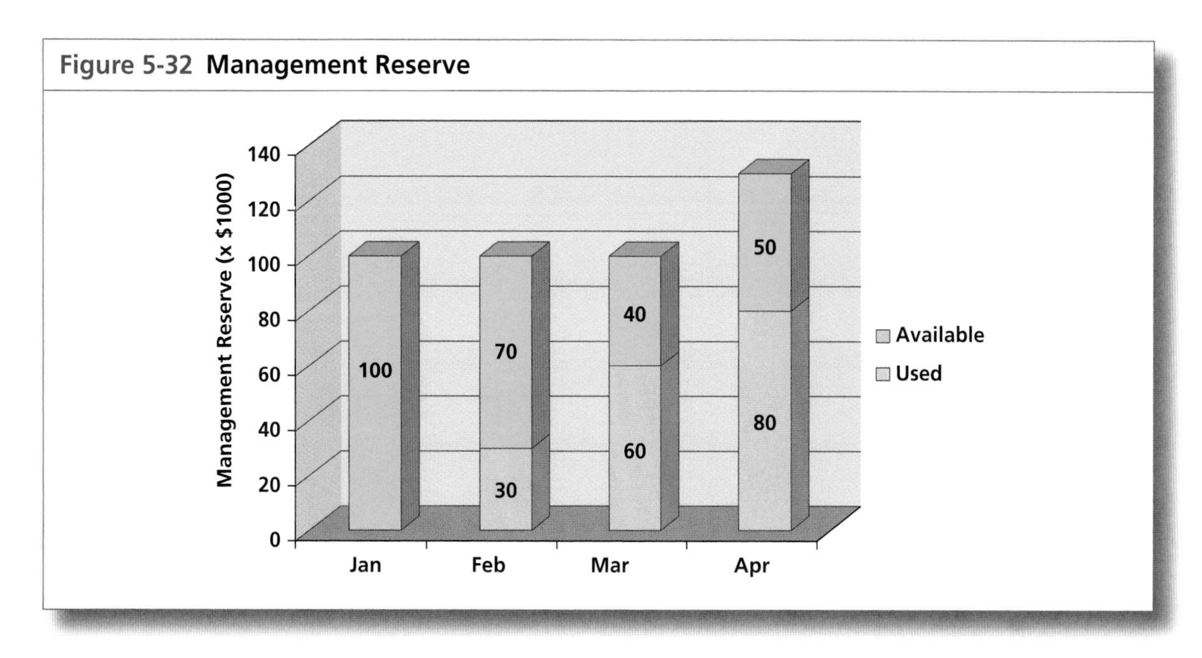

Figure 5-32 Management Reserve

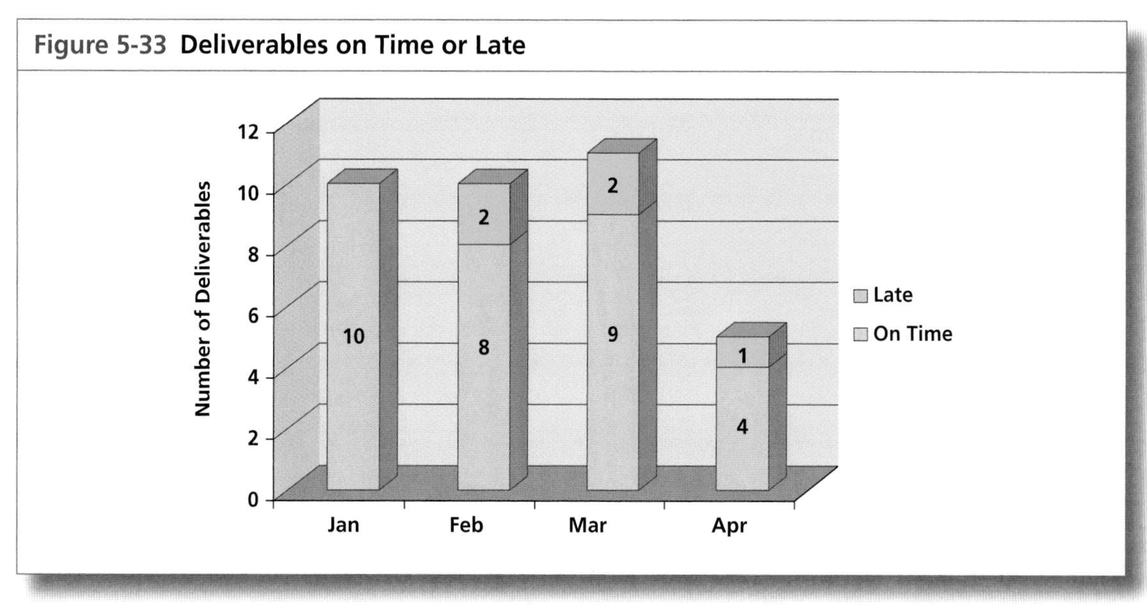

Figure 5-33 Deliverables on Time or Late

metric, as shown in Figure 5–34, which shows how many of the deliverables were accepted and rejected.

Figure 5–35 shows the trend on the cost performance index (CPI) and the schedule performance index (SPI). Since they represent trends, it is better to show them on the trend chart on the left rather than using a gauge

Figure 5-34 Deliverables Accepted or Rejected

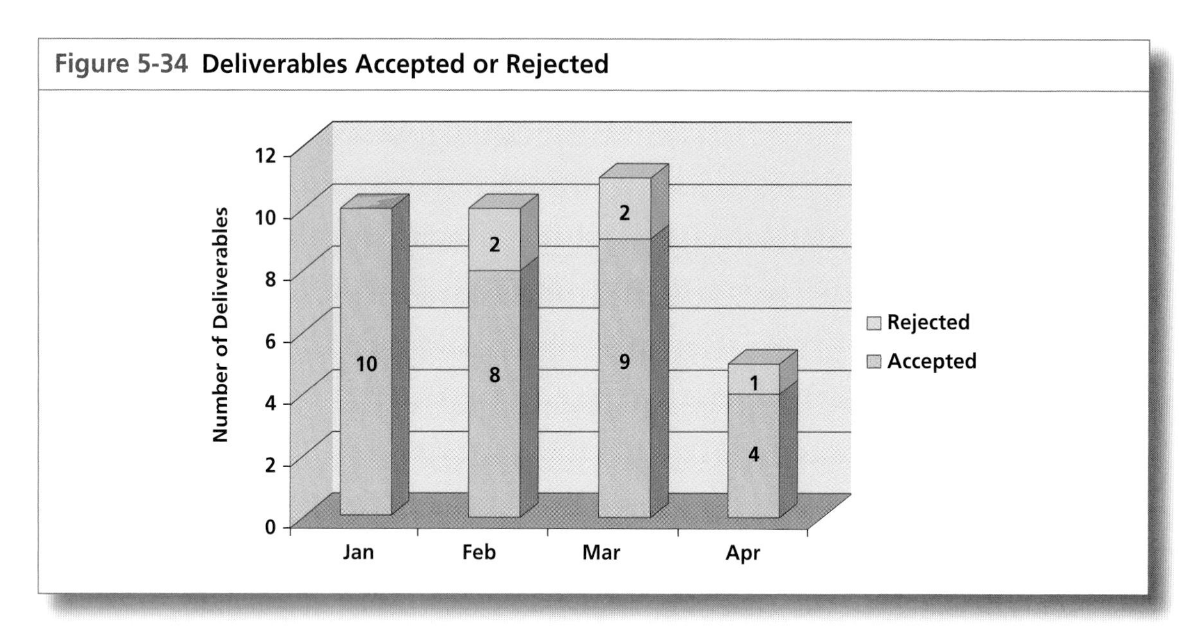

Figure 5-35 Cumulative Month-End CPI and SPI Data

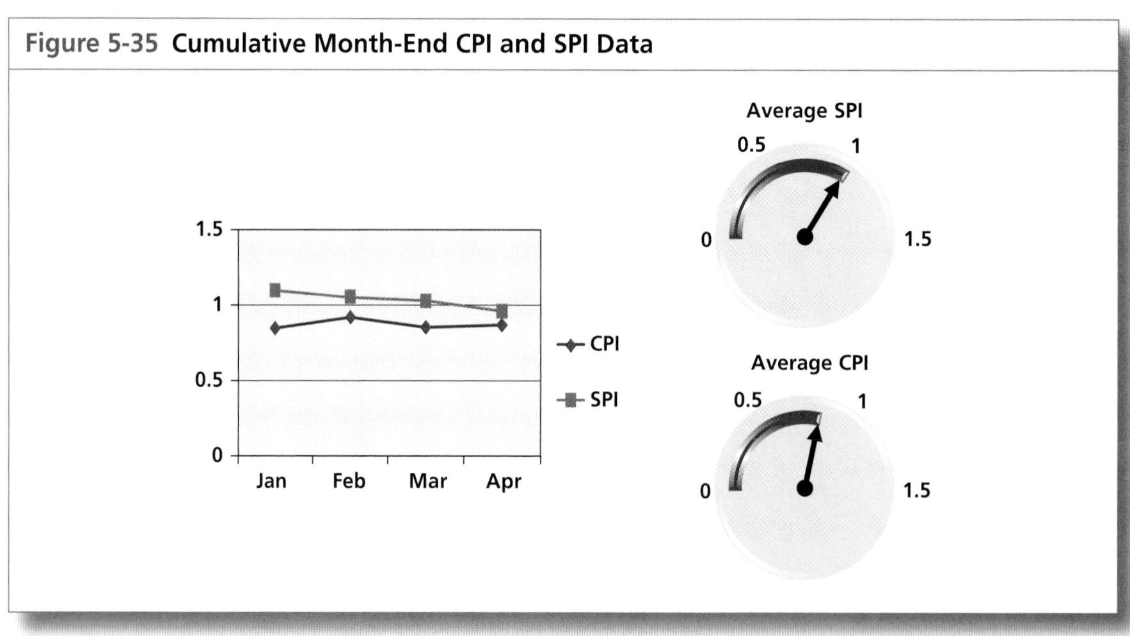

that shows the average value. Under normal performance, a value of 1.0 for SPI would indicate we are meeting our schedule., but the gauge shows an "average" SPI value. If we look at the trend chart on the left, we can see that the trend for SPI is unfavorable even though the average value is 1.0. Care must be taken to use the correct image when displaying the metric. Since this chart is used to show trends, we often do not place it on the dashboard until we have a sufficient number of data points to indicate a trend.

The cost performance index is generally more meaningful than the cost variance because it shows trends. However, if the cost variance is to be provided in metric form, it may be best to show the difference between last month and this month. This is shown in Figure 5-36. Work package #8 is significantly worse than last month. Work package #1 is still unfavorable, but there has been a favorable decrease in the magnitude of the unfavorable variance. In other words, there can be improvements in the magnitude of a negative variance. Likewise, in Figure 5-37, there can be unfavorable decreases in the magnitude of a positive variance.

Trends are normally used to predict the estimated cost at completion (EAC). This is shown in Figure 5-38. The original budget is $800,000 and we can see the fluctuations in EAC. There are several formulas that can be used for EAC and sometimes the fluctuations occur when we change the formulas we are using.

Figure 5-36 Color-Coded Unfavorable Variances (Monthly)

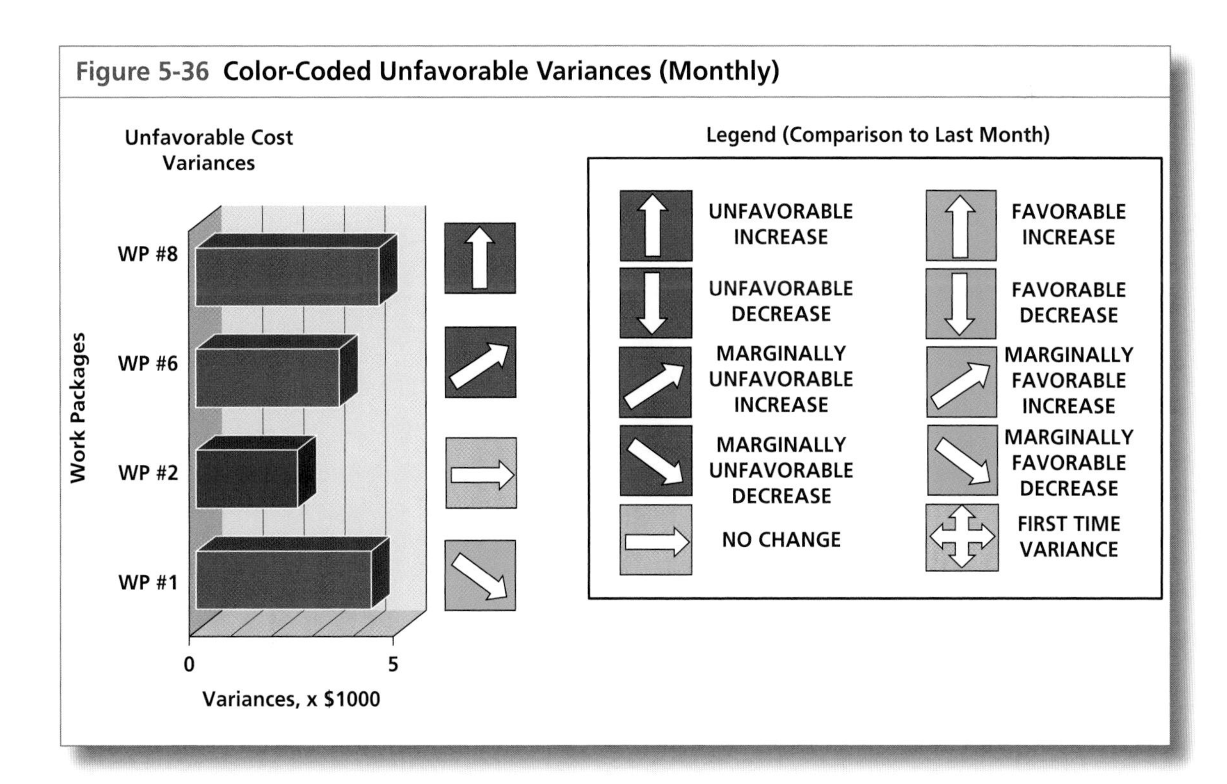

Figure 5-37 Color-Coded Favorable Variances (Monthly)

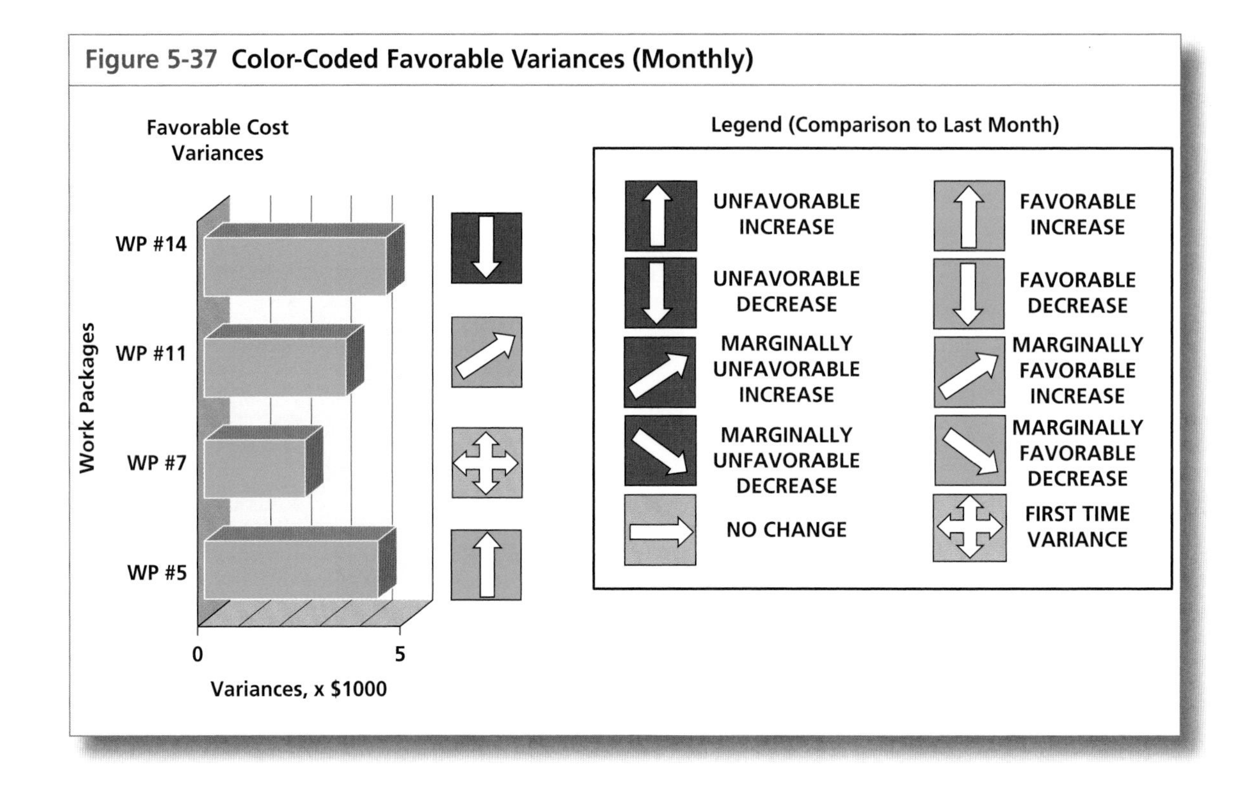

Figure 5-38 Estimate at Completion

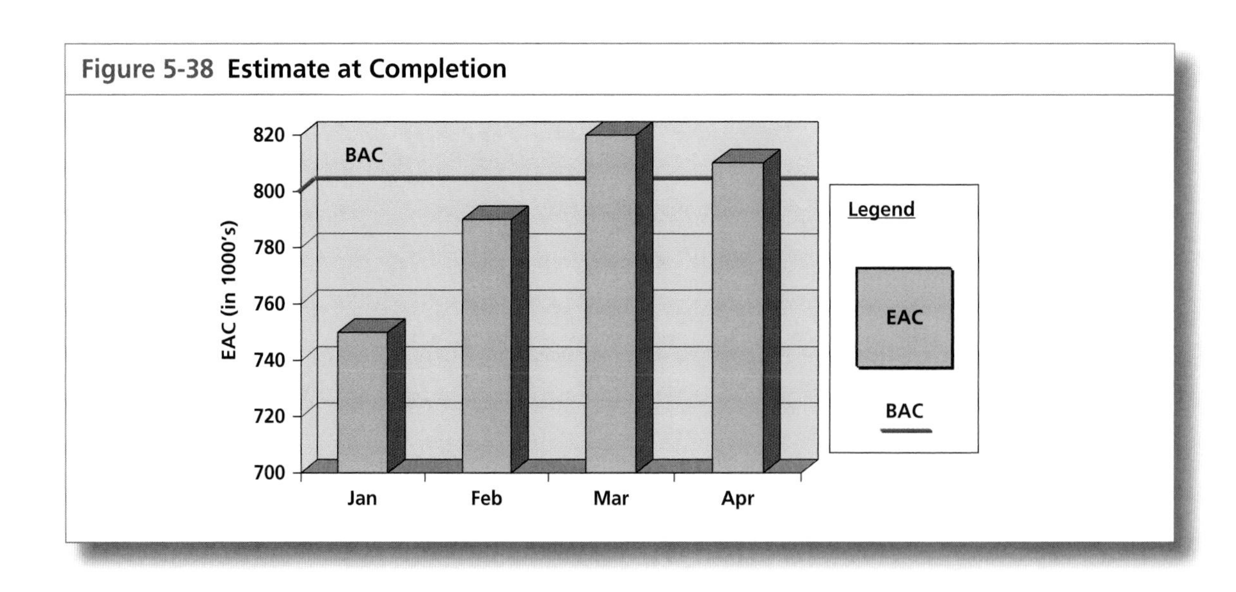

Figure 5-39 Risks Including Aging

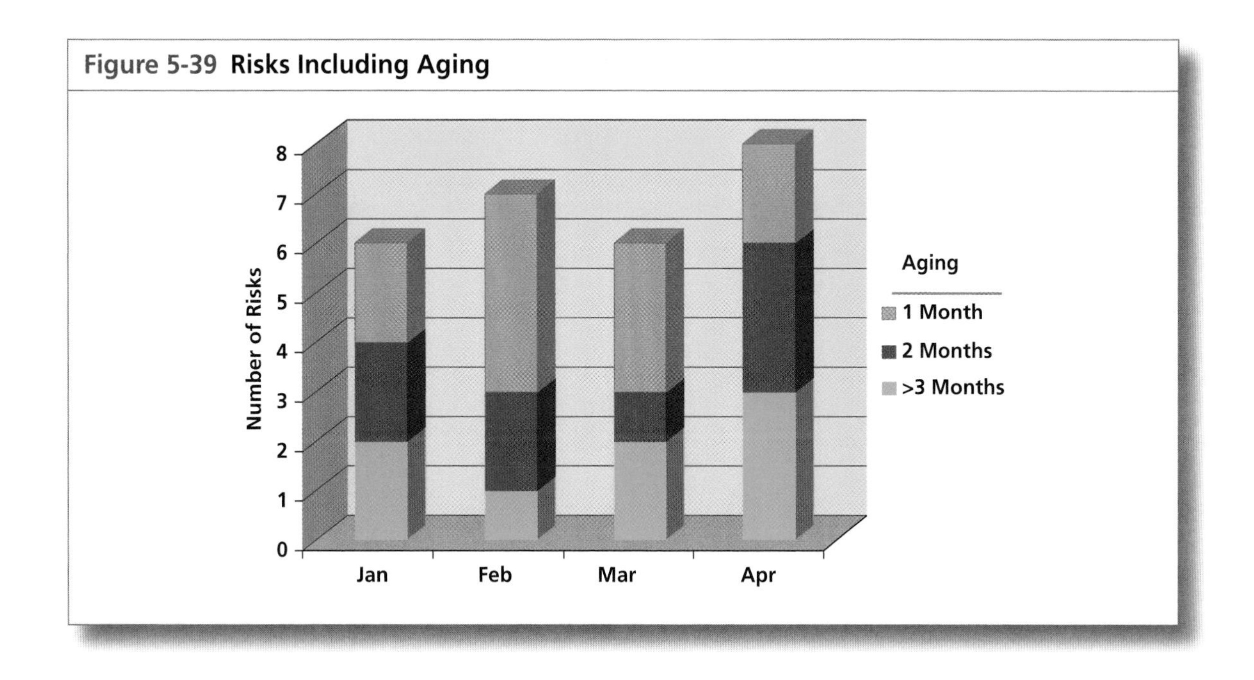

Sometimes, the risks on a project are like a chronic; they simply do not go away even with mitigation attempts. Figure 5–39 shows how a metric can be used to track the aging of a risk. One metric alone cannot always show the status of the project. A combination of metrics may be needed.

5.21 CREATING A PROJECT VALUE BASELINE

All of the information covered thus far in this chapter makes it clear that we need a baseline for measuring value. Executives and clients expect project managers to effectively monitor and control projects. As part of monitoring and control, project managers must prepare progress, status, and forecast reports that clearly articulate the performance of the project. But to measure performance, one needs a reference point or baseline from which measurements can be made. The necessity for a baseline is clear:

- Without a baseline, performance cannot be measured.
- If performance cannot be measure, it cannot be managed.
- Performance that can be measured gets watched.
- What gets watched gets done.

For a project to be able to be controlled, it must be organized as a closed system. This requires that baselines be established for possibly all constraints, including scope, time, and cost at a minimum. Without such baselines, a project is considered out of control, and it may be impossible to track what has changed without knowing where you started. It may also be impossible to identify value added opportunities.

The Performance Measurement Baseline

The reference point for measuring performance is the performance measurement baseline (PMB). It serves as the metric benchmark against which performance can be measured in terms of the triple constraints, namely time, cost, and scope. It can also be used as the basis for business value tracking provided that value metrics and value constraints be established.

The primary reasons for establishing, approving, controlling, and documenting the PMB were to:

- Ensure achievement of project objectives
- Manage and monitor progress during project execution
- Ensure accurate information on the accomplishment of the deliverables and requirements
- Establish performance measurement criteria

The PMB is finalized at the end of the planning phase once the requirements have been defined, the initial costs have been developed and approved, and the schedule has been set. Once established, the PMB serves as the benchmark from which to measure and gauge the project's progress. The baseline is then used to measure how actual progress compares to planned performance. Performance measurement may be meaningless without an accurate baseline as a starting point. Unfortunately, project managers in the past created baselines based upon just those elements of work they felt were important and this may or may not have been in full alignment with customer requirements or the customer's need for identification of value. *The baseline was what the project manager planned to do, not necessarily what the customer had asked for.*

Project Value Management

The problem with the PMB the way it was traditionally used is that it did not account for the value that was expected from the project by people outside of the company. The importance of value has been known for some time. For years, many companies established value engineering (VE) departments that focused on internal value achieved in engineering and manufacturing activities. Later, VE was expanded and called value analysis (VA), and included consideration of internal business value. We have now combined VE and VA, and call it Value Management (VM) where VM = VE + VA.

Today, we are asking project managers to understand the importance of project value management and its relationship to both the customer's and their consumers' understanding of value. Project value management is a mind-set and its principles should be employed at the onset of the project and used throughout the project's full life cycle. Project value management is a combination of attributes, people, requirements, enterprise environmental factors and circumstances. The attributes of project value may include time, cost, quality, risk as well as other such factors. Project value is more than just obtaining customer satisfaction at the lowest price or with a minimization of expended resources. Project value cannot be determined exclusively in terms of the traditional triple constraints. Focusing on the triple constraints is not a mindset to create project value. Establishing good value-reflective metrics allow the definition of project success to be made in terms of factors other than the traditional triple constraints. For example, let's assume you manufacture a component for a customer who, in turn, uses the component in a product they sell to consumers (i.e., their customers). Each can have a completely different definition of success such as:

- **Consumer:** A deliverable or solution that removes obstacles from or improves the consumer's way of life
- **Customer:** A deliverable or solution that is aligned to the customer's corporate strategic goals and objectives
- **Project manager:** Providing sustainable business value within the competing constraints

Simply stated, VOC is no longer just the voice of the customer. VOC can also be the voice of the consumers, who are your customer's customers. To adequately listen to these voices, the project manager must understand the customer's business model and strategy for reaching out to consumers as well as the customer's value management program.

Project value management must begin with a clear understanding of the customer's definition of value. Value mismatches generally lead to bad results. However, it should be understood that on long term projects the customer's definition of value may change.

Agreeing with everyone's definition of value works only if the project staffing function provides resources with the capability to equal or exceed the desired project value. Figure 5–40 shows that the best resources should be assigned to projects that provide high value to customers and consumers, provide high strategic value to the project manager's company, and require a small investment by the project manager's company.

The Value Management Baseline

Configuration management is the process of managing changes in the deliverables, hardware, software, documentation, and even measurements. The value baseline supports configuration management processes. Traditionally,

Figure 5-40 Value-Based Resource Application Model

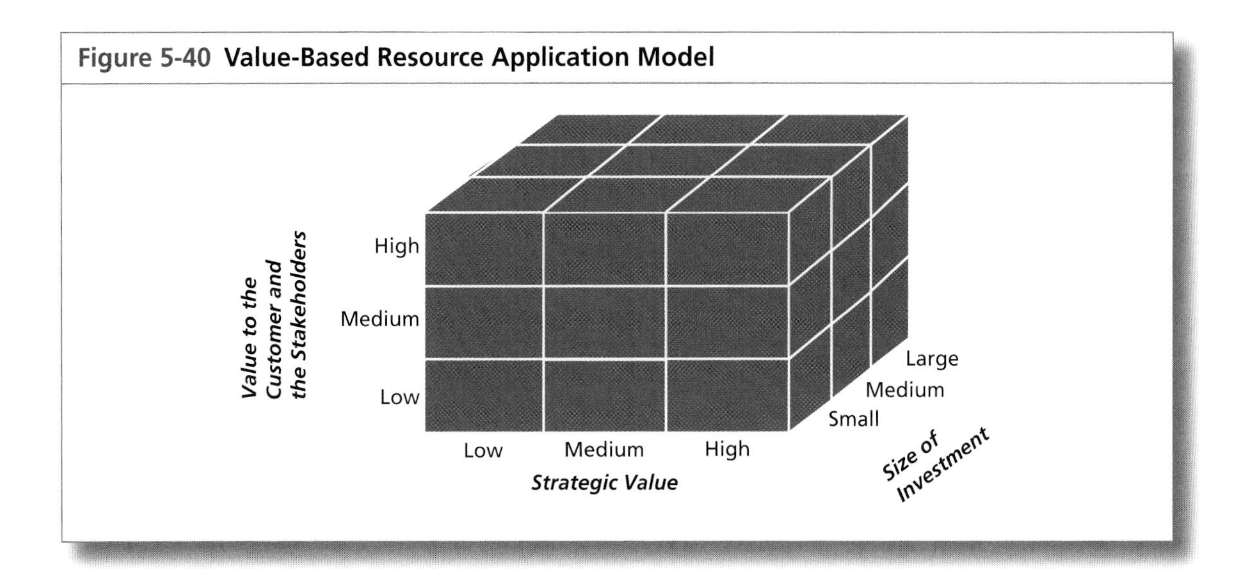

configuration management included input from multiple baselines but not from a value baseline. Therefore, at change control board meetings it was somewhat difficult to quantitatively define exactly what added value would be achieved. Today, the answers to the critical questions to be addressed at the change control board meetings come from four baselines:

- **Cost baseline:** Cost of the change
- **Schedule baseline:** Impact on the schedule
- **Risk baseline:** Risk effects
- **Value baseline:** Value-added opportunities

All baselines are reference points against which a comparison is made between planned and actual progress and are established at a fixed point in time. However, some value baselines must evolve over time. Unlike the traditional baselines for time, cost, and scope, value baselines are highly dependent on when the measurements can be made, the measurement intervals, and the fact that value baselines are often displayed as step functions rather than linear functions.

Value baselines have several characteristics:

- The value baseline can be composed of attributes from other baselines.
- The value baseline should be shown to the customer. Other baselines may or may not be shown to the customers.
- The value baseline may or may not be a contractual obligation.

- Unlike other baselines, the value baseline can change from life cycle phase to phase.
- Stakeholders must understand the differences between actual and planned value, whether favorable or unfavorable.
- Baseline changes may require modifying or reworking the project plan, or even result in project cancellation.
- The value baseline can change without any changes occurring in other baselines.
- On some projects, monitoring the value baseline in the early life cycle phases may provide no meaningful data.
- It is important to determine how often you need to update the value baseline.
- Value baselines may need to be continuously explained to the viewer. This may not be the case with other baselines.
- An increase of say 5% in the value baseline does not necessarily mean that the actual value has increased by 5%.
- Performance measurements for value may need to be customized rather than off-the-shelf techniques. This may need to be agreed upon upfront on the project.

Getting agreement on project value at the beginning of a project may be difficult. Even with an agreement, each stakeholder can still have their own subjective agenda on what value means to them:

- **Project manager:** Lowest cost
- **Project design team:** Functionality
- **Client:** Market share
- **Consumers:** Best buy

Stakeholder involvement is essential when establishing the value baseline. Stakeholders must clearly understand what is meant by:

- Normal value performance
- An increase in value
- A decrease in value

Factors that can cause the value baseline to deteriorate include:

- Faulty initial expectations of value
- Unrealistic expectations of value
- Unattainable expectations of value
- Poor alignment with business objectives
- Recessions
- Changing customer/consumer requirements and habits
- Unexpected crises

Figure 5-41 Value Metric Attributes

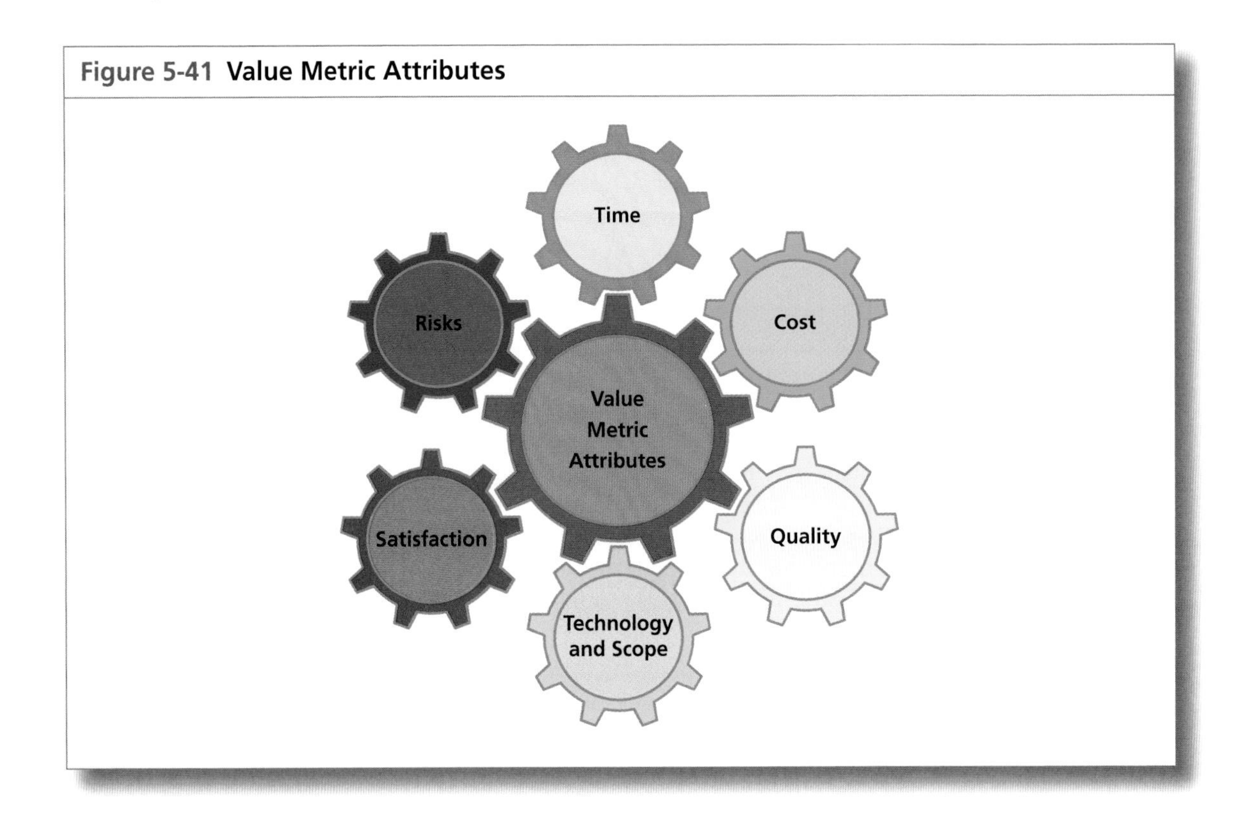

Selecting the Value Baseline Attributes

The value baseline is composed of a collection of attributes taken from other baselines. Each project's value baseline can be different. Figure 5-41 shows the six generic categories for the selection of value attributes. The attributes in Figure 5–41 are represented as gears because changes in one attribute could result in changes to all of the other attributes. Each of the value attributes can have a different meaning to each of the stakeholders. This is shown in Table 5–19.

Sometimes, metrics such as time and cost are not treated as value attributes but are still important to the client. As an example, let's look at a project where the life cycle cost of the deliverable is more important to the client than the original contract cost. Life cycle costing is the total cost to the consumer for the acquisition, ownership, operations, support, and disposal of the deliverable. What if the project's cost increases by $100,000 but life cycle costing indicates a tenfold savings in costs associated with safety, reliability, operability, maintenance, and environmental factors? In this example, the cost metric is still important during project execution but is not regarded as critical so as to be used as a value attribute. The same

TABLE 5-19 Interpretation of Attributes

GENERIC VALUE ATTRIBUTE	PROJECT MANAGER'S VALUE ATTRIBUTE	CUSTOMER'S VALUE ATTRIBUTE	CONSUMER'S VALUE ATTRIBUTE
Time	Project Duration	Time-to-Market	Delivery Date
Cost	Project Cost	Selling price	Purchase rice
Quality	Performance	Functionality	Usability
Technology and Scope	Meeting specifications	Strategic alignment	Safe buy and reliable
Satisfaction	Customer satisfaction	Consumer satisfaction	Esteem in ownership
Risks	No future business from this customer	Loss of profits and market share	Need for support and risk of obsolescence

could occur with the time metric where getting to the marketplace with a quality product is more important than simply time-to-market.

Life cycle costing decisions can add significant value to customers and consumers. Project managers should seek out project value opportunities but not necessarily at a significant expense to the project. Stakeholders may need to approve project value opportunity initiatives. Project value initiatives can lead to overachievement actions by the project team:

Overachievement trends:

- Exceeding specifications rather than meet them
- Giving the client more than they ask for or need unnecessarily
- Identifying opportunities for the future at the expense of the project

Risks of overachievement

- Increased complexities
- Greater overall uncertainty and risk
- Possible conflicting priorities
- Inability to meet time compression goals

We can now summarize the skills needed for effective project value management:

- Innovation skills
- Brainstorming skills
- Problem solving skills
- Process skills
- Life cycle costing skills
- Risk management skills

6 DASHBOARDS

CHAPTER OVERVIEW

Dashboards are an attempt to go to paperless project management and yet convey the most critical information to the stakeholders the fastest way. Dashboards are communication tools for providing data to viewers. If your goal is to communicate the big picture quickly, perhaps in summary format, dashboards are the way to go. Other project management tools, such as written reports, may be necessary to provide supporting information and details.

Designing the dashboard is not always easy, however. Multiple dashboards may be required on the same project. There are rules that can be followed to make the design effort easier. Included in this chapter are white papers from companies that assist clients in dashboard design efforts.

CHAPTER OBJECTIVES

- To understand the characteristics of a dashboard
- To understand the differences between dashboards and scorecards
- To understand the different types of dashboards
- To understand the benefits of using dashboards
- To understand and be able to apply the dashboard rules

KEY WORDS

- Dashboards
- Rules for dashboard design
- Scorecards
- Traffic light reporting

6.0 INTRODUCTION

The idea behind digital dashboards was an outgrowth of decision support systems in the 1970s. With the surge of the web in the late 1990s, business-related digital dashboards began to appear. Some dashboards were laid out to track the flows inherent in business processes, while others were use to track how well the business strategy was being executed. Dashboards were constructed to represent financial measures that even executives could understand. Figure 6–1 shows what a typical dashboard might look like.

Figure 6-1 The Framework for a Typical Dashboard

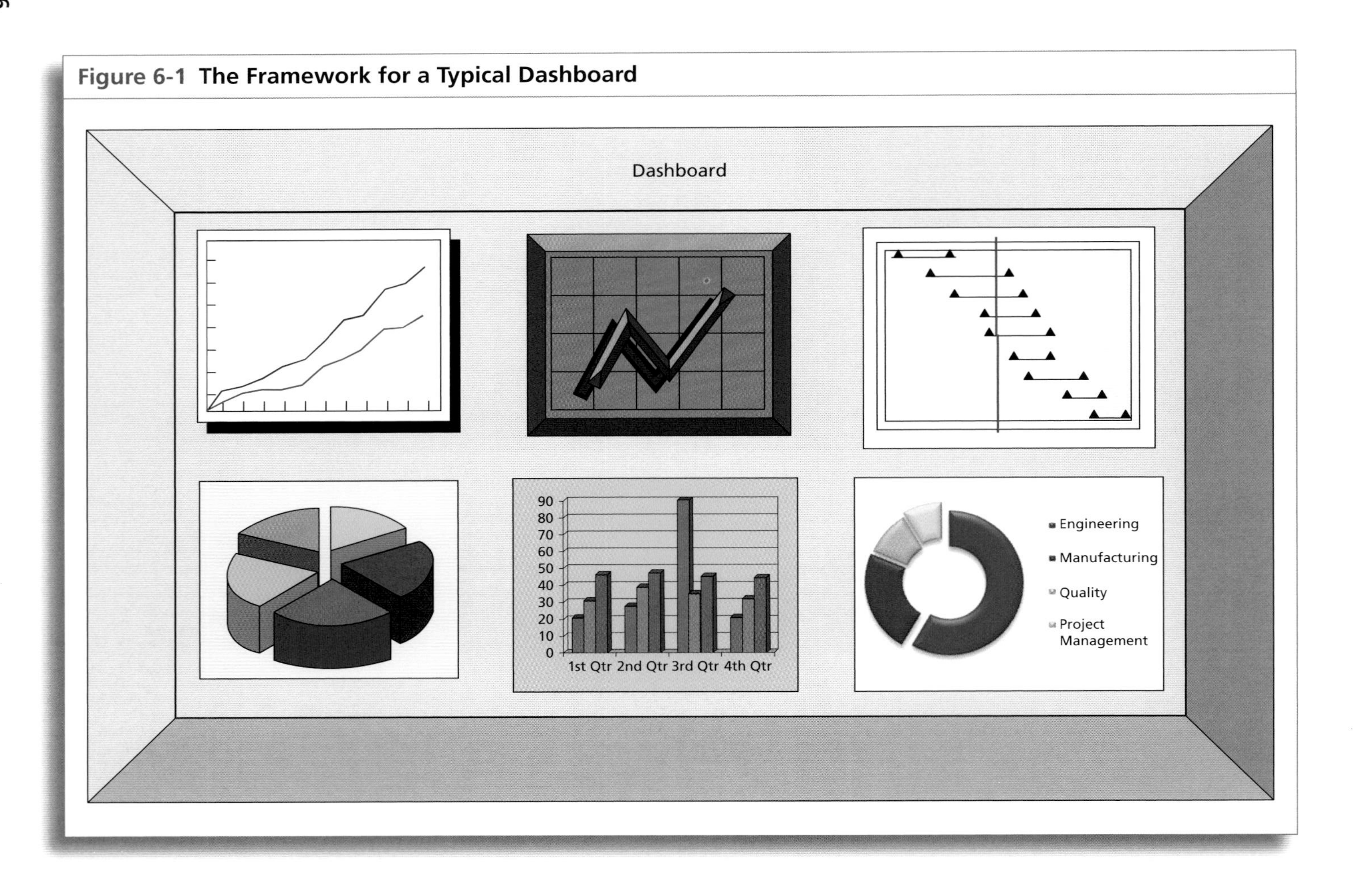

Perhaps the single most important event affecting dashboards was the introduction of the importance of key performance indicators as part of the Balanced Scorecard approach published by Robert S. Kaplan and David P. Norton in the mid-1990s.[1]

Later, in an article by Mark Leon ("Dashboard Democracy," in *Computerworld*, June 16, 2003), 135 companies were surveyed and more than half were using dashboards.

Even though dashboards are quite common in certain industries, the uses for the dashboards can vary significantly. As an example, here are some specific types of dashboards:

- Hospital workflows and bed management dashboard
- Museums dashboard
- Best places to work dashboard
- Casino management dashboard
- Dentist dashboard
- Energy dashboard
- Federal government dashboards to improve performance and cut waste
- Flex dashboard at the Federal Reserve Bank
- Food quality control dashboard
- Investor risk dashboard
- Sales compensation dashboard

The web site www.dashboardspy.com has archives that describe a multitude of dashboards for almost every industry.

In today's business environment, the ability to make dashboard presentations is almost as indispensable as writing skills. People tend to take graphics for granted, but do not realize what's wrong with various types of information graphics because this is not traditionally taught in schools. There are specialized seminars and webinars today on dashboard design to fill this gap.

Many dashboards fail to provide value because of design issues, not technology. Effective dashboards are not bells and whistles, glitter, and bright lights. Dashboard design is effective communication. Most people fail to understand that information visualization is a science, not an art.

According to Stephen Few,[2]

1. Robert S. Kaplan and David P. Norton, *The Balanced Scorecard: Translating Strategy into Action*, Cambridge, MA: Harvard Business School Press, 1996.

2. Stephen Few, "Dashboard Design: Beyond Meters, Gauges and Traffic Lights," *Business Intelligence Journal*, 2005. Stephen Few has also written excellent books on dashboards: Stephen Few, *Information Dashboard Design*, Sebastopol, CA: O'Reilly, 2006; Stephen Few, *Show Me the Numbers*, Oakland, CA: Analytics Press 2004; and Stephen Few, *Now You See It*, Oakland, CA: Analytics Press, 2009.

> The primary purpose of a dashboard is to display all of the required information on a single screen, clearly and without distraction, in a manner that can be assimilated quickly. If this objective is hard to meet in practice, it is because dashboards often require a dense display of information. You must pack a lot of information into a very limited space, and the entire display must fit on a single screen, without clutter. This is a tall order that requires a specific set of design principles.

We can modify Stephen Few's comments to provide a definition of a dashboard as it relates to project management:

> A project management dashboard is a visual display of a small number of critical metrics or key performance indicators such that stakeholders and all project personnel can see the necessary information at a glance in order to make an informed decision. Raw data is converted into meaningful information. All of the information should be clearly visible on one computer screen.

There are some simple facts related to dashboards:

- Dashboards are communication tools.
- Dashboards provide the viewers with situational awareness of what the information means now and what it might mean in the future.
- Dashboards are not detailed reports.
- Some dashboards simply may not work.
- Some dashboards may be inappropriate for a particular application and should not be forced upon the stakeholders.
- More than one dashboard may be required to convey the necessary information. This should be done by providing the right data and without overwhelming people with information.
- It is important that the information displayed focus on the future. Otherwise, viewers will get bogged down analyzing the past rather than thinking about what's ahead.
- With the growth in the use of KPIs, it is extremely important that the stakeholders and other viewers of the dashboards have a good understanding of what is being measured. Deciding what to track is critical. Many projects fail because the dashboard designers insert too many bells and whistles, which can become distractions. Also, simply because an indicator on a dashboard is not flashing does not mean that things are going well.

TIP The project manager must explain to the stakeholders how to identify when things are going well and when things are going poorly using the dashboard KPIs.

There are also traps that dashboards designers must understand. Two such traps involve dashboard security and the use of branding. Dashboard security refers to the process of restricting the information presented in the dashboard only to those that have the right to that information.

The security system may be quite complex since each of the stakeholders may receive a different dashboard.

Also, because the space is limited on dashboards, care must be taken to avoid the heavy usage of company or project logos and other branding information. Company branding is always nice to have, but screen real estate is limited and expensive. Cluttering up a dashboard with too much information can lead to information overload.

Dashboard designers must understand:

- The end user's needs
- How the dashboard will be used
- How the measurements will be made
- How often the measurements will be made
- How the dashboard will be updated
- How to maintain uniformity in design, if possible

Unlike business dashboards, which can be updated quarterly, project management dashboards focus on month-to-date and cumulative-to-date comparisons, or the proximity to the target. Project management dashboards might also possess real-time reporting.

6.1 TRAFFIC LIGHT DASHBOARD REPORTING

In our attempt to go to paperless project management, emphasis is being given to visual displays such as dashboard and scorecards. Executives and customers desire a visual display of the most critical project performance information in the least amount of space. Simple dashboard techniques, such as traffic light reporting, shown in Figure 6–2, can convey critical performance information.

The following are examples of the meaning of the indicators in Figure 6–2:

- **Red traffic light:** A problem exists which may affect time, cost, quality, or scope. Sponsorship or stakeholder involvement may be necessary.
- **Yellow or amber light:** This is a caution. A potential problem may exist, perhaps in the future if the situation is not monitored. The sponsors/stakeholders are informed, but no action is necessary at this time. If some action has been taken, the problem has not been resolved as yet.
- **Green light:** Work is progressing as planned. No involvement by the sponsors or stakeholders is necessary.

The colors, red, yellow, and green, may be interpreted differently by each viewer. Therefore, an explanation of the colors may be necessary. For example:

- When discussing project risks:
  - **Red:** Some risk events exists, and there are no workable mitigation strategies.

Figure 6-2 **Traffic Light Dashboard Indicators**

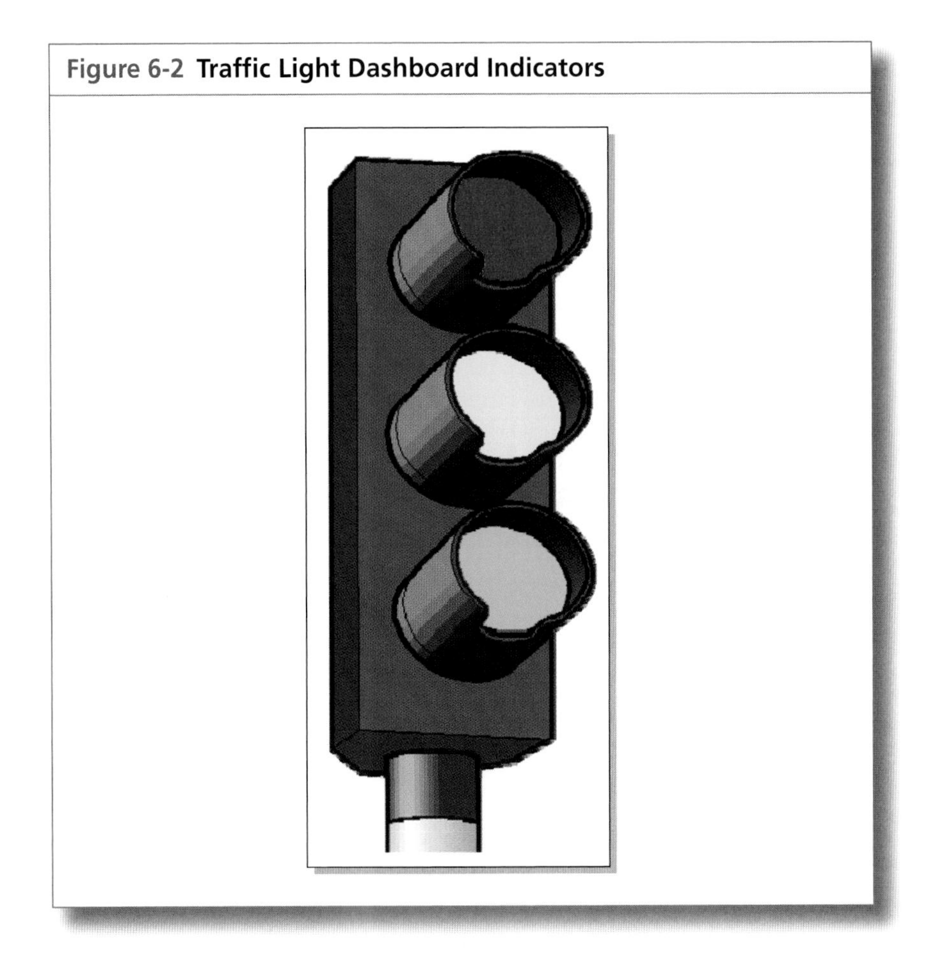

- **Yellow:** Some risk events have been identified, and mitigation strategies are being developed.
- **Green:** There are no risks.

- When discussing project staffing:
  - **Red:** We either lack sufficient resources or the assigned resources have questionable credentials.
  - **Yellow:** We have resource staffing issues but are ramping up.
  - **Green:** Sufficient and qualified resources are available.

While a traffic light dashboard with just three colors is most common, some companies use many more colors. The IT group of a retailer had an eight-color dashboard for IT projects. An amber color meant that the targeted end date had past and the project was still not complete. A purple color meant that this work package was undergoing a scope change that could have an impact on the triple constraints.

Although dashboards for project management applications are just in the infancy stages, companies have been using traffic light reporting for some time. It is common for project stakeholders to be briefed by the project manager without paperwork exchanging hands. The project manager displays the status of a project on a screen, using a computer and an LCD projector. Beside all of the work packages in the WBS is a traffic light. Senior management then takes keen interest in all of the work packages indicated in red. One Detroit-based company believes that in the first year of using this technique and going to paperless meetings they saved $1 million and expect the savings to increase each year.

6.2 DASHBOARDS AND SCORECARDS

Some people confuse dashboards with scorecards. There is a difference between dashboards and scorecards. According to Wayne W. Eckerson,[3]

> Dashboards are visual display mechanisms used in an **operationally** oriented performance measurement system that measure performance against targets and thresholds using right-time data.
>
> Scorecards are visual displays used in a **strategically** oriented performance measurement system that chart progress towards achieving strategic goals and objectives by comparing performance against targets and thresholds.
>
> Both dashboards and scorecards are visual display mechanisms within a performance measurement system that convey critical information. The primary difference between dashboards and scorecards is that dashboards monitor operational processes such as those used in project management, whereas scorecards chart the progress of tactical goals. Table 6–1 and the description following it show how Eckerson compares the features of dashboards to scorecards.[4]

Dashboards

Dashboards are more like automobile dashboards. They let operational specialists and their supervisors monitor events generated by key business processes. Unlike automobiles, however, most business dashboards do not display events in "real time" as they occur; they display them in "right time" as users need to view them. This could be every second, minute, hour, day, week, or

3. Wayne W. Eckerson, *Performance Dashboards: Measuring, Monitoring and Managing Your Business*, Hoboken, NJ: John Wiley and Sons, 2006; pp. 293, 295. Chapter 12 provides an excellent approach to designing dashboard screens.

4. Ibid., p. 13.

TABLE 6-1 Comparing Features

FEATURE	DASHBOARD	SCORECARD
Purpose	Measures performance	Charts progress
Users	Supervisors, specialists	Executives, managers, and staff
Updates	Right-time feeds	Periodic snapshots
Data	Events	Summaries
Display	Visual graphs, raw data	Visual graphs, comments

month, depending on the business process, its volatility, and how critical it is to the business. However, most elements on a dashboard are updated on an intraday basis, with latency measured in either in minutes or hours.

Dashboards often display performance visually, using charts or simple graphs, such as gauges and meters. However, dashboard graphs are often updated in place, causing the graph to "flicker" or change dynamically. Ironically, people who monitor operational processes often find the visual glitz distracting and prefer to view the data in its original form, as numbers or text, perhaps accompanied by visual graphs.

Scorecards

Scorecards, on the other hand, look more like performance charts used to track progress toward achieving goals. Scorecards usually display monthly snapshots of summarized data for business executives who track strategic and long-term objectives, or daily and weekly snapshots of data for managers who need to chart the progress of their group of projects toward achieving goals. In both cases, the data are summarized, so users can view their performance status at a glance.

Like dashboards, scorecards also make use of charts and visual graphs to indicate performance state, trends, and variance from goals. The higher up the users are in the organization, the more they prefer to see performance encoded visually. However, most scorecards also contain (or should contain) a great deal of textual commentary that interprets performance results, describes action taken, and forecasts future results.

Summary

In the end, it does not really matter whether you use the term "dashboard" or "scorecard" as long as the tool helps to focus users and organizations on what really matters. Both dashboards and scorecards need to display critical performance information on a single screen so that users can monitor results at a glance.

TABLE 6-2 Comparison of Dashboards and Scorecards

FACTOR	DASHBOARDS	SCORECARDS
Performance	Operational issues	Strategic issues
WBS level for measurement	Work package level	Summary level
Frequency of update	Real time data	Periodic data
Target audience	Working levels	Executive levels

TABLE 6-3 Three Types of Performance Dashboards

	OPERATIONAL	TACTICAL	STRATEGIC
Purpose	Monitor operations	Measure progress	Execute strategy
Users	Supervisors, specialists	Managers, analysts	Executives, managers, staff
Scope	Operational	Departmental	Enterprise
Information	Detailed	Detailed/summary	Detailed/summary
Updates	Intraday	Daily/weekly	Monthly/quarterly
Emphasis	Monitoring	Analysis	Management

Dashboards appear to be more appropriate for project management than scorecards. Table 6–2 shows some of the factors relating to this.

Although the terms are used interchangeably, most project managers prefer to use dashboards and/or dashboard reporting rather than scorecards. Eckerson defines three types of dashboards as shown in Table 6–3 and the description that follows:[5]

> **Operational dashboards** monitor core operational processes and are used primarily by front-line workers and their supervisors who deal directly with customers or manage the creation or delivery of organizational products and services. Operational dashboards primarily deliver detailed information that is only lightly summarized. For example, an online Web merchant may track transactions at the product level rather than the customer level. In addition, most metrics in an operational dashboard are updated on an intraday basis, ranging from minutes to hours, depending on the application. As a result, operational dashboards emphasize monitoring more than analysis and management.
>
> **Tactical dashboards** track departmental processes and projects that are of interest to a segment of the organization or a limited group of people.

5. Ibid., pp. 17–18.

Managers and business analysts use tactical dashboards to compare performance of their areas or projects, to budget plans, forecasts, or last period's results. For example, a project to reduce the number of errors in a customer database might use a tactical dashboard to display, monitor, and analyze progress during the previous 12 months toward achieving 99.9 percent defect-free customer data by 2007.

Strategic dashboards monitor the execution of strategic objectives and are frequently implemented using a Balanced Scorecard approach, although Total Quality Management, Six Sigma, and other methodologies are used as well. The goal of a strategic dashboard is to align the organization around strategic objectives and get every group marching in the same direction. To do this, organizations roll out customized scorecards to every group in the organization and sometimes to every individual as well. These "cascading" scorecards, which are usually updated weekly or monthly, give executives a powerful tool to communicate strategy, gain visibility into operations, and identify the key drivers of performance and business value. Strategic dashboards emphasize management more than monitoring and analysis.

There are three critical steps that must be considered when using dashboards; (1) the target audience for the dashboard, (2) the type of dashboard to be used, and (3) the frequency in which the data will be updated. Some project dashboards focus on the key performance indicators that are part of earned value measurement. These dashboards may need to be updated daily or weekly. Dashboards related to the financial health of the company may be updated monthly or quarterly.

6.3 BENEFITS OF DASHBOARDS

Digital dashboards allow viewers to gauge exactly how well the project is performing overall, and allow the viewers to capture specific data. The benefits of using dashboards include:[6]

- Visual representation of performance measures
- Ability to identify and correct negative trends
- Measure efficiencies/inefficiencies
- Ability to generate detailed reports showing new trends
- Ability to make more informed decisions based upon collected intelligence
- Align strategies and overall goals
- Save time over running multiple reports
- Gain total visibility of all systems instantly

6. Wikipedia, www.wikipedia.com.

In order for a project to continuously improve, four steps are required:

- Measure performance and turn it into data
- Turn data into knowledge
- Turn knowledge into action
- Turn action into improvements

Dashboards are tools that allow this to happen.

6.4 RULES FOR DASHBOARDS

Throughout this chapter, we will be discussing certain rules for dashboards, such as rules for colors, rules for metaphor selection, and rules for positioning the metaphors. However, there are certain overall rules that we should consider as well. Some of these include:

- Dashboards are communication tools to provide information at a glance.
- Be sure to understand aesthetics, especially in symmetry and proportions.
- Be sure to understand computer resolution considerations involving readability.
- Dashboard design begins with an understanding of the user's needs.
- Be sure to understand content selection and the accuracy of the information.
- Dashboard design can be done with simple displays.
- Dashboard design can be done with simple tools.
- Use the fewest metrics necessary.
- Determine the fewest metrics that can be retained in short term memory.
- Using too many colors or sophisticated, complex metaphors leads to distractions.
- Limit metrics to a single screen.
- Be sure to understand the available space; know the number of windows/frames available within the dashboard.
- Perfection in design can never be achieved.
- Asking for assistance with the design effort is not an embarrassment.
- Monitor the health and user friendliness of the dashboard.

6.5 BITWORK, INC.: TEN QUESTIONS TO ASK BEFORE IMPLEMENTING A DASHBOARD OR REPORTING SYSTEM[7]

Implementing an enterprise dashboard, a balanced scorecard, or another reporting solution is a big step and a big investment for any size of organization. We've put together the following ten questions for use in fleshing

7. This entire section is copyrighted material reproduced with permission of Bitwork, Inc. All rights reserved. For more information, visit www.bitwork.com or call (877) 724-8967.

out project plans, avoiding common challenges, and maximizing the success of your project.

1. What Are Your Needs?

Before starting any project, it's important to know what you want to get out of it and how that will help your organization. Many projects go awry when technology is implemented for technology's sake or when the focus is on the latest bells and whistles rather than what's needed to meet the project requirements.

There are many companies that sell dashboard systems. Choosing a vendor that is not only familiar with your industry but that also focuses specifically on your industry will give you more assurance that certain pitfalls are avoided as the project moves along.

2. What Do You Have in Place Already?

Many organizations have some sort of reporting in place already. Sometimes it's from a third party, and sometimes it is a piecemeal solution that's been developed internally over time. Often different departments will have had put together their own reports without much communication with others.

Whatever the case, the next question to ask is what the relationship will be between the new solution and the old. One option is for the new product to exist alongside the other services, with each one generating their its own reports. Another option is to integrate all of the packages so that they feed data into one central reporting platform. And the third option is for the new solution to replace all of the others.

3. What Is Involved in Integration?

The new reporting solution will need to integrate with your organization's software applications and with any existing reporting applications. Vendors generally provide out-of-the-box integration with many databases and some common applications. Your business critical applications will generally be the least supported and the most difficult to connect to. Any applications that you've developed in house will also need custom integration.

Ask your vendor how they will connect to the important applications. Assess whether they have the necessary skill sets available for this and whether they can efficiently develop the necessary connectors in time to fit your schedule.

4. How Long Does Installation Take?

It is very common to hear of dashboard projects that have gotten a year along and still aren't displaying information. And still others require huge IT commitments with new systems and databases.

With a software appliance solution, the hardware can generally be installed in a matter of hours. You then don't have to worry about installing an operating system and patches, loading software, or setting up external storage. Make sure your vendor's project plan shows not only how long the entire project will take but at what points along the way you'll see specific information reported.

5. How Easy Is the System to Use?

Some organizations employ a small team of people who are experts in writing SQL queries for databases. They design the queries that run overnight and create the reports that have been requested from around the organization. This may work for them, but it is expensive, and it isn't very scalable. Some products wrap graphics around text forms where users still enter SQL queries, but this still doesn't remove the requirement for specialized database knowledge.

Look for solutions that provide a graphical interface that is both flexible and requires minimal training to use. A web interface is easy to access from the different operating systems, which will support your users in different departments. Make sure that it's easy to find your way around the application since people will probably be using it frequently, and you want to minimize training and support calls. Also check how long it takes to get requested information from the system. Most requests should be fulfilled in just seconds with very few taking over a minute.

6. Who Will Use the System?

Another complication that comes up in dashboard and other reporting projects is the expansion of scope. One department may start a reporting initiative, and then once it gains momentum, other departments want to get onboard. Though it's great to have everyone moving in a common direction and using a common system, care needs to be taken to keep the project organized, make sure everyone's needs are met, and make sure everyone can use the system.

Plan ahead on what it will take to scale the system inside your organization. Estimate how many people might be using the new dashboard at once, and check that the proposed solution will support that number. Then work with the different departments to determine in what order and at what point in the project their information will be added to the dashboard.

7. Can You Get Customizations?

Every industry is different, and within each industry, every organization is unique. An out-of-the-box solution can cover many of your needs, but with customizations specific to your environment, your solution can truly become part of and improve your organization.

Discuss the uniqueness of your environment and the specific customizations you're interested in with the vendor. See what is both possible and efficient for them to add in your installation. Some vendors will give you a certain amount of customization at no charge.

8. What's Involved in Operations and Maintenance?

There is no standard architecture for reporting systems, so you'll have to look closely at what's involved in running a specific vendor's product. The first thing to look at is the hardware. After counting the main system, look to see what else is required, from database servers to auxiliary reporting servers or a front end web server for the user interface. Next is software, which includes the reporting software, operating systems, any database, and other third-party software. And last on the list is supporting your user community through training, support calls, and configuring new charts and reports.

All of these things require time. Look closely at how a vendor's solution will extend your IT infrastructure, and calculate how many additional full-time employees you will need to support the product on an ongoing basis. Aim to work with a vendor's product that requires little or no support.

9. What Does the System Cost?

No matter what a vendor charges you for a dashboard or reporting system, it is important to look beyond the sticker price to the total cost of ownership. This can be divided into two parts.

The first is to assess the cost of the installation. This includes the hardware and software, as well as any vendor services for the installation, including integration and customization. The other part is the cost involved in running and maintaining the system. This includes the vendor's support and maintenance contract, the support and maintenance on any new hardware and software for the installation, and finally the human resources.

When selecting a vendor, make sure you're getting the functionality you need and that you're not also buying a lot of features you're realistically not going to use. Then make sure that the system footprint is small in order to minimize both installation cost and ongoing expense.

10. How Long Will It Last?

The final question to ask is how long this new dashboard or reporting solution will last once you've spent the time and money to implement it. Consider what your needs will be tomorrow, how well the product will grow with you, and how easy it is to work with your vendor. Discuss what information you might want, and see where that fits in on your vendor's roadmap. Make sure you can start using the product as soon as possible in order to maximize your project's useful lifetime and return on investment.

6.6 BRIGHTPOINT CONSULTING, INC.: DESIGNING EXECUTIVE DASHBOARDS[8]

Introduction

Corporate dashboards are becoming the "must have" business intelligence technology for executives and business users across corporate America. Dashboard solutions have been around for over a decade but have recently seen resurgence in popularity because of the advance of enabling business intelligence and integration technologies.

Designing an effective business dashboard is more challenging than it might appear because you are compressing large amounts of business information into a small visual area. Every dashboard component must effectively balance its share of screen real estate with the importance of the information it is imparting to the viewer.

This article will discuss how to create an effective operational dashboard and some of the associated design best practices.

Dashboard Design Goals

Dashboards can take many formats, from glorified reports to highly strategic business scorecards. This article refers to operational or tactical dashboards employed by business users in performing their daily work; these dashboards may directly support higher-level strategic objectives or be tied to a very specific business function. The goal of an operational dashboard is to provide business users with *relevant* and *actionable* information that empowers them to make effective decisions in a more efficient manner than they could without a dashboard. In this context, "relevant" means information that is directly tied to the user's role and level within the organization. For instance, it would be inappropriate to provide the CFO with detailed metrics about web site traffic but appropriate to present usage costs as they

---

8. Material in this section has been taken from BrightPoint Consulting white paper, "Designing Executive Dashboards" by Tom Gonzalez, Managing Director, BrightPoint Consulting, Inc., © 2005 by BrightPoint Consulting, Inc. Reproduced by permission. All rights reserved. Mr. Gonzalez is the founder and Managing Director of BrightPoint Consulting, Inc., serving as a consultant to both Fortune 500 companies and small-medium businesses alike. With over 20 years experience in developing business software applications, Mr. Gonzalez is a recognized expert in the fields of business intelligence and enterprise application integration within the Microsoft technology stack. BrightPoint Consulting, Inc. is a leading technology services firm that delivers corporate dashboard and business intelligence solutions to organizations across the world. BrightPoint Consulting leverages best of breed technologies in data visualization, business intelligence and application integration to deliver powerful dashboard and business performance solutions that allow executives and managers to monitor and manage their business with precision and agility. For further company information, visit BrightPoint's web site at www.brightpointinc.com. To contact Mr. Gonzalez, e-mail him at tgonzalez@brightpointinc.com.

relate to bandwidth consumption. "Actionable" information refers to data that will alert the user as to when and what type of action needs to be taken in order to meet operational or strategic targets. Effective dashboards require an extremely efficient design that takes into account the role a user plays within the organization and the specific tasks and responsibilities that user performs on a daily/weekly basis.

Defining Key Performance Indicators

The first step in designing a dashboard is to understand what key performance indicators (KPI) users are responsible for and which KPIs they wish to manage through their dashboard solution. A KPI can be defined as a measure (real or abstract) that indicates relative performance in relationship to a target goal. For instance, we might have a KPI that measures a specific number, such as daily Internet sales with a target goal of $10,000. In another instance we might have a more abstract KPI that measures "financial health" as a composite of several other KPIs, such as outstanding receivables, available credit and earnings before tax and depreciation. Within this scenario the higher-level "financial" KPI would be a composite of three disparate measures and their relative performance to specific targets. Defining the correct KPIs specific to the intended user is one of the most important design steps, as it sets the foundation and context for the information that will be subsequently visualized within the dashboard.

Defining Supporting Analytics

In addition to defining your KPIs, it is helpful to identify the information a user will want to see in order to diagnose the condition of a given KPI. We refer to this non-KPI information as "supporting analytics" because it provides context and diagnostic information for end users in helping to understand why a KPI is in a given state. Often, these supporting analytics take the form of more traditional data visualization representations such as charts, graphs, tables and, with more advanced data visualization packages, animated what-if or predictive analysis scenarios.

For each KPI on a given dashboard you should decide if you want to provide supporting analytics and, if so, what type of information would be needed to support analysis of that KPI. For instance, in the case of a KPI reporting on aging receivables, you might want to provide the user a list of accounts due with balances past 90 days. In this case when a user sees that the aging KPI is trending in the wrong direction he/she could click on a supporting analytics icon to bring up a table of accounts due sorted by balance outstanding. This information would then support the user in his/her ability to decide what, if any, action needed to be taken in relationship to the condition of the KPI.

Choosing the Correct KPI Visualization Components

Dashboard visualization components fall into two main categories: key performance indicators and supporting analytics. In either case, it is important to choose the visualization that best meets the end users' need in relationship to the information they are monitoring or analyzing.

For KPIs there are five common visualizations used in most dashboard solutions. The following list describes each component's relative merits and common usage scenario.

1. **Alert Icons:** The simplest visualization is perhaps an alert icon, which can be a geometric shape that is either color-coded or shaded various patterns based on its state. Potentially, the most recognizable alert icon is a green, yellow, or red circle, whereby the color represents a more or less desirable condition for the KPI.

   When to use: These types of visualizations are best used when they are placed in the context of other supporting information, or when you need a dense cluster of indicators that are clearly labeled. Traditional business scorecard dashboards that are laid out in table-like format can benefit from this visualization in which other adjacent columns of information can be analyzed, depending on the state of the alert icon. These types of icons are also useful in reporting on system state, such as whether a machine or application is online or not. Be cautious of using icons that depend exclusively on color to differentiate state, as 10 percent of the male population and 1 percent of the female population is color-blind; consider using shapes in conjunction with color to differentiate state.
2. **Traffic Light Icons:** The traffic light is a simple extension of the alert icon, and has little advantage over the alert icon in terms of data visualization. Like the alert icon, this component only offers one dimension of information, but it requires 100 percent of the screen real estate. The one advantage of the traffic light icon is that it is a more widely recognized symbol of communicating a "good state," "warning state" or "bad state."

   When to use: In most cases a simple alert icon is a more efficient visualization, but in situations where your dashboard is being used by a wide audience on a less frequent basis, a traffic light component will allow users to more quickly assimilate the alert information because of their familiarity with the traffic light symbol from real-world experience.
3. **Trend Icons:** A trend icon represents how a key performance indicator or metric is behaving over a period of time. It can be in one of three states: moving toward a target, [moving] away from a target, or static. Various symbols may be used to represent these states, including arrows or numbers. Trend icons can be combined with alert icons to display two dimensions of information within the same visual space. This can

be accomplished by placing the trend icon within a color- or shape-coded alert icon.

When to use: Trend icons can be used by themselves in the same situation you would use an alert icon, or to supplement another more complex KPI visualization when you want to provide a reference to the KPI's movement over time.

4. **Progress Bars:** A progress bar represents more than one dimension of information about a KPI via its scale, color, and limits. At its most basic level, a progress bar can provide a visual representation of progress along a one-dimensional axis. With the addition of color and alert levels, you can also indicate when you have crossed specific target thresholds as well as how close you are to a specific limit.

   When to use: Progress bars are primarily used to represent relative progress toward a positive quantity of a real number. They do not work well when the measure you want to represent can have negative values: The use of shading within a "bar" to represent a negative value can be confusing to the viewer because any shading is seen to represent some value above zero, regardless of the label on the axis. Progress bars also work well when you have KPIs or metrics that share a common measure along an axis (similar to a bar chart), and you want to see relative performance across those KPIs/metrics.

5. **Gauges:** A gauge is an excellent mechanism by which to quickly assess both positive and negative values along a relative scale. Gauges lend themselves to dynamic data that can change over time in relationship to underlying variables. Additionally, the use of embedded alert levels allows you to quickly see how close or far away you are from a specific threshold.

   When to use: Gauges should be reserved for the highest level and most critical metrics or KPIs on a dashboard because of their visual density and tendency to focus user attention. Most of these critical operational metrics/KPIs will be more dynamic values that change on a frequent basis throughout the day. One of the most important considerations in using gauges is their size: too small and it is difficult for the viewer to discern relative values because of the density of the "ink" used to represent the various gauge components; too large and you end up wasting valuable screen space.

   With more sophisticated data visualization packages, gauges also serve as excellent context-sensitive navigation elements because of their visual predominance within the dashboard.

Supporting Analytics

Supporting analytics are additional data visualizations that a user can view to help diagnose the condition of a given KPI or set of KPI's. In most business cases, these supporting analytics take the form of traditional charts and tables or lists. While the scope of this article is not intended to cover the myriad of best practices in designing traditional charting visualizations, we will discuss some of the basics as they relate to dashboard design.

When creating supporting analytics, it is paramount that you take into account the typical end user who will be viewing the dashboard. The more specialized and specific the dashboard will be, the more complexity and detail you can have in your supporting analytics. Conversely, if you have a very high-level dashboard, your supporting analytics will generally represent higher-level summary information with less complex detail.

In the following list, we will discuss some of the most common visualizations used for designing supporting analytics.

1. **Pie charts:** Pie charts are generally considered a poor form of data visualization for any data set with more than half a dozen elements. The problem with pie charts is that it is very difficult to discern proportional differences with a radially divided circle, except in the case of a small data set that has large value differences within it. Pie charts also pose a problem for labeling, because they are either dependent on a color or pattern to describe the different data elements, or the labels need to be arranged around the perimeter of the pie, creating a visual distraction.

   When to use: Pie charts should be used to represent very small data sets that are geared to high-level relationships between data elements. Usually pie charts can work for summary-level relationships but should not be used for detailed analysis.
2. **Bar charts:** Bar charts are an ideal visualization for showing the relationship of data elements within a series or multiple series. Bar charts allow for easy comparison of values because of the fact that the "bars" of data share a common measure and can be easily visually compared to one another.

   When to use: Bar charts are best suited for categorical analysis but can also be used for small time series analysis (e.g., the months of a year.) An example of categorical analysis is examining sales of products broken down by product or product group, with sales in dollars being the measure and product or product group being the category. Be careful in using bar charts if you have a data set that can have one element with a large outlier value; this will render the visualization for the remaining data elements unusable. This is because of the fact that the chart scale is linear and will not clearly represent the relationships between the remaining data elements. An example is shown in Figure 6–3. Notice that, because widget 2 has sales of $1.2MM, you cannot easily discern that widget 3 has twice as many sales ($46,000) as widget 1 ($23,000).
3. **Line charts:** Line charts are ideal for time series analysis where you want to see the progress of one or more measures over time. Line charts

Figure 6-3 Typical Bar Chart *Material in this section has been taken from BrightPoint Consulting white paper, "Designing Executive Dashboards" by Tom Gonzalez, managing director, BrightPoint Consulting, Inc., © 2005 by BrightPoint Consulting, Inc. Reproduced by permission. All rights reserved.*

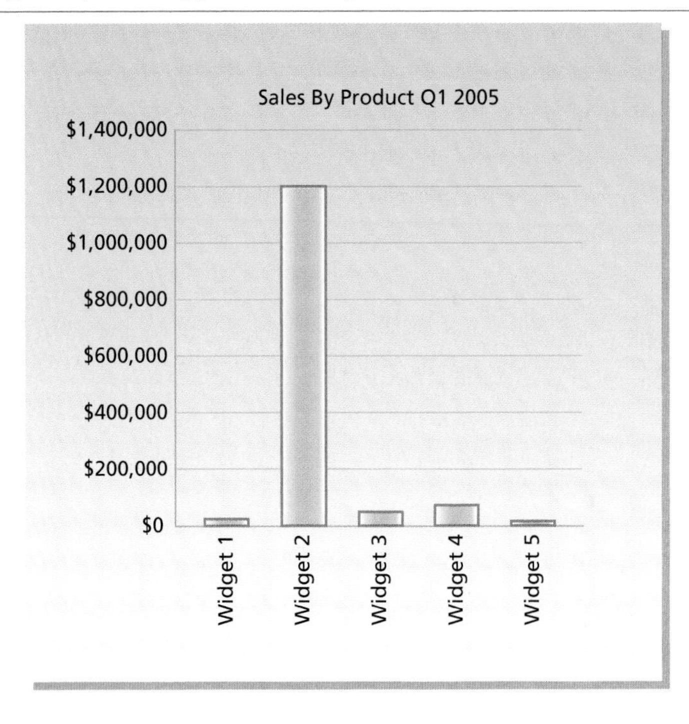

also allow for comparative trend analysis because you can stack multiple series of data into one chart.

When to use: Use line charts when you would like to see trends over time in a measure versus a side-by-side detailed comparison of data points. Time series line charts are most commonly used with the time dimension along the X-axis and the data being measured along the Y-axis.

4. **Area charts:** Area charts can be considered a subset of the line chart, where the area under or above the line is shaded or colored.

    When to use: Area charts are good for simple comparisons with multiple series of data. By setting contrasting color hues, you can easily compare the trends over time between two or more series.

5. **Tables and lists:** Tables and lists are best used for information that either contains large lists of non-numeric data, or data that has relationships not easily visualized.

    When to use: You will want to use tables or lists when the information you need to present does not lend itself to easy numeric analysis. One example is a financial KPI that measures a company's current liquidity ratio. In this case, there can be a complex interrelationship of line items within the company's balance sheet, where a simple table of balance sheet line items would provide a more comprehensive supporting analytic than a series of detailed charts and graphs.

A Word about Labeling Your Charts and Graphs

Chart labels are used to give the users context for the data they are looking at, in terms of both scale and content. The challenge with labeling is that the more labels you use and the more distinctive you make them, the more it will distract the user's attention from the actual data being represented within the chart.

When using labels, there are some important considerations to take into account. Foremost among these is how often your user will be viewing these charts. For charts being viewed on a more frequent basis, the user will form a memory of relevant labels and context. In these scenarios, you can be more conservative in your labeling by using smaller fonts and less color contrast. Conversely, if a user will only be seeing the chart occasionally, you will want to make sure everything is labeled clearly so that the user does not have to decipher the meaning of the chart.

Putting It All Together: Using Size, Contrast, and Position

The goal in laying out an effective dashboard is to have the most important business information be the first thing to grab your user's visual attention. In your earlier design stages, you already determined the important KPI's and supporting analytics, so you can use this as your layout design guide. Size, contrast, and position all play a direct role in determining which visual elements will grab the user's eye first.

Size: In most situations, the size of a visual element will play the largest role in how quickly the user will focus their attention upon it. In laying out your dashboard, figure out which element or group of elements will be the most important to the user and make their size proportionally larger than the rest of the elements on the dashboard.

This principle holds true for single element or groups of common elements that have equal importance.

Contrast: After size, the color or shade contrast of a given element in relationship to its background will help determine the order in which the user focuses attention on that element. In some situations, contrast alone will become the primary factor, even more so than size, for where the user's eye will gravitate. Contrast can be achieved by using different colors or saturation levels to distinguish a visual element from its background. A simple example of this can be seen in the Figure 6–4.

As you can see, the black circle instantly grabs the user's attention because of the sharp contrast against the white background. In this example,

Figure 6-4 Contrasting Colors *Material in this section has been taken from BrightPoint Consulting white paper "Designing Executive Dashboards" by Tom Gonzalez, managing director, BrightPoint Consulting, Inc., © 2005 by BrightPoint Consulting, Inc. Reproduced by permission. All rights reserved.*

Figure 6-5 Positioning of Icons *Material in this section has been taken from BrightPoint Consulting white paper "Designing Executive Dashboards" by Tom Gonzalez, managing director, BrightPoint Consulting, Inc., © 2005 by BrightPoint Consulting, Inc. Reproduced by permission. All rights reserved.*

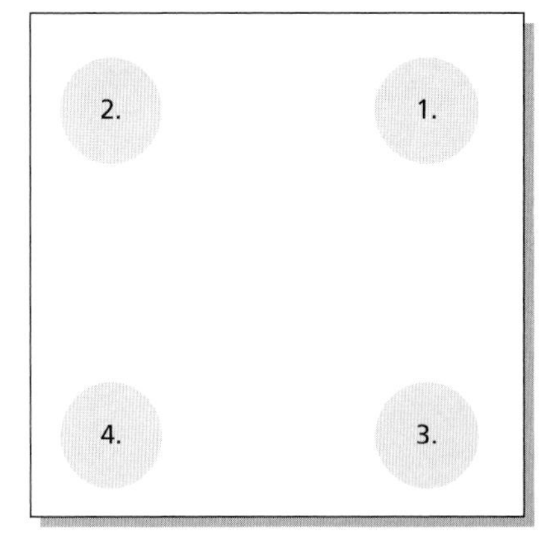

the contrast even overrides the size of the larger circle in its ability to focus the user's visual awareness.

Position: Visual position also plays a role in where a user will focus his or her attention. All other factors being equal, the top-right side of a rectangular area will be the user's first focal point, as seen in Figure 6–5. The next area a user will focus is the top-left side, followed by the bottom right, and finally the bottom left. Therefore, if you need to put an element on the dashboard that you don't want the user to have to hunt around for, the top-right quadrant is generally the best place for it.

Position is also important when you want to create an association between visual elements. By placing elements in visual proximity to each other, and grouping them by color or lines, you can create an implied context and relationship between those elements. This is important in instances when you want to associate a given supporting analytic with a KPI or group together related supporting analytics.

Validating Your Design

You will want to make sure that your incorporation of the preceding design techniques achieves the desired effect of focusing the user's attention on the most important business information and in the proper order. One way to see if you have achieved this successfully is to view your dashboard with an out-of-focus perspective. This can be done by stepping back from your dashboard and relaxing your focus until the dashboard becomes blurry and you can no longer read words or distinguish finer details. Your visual cortex will still recognize the overall visual patterns, and you will easily see the most attention-grabbing elements of your design. You want to validate that the elements attracting the most visual attention correspond with the KPIs and supporting analytics that you had previously identified as being most critical to the business purpose of your dashboard.

Please bear in mind, the design guidelines presented in this article should be used as general rules of thumb, but it is important to note that these are not hard-and-fast rules that must be followed in every instance. Every dashboard has its own unique requirements, and in certain cases you will want to deviate from these guidelines and even contradict them to accomplish a specific visual effect or purpose.

REFERENCES

Fitts, P. M. (1954). "The information capacity of the human motor system in controlling the amplitude of movement," *Journal of Experimental Psychology*, 47, 381–391.

Tractinsky, Noam "Aesthetics and Apparent Usability: Empirically Assessing Cultural and Methodological Issues," Association for Computing Machinery, Inc., 1997.

Tufte, Edward, *Envisioning Information*, Graphics Press, 1990.

6.7 ALL THAT GLITTERS IS NOT GOLD

In the previous sections, we discussed the use of various metaphors and icons to display information in dashboards. Designing a perfect dashboard may be impossible. An image that works well for one dashboard may be inappropriate for another. Also, there are both advantages and disadvantages to all images and colors. My belief is that, if the image works and provides the necessary and correct information for the stakeholder and the stakeholders understands the image, continue using it.

There are several rules that can be used for dashboard designs. In-depth explanations of each of these can be found on the Internet. As an example:

- **Rules for selecting the right artwork:** Selecting the right image is critical. As an example, gauges cannot show trends. Options for images include:
  - Gauges
  - Thermometers
  - Traffic lights
  - Area charts
  - Bar charts
  - Stacked charts
  - Bubble charts
  - Clustered charts
  - Performance trends
  - Performance variances
  - Histograms
  - Pie charts
  - Rectangles with quadrants
  - Alert buttons
  - Cylinders
  - Composites
- **Rules for positioning the artwork (Number of windows and frames):** There must be a speed of perception. Also, the upper left (or upper right, based on the designer's preference) is usually considered to be more important than the lower-right corner.
- **Rules for visualization (Readability and symmetry and proportions):** The image and information should be easy to read and aesthetically pleasing to the eye.
- **Rules for accuracy of the information (Context selection):** The image must provide reasonably accurate information for informed decision making without requiring an interpretation by the viewer. However, some stakeholders are more interested in trends than absolute performance.
- **Rules for color selection:** Factors that must be considered include:
  - Colors
  - Positioning of the colors
  - Brightness

- Orientation
- Saturation
- Size
- Texture
- Shape

As mentioned before, perfection in dashboard design may not be possible. Even the simplest designs can have possible flaws for the viewer. As an example, consider the following area charts:[9]

- **Traditional area chart (Figure 6–6):** This displays the trend over time or categories.
- **Stacked area chart (Figure 6–7):** This displays the trend of the contribution of each value over time or categories.
- **100% stacked area chart (Figure 6–8):** This displays the trend of the percentage each value contributes over time or categories.

Figure 6-6 Area Chart *From Rasmussen,* Business Dashboards, *Wiley, 2009*

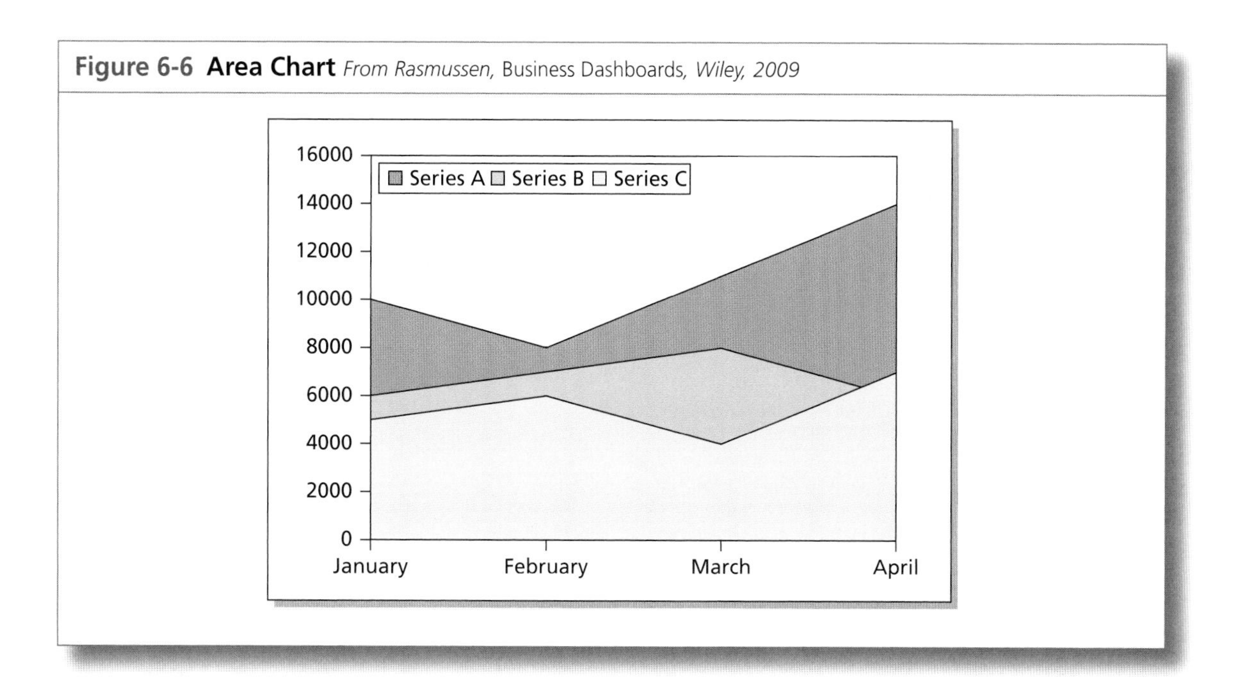

9. Figures 6–6 through 6–20 were taken from Nils Rasmussen, Claire Y. Chen, and Manish Bansal, *Business Dashboards*, Hoboken, NJ: John Wiley and Sons, 2009, pp. 94–100, Exhibits 14–1 through 14–15. The book has excellent examples of business dashboards, and Appendix E in the book identifies metrics and KPIs for a variety of industries and applications.

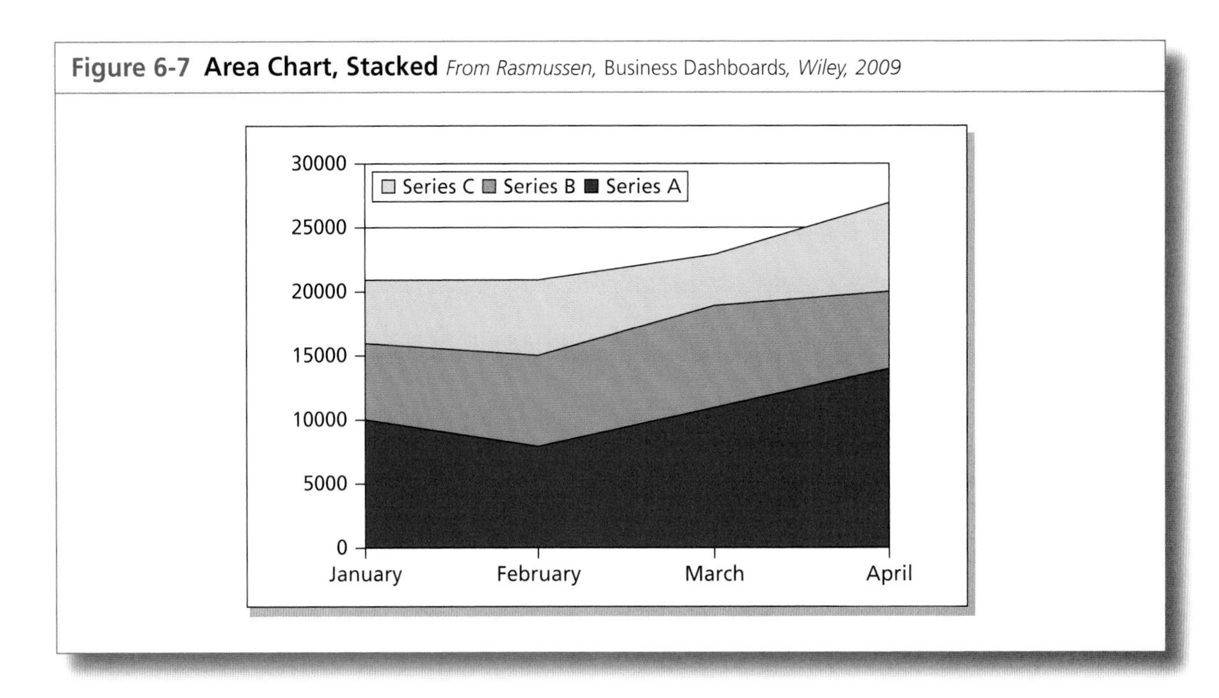

Figure 6-7 Area Chart, Stacked *From Rasmussen,* Business Dashboards, *Wiley, 2009*

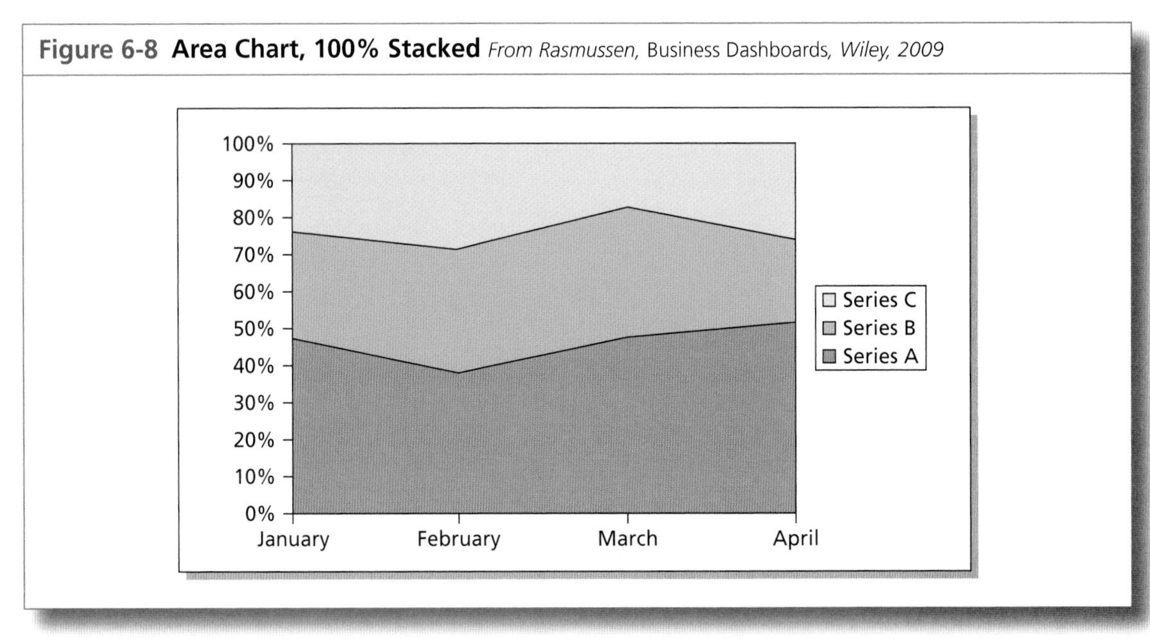

Figure 6-8 Area Chart, 100% Stacked *From Rasmussen,* Business Dashboards, *Wiley, 2009*

These charts are good for looking at trends. Some stakeholders may be interested more in trends than hard numbers. However, to get a precise value at a specific time period would require detailed measurements from the charts, and this would introduce the opportunity for error.

Another commonly used image is the bar chart. As an example, consider the following Bar Charts:

- **Clustered bar chart (Figure 6–9):** This compares values across categories.
- **Stacked bar chart (Figure 6–10):** This compares the contribution of each value to a total across categories.

Figure 6-9 Bar Chart, Clustered *From Rasmussen,* Business Dashboards, *Wiley, 2009*

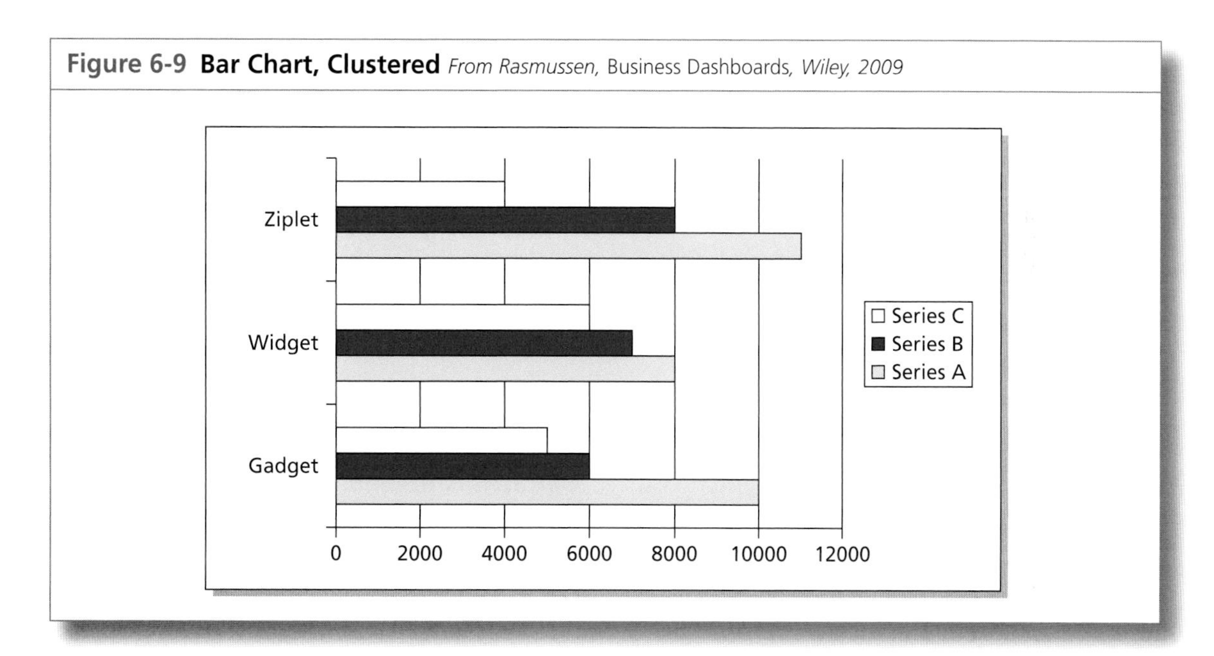

Figure 6-10 Bar Chart, Stacked *From Rasmussen,* Business Dashboards, *Wiley, 2009*

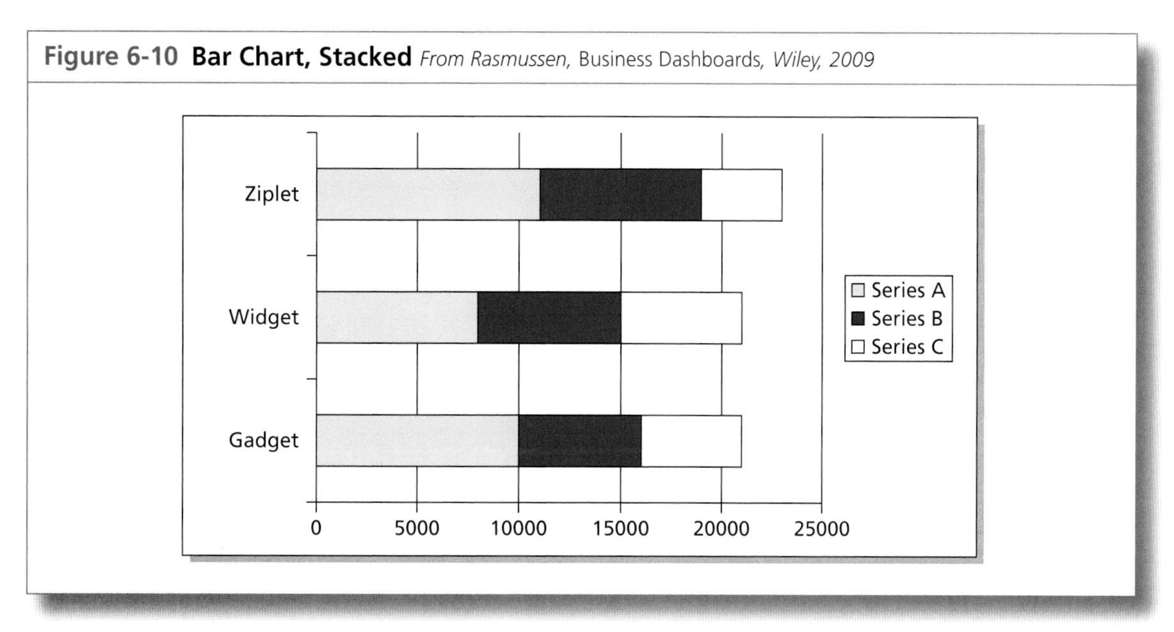

- **100% stacked bar chart (Figure 6–11):** This compares the percentage each value contributes to a value across categories.

In these figures, viewers can be distracted if part of the bar appears in the same color as the background of the image. Also, in the stacked bar charts, getting the exact value of Series B and Series C may require measurement that can lead to error.

Bubble charts, as shown in Figure 6–12, are more appropriate to business dashboards than project dashboards. The chart compares three sets of values, similarly to line charts, but with a third value displayed as the size of the bubble marker.

Column charts are similar to bar charts. As an example, consider the following three column charts:

- **Clustered column chart (Figure 6–13):** This compares values across categories.
- **Stacked column chart (Figure 6–14):** This compares the contribution of each value to a total across categories.
- **100% stacked column chart (Figure 6–15):** This chart compares the percentage each value contributes to a value across categories.

Some form of column chart appears in almost all dashboards. However, care must be taken in the selection of the colors. In Figures 6–13 and 6–14, the shades of the colors on the columns may create a visual problem. The

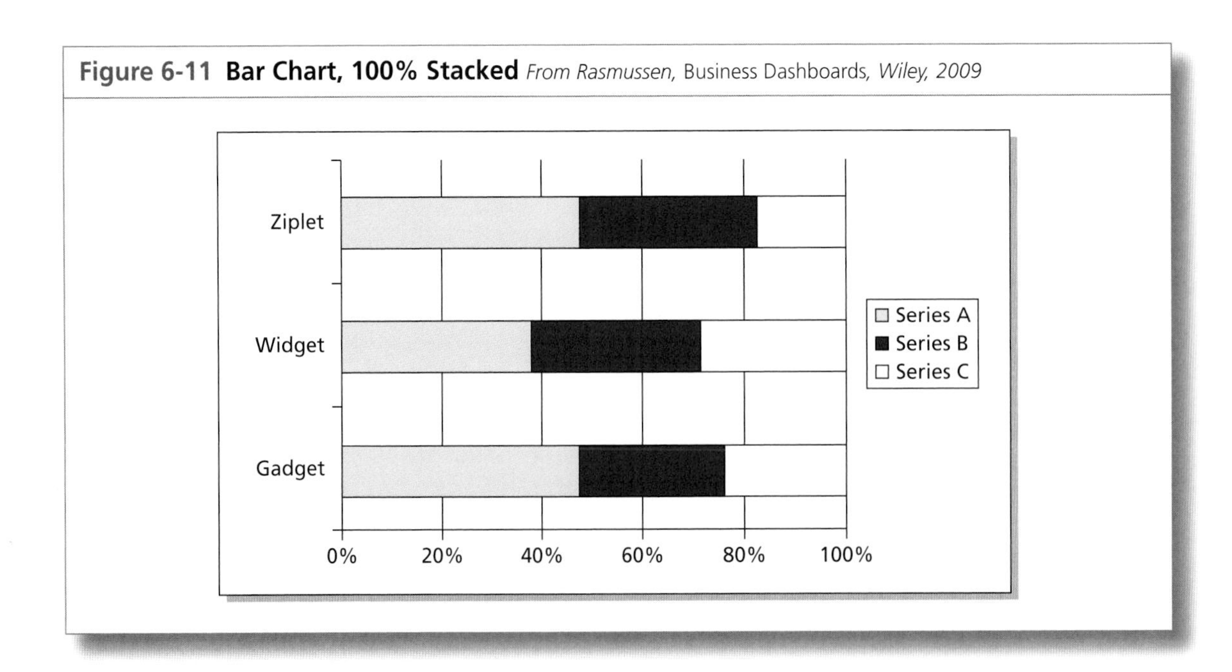

Figure 6-11 Bar Chart, 100% Stacked *From Rasmussen,* Business Dashboards, *Wiley, 2009*

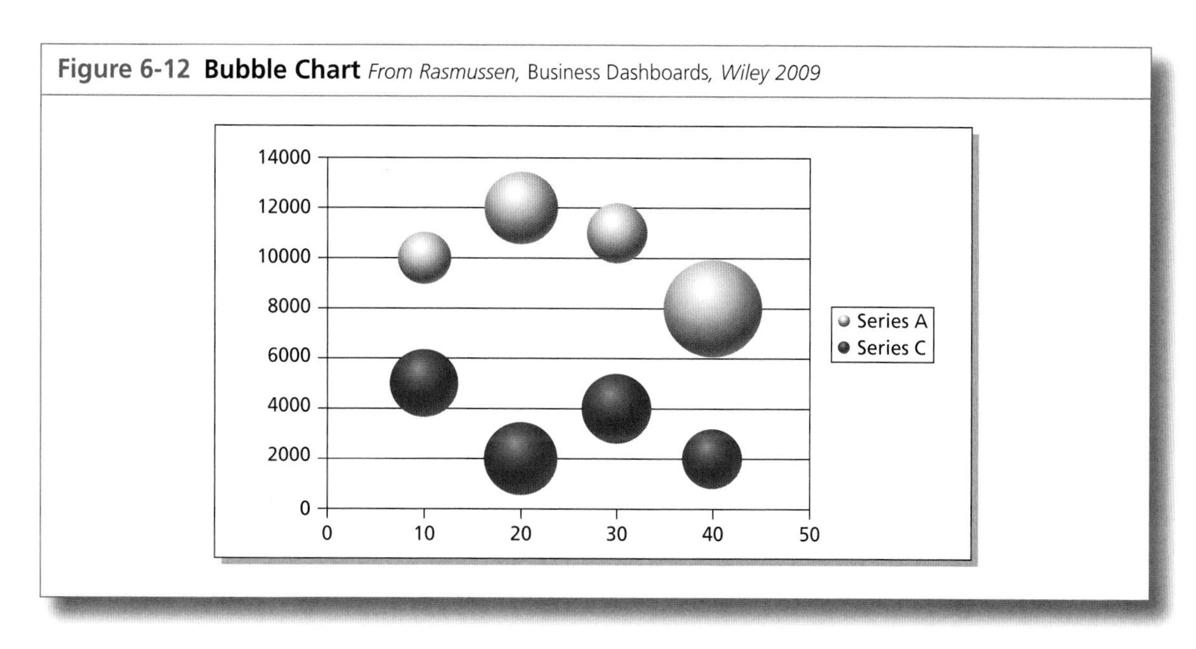

Figure 6-12 Bubble Chart *From Rasmussen,* Business Dashboards, *Wiley 2009*

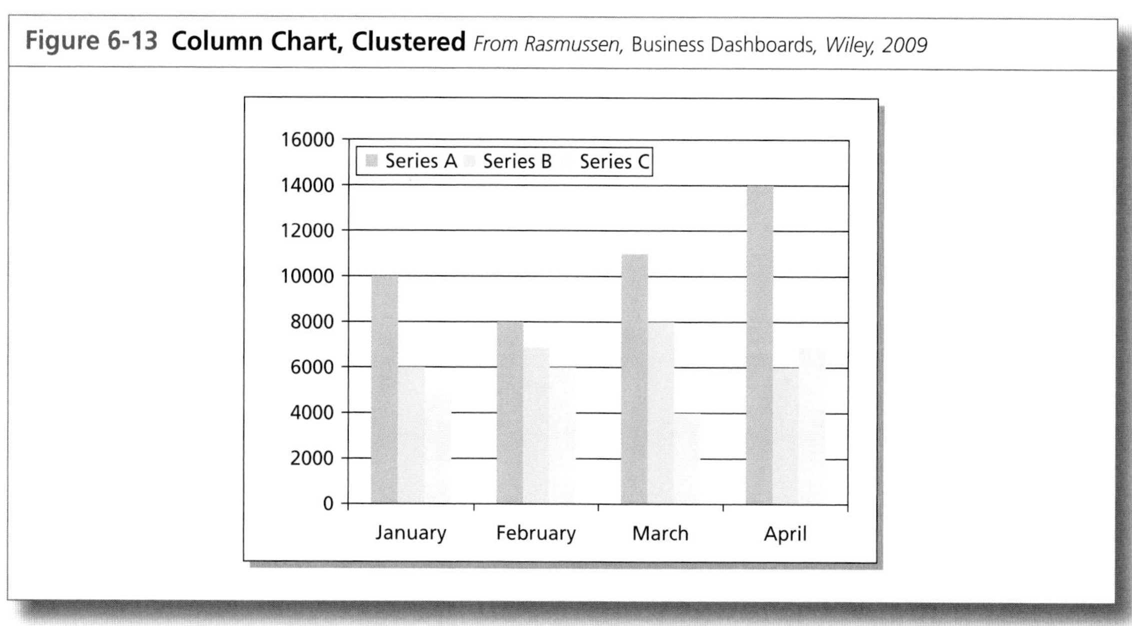

Figure 6-13 Column Chart, Clustered *From Rasmussen,* Business Dashboards, *Wiley, 2009*

shades in Figure 6–15 are easier to read, provided that the individual is not color blind.

Gauges are used to show a single value. Typically, gauges, such as those shown in Figure 6–16, will also use colors to indicate whether the value that is displayed is "good," "acceptable," or "bad."

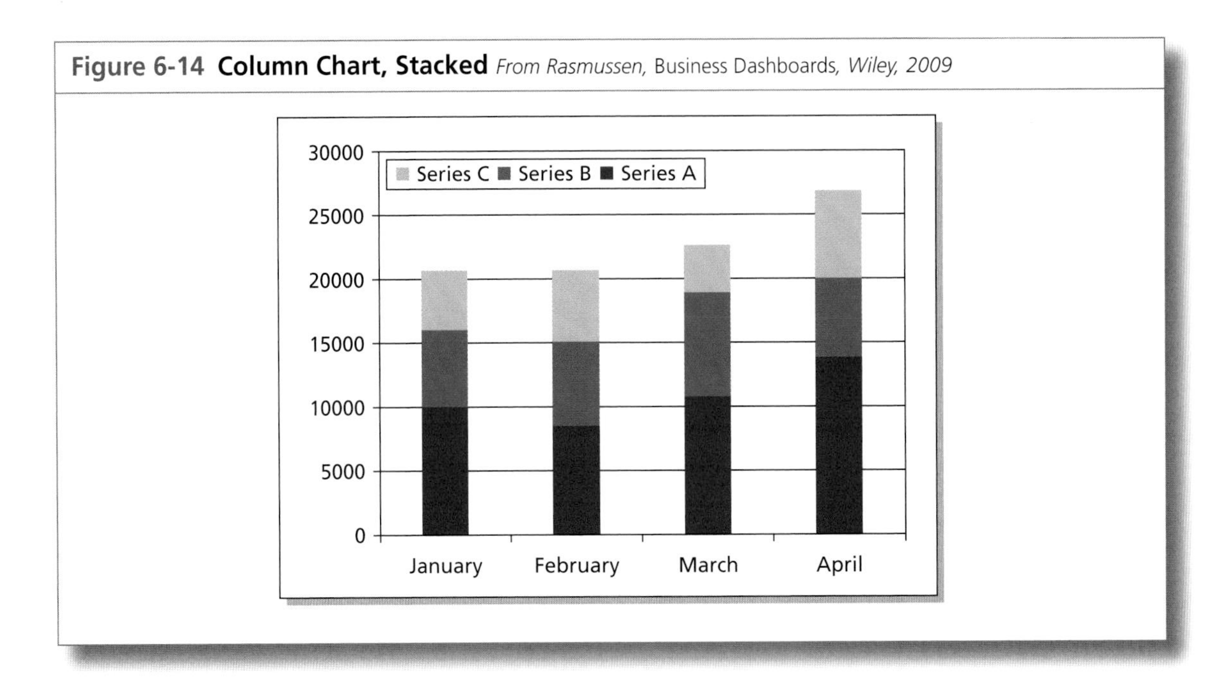

Figure 6-14 Column Chart, Stacked *From Rasmussen,* Business Dashboards, *Wiley, 2009*

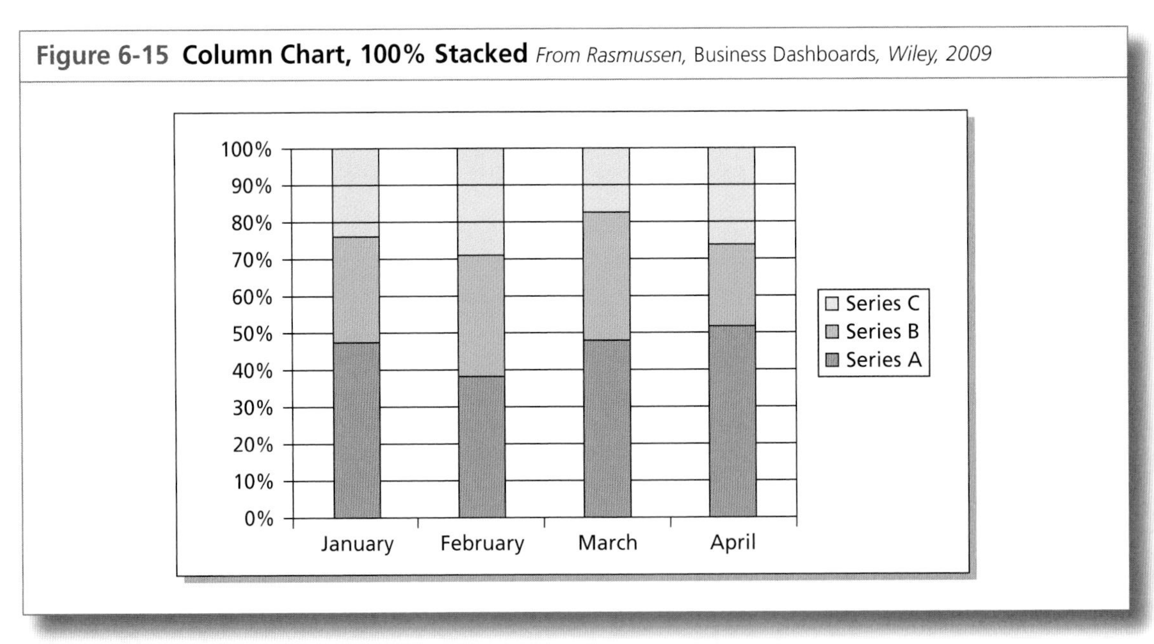

Figure 6-15 Column Chart, 100% Stacked *From Rasmussen,* Business Dashboards, *Wiley, 2009*

Icons can be found in a variety of shapes. Most popular are traffic lights (oval circles) or arrows used in conjunction with dashboards or scorecards to visualize and highlight variances. This is shown in Figure 6–17. Colors like green, yellow, and red are used to indicate values as "good,"

Figure 6-16 Gauges *From Rasmussen,* Business Dashboards, *Wiley, 2009*

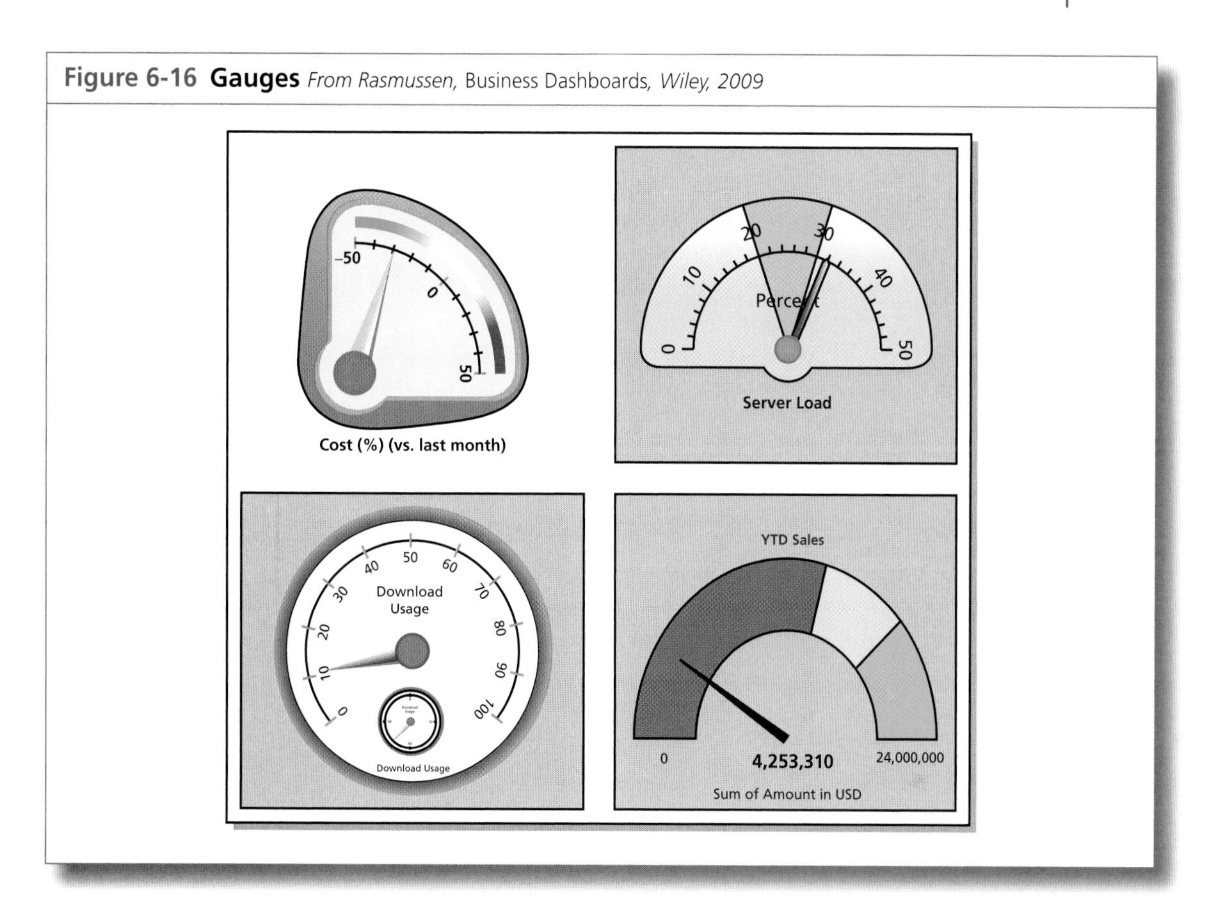

Figure 6-17 Icons *From Rasmussen,* Business Dashboards, *Wiley, 2009*

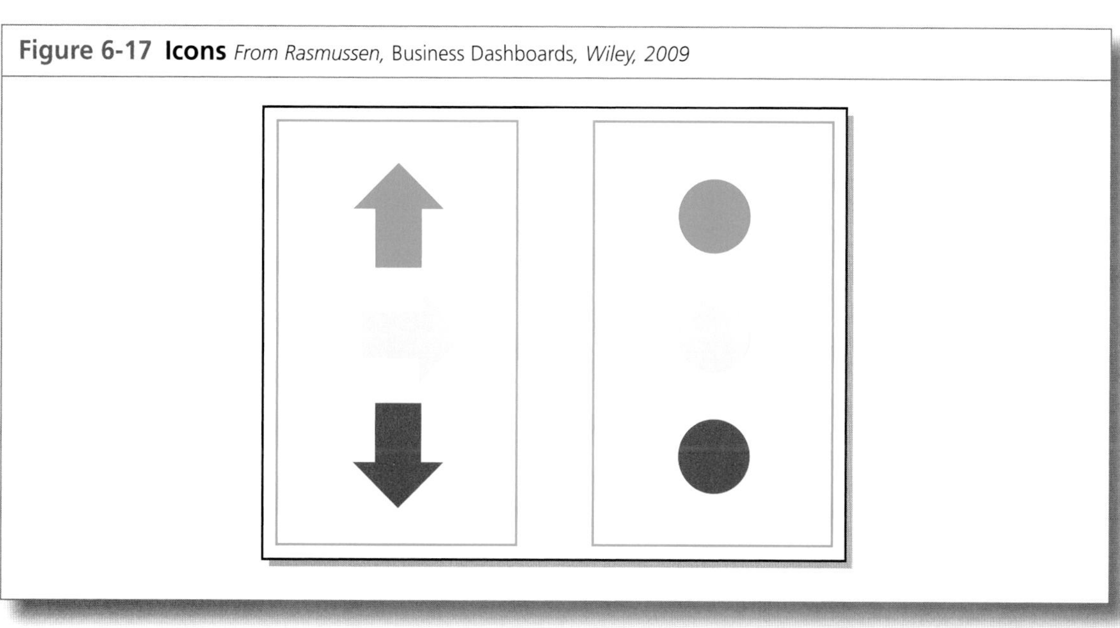

"acceptable," and "bad." The color green can have more than one meaning. For example, in some icons, green may indicate that a change is need rather than no change is necessary.

Line charts are also images that can be used to show trends. However, no more than three or four lines should appear on a chart. Examples of line charts are:

- **Traditional line chart (Figure 6–18):** This displays trends over time or categories.
- **Stacked line chart (Figure 6–19):** This displays the trend of the contribution of each value over time or categories.
- **100% stacked line chart (Figure 6–20):** This displays the trend of the percentage each value contributes over time or categories.

Perhaps the most important word in dashboard design is *simplicity*. Colorful graphics, intricate designs, and three-dimensional (3-D) artwork can distract the viewer from the more critical information. Figure 6–21 shows primary and secondary stakeholders. When you first look at the figure, you are intrigued by the 3-D effect, which adds nothing to the information you want to convey. Putting the information in a table or line chart would have achieved the same effect and might have been easier to understand. Also, there are no numbers in the figure, so the viewer may not be sure exactly how many stakeholders are in each category.

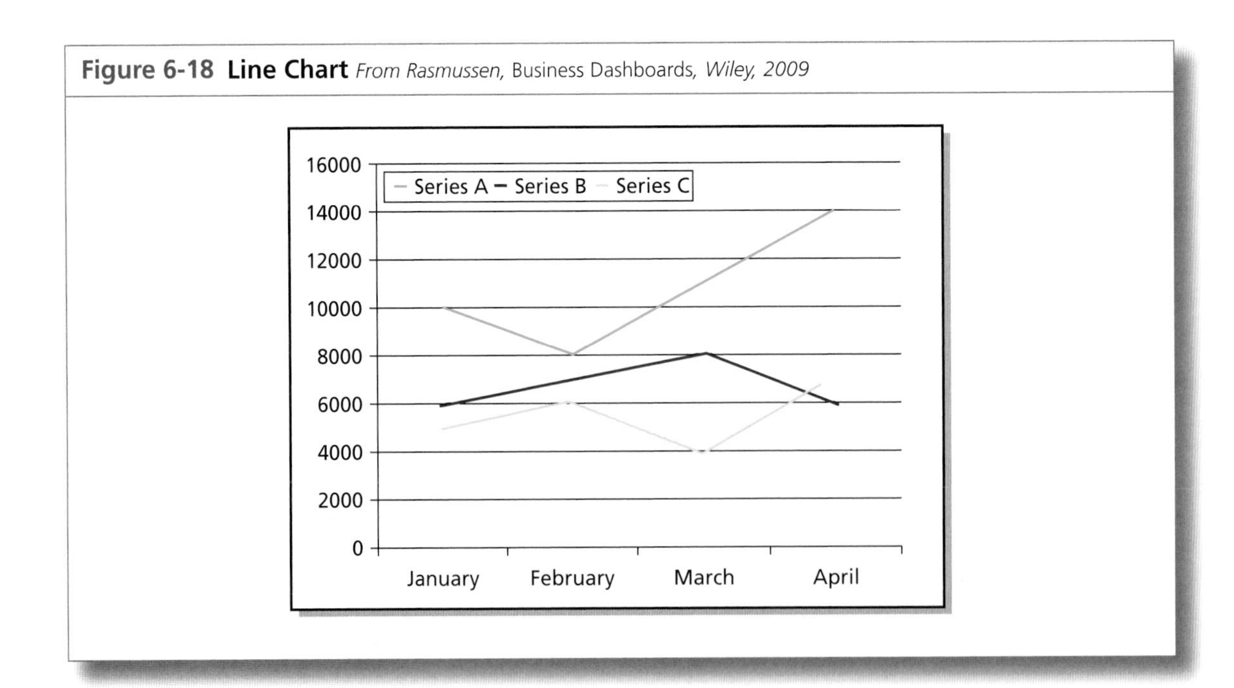

Figure 6-18 Line Chart *From Rasmussen,* Business Dashboards, *Wiley, 2009*

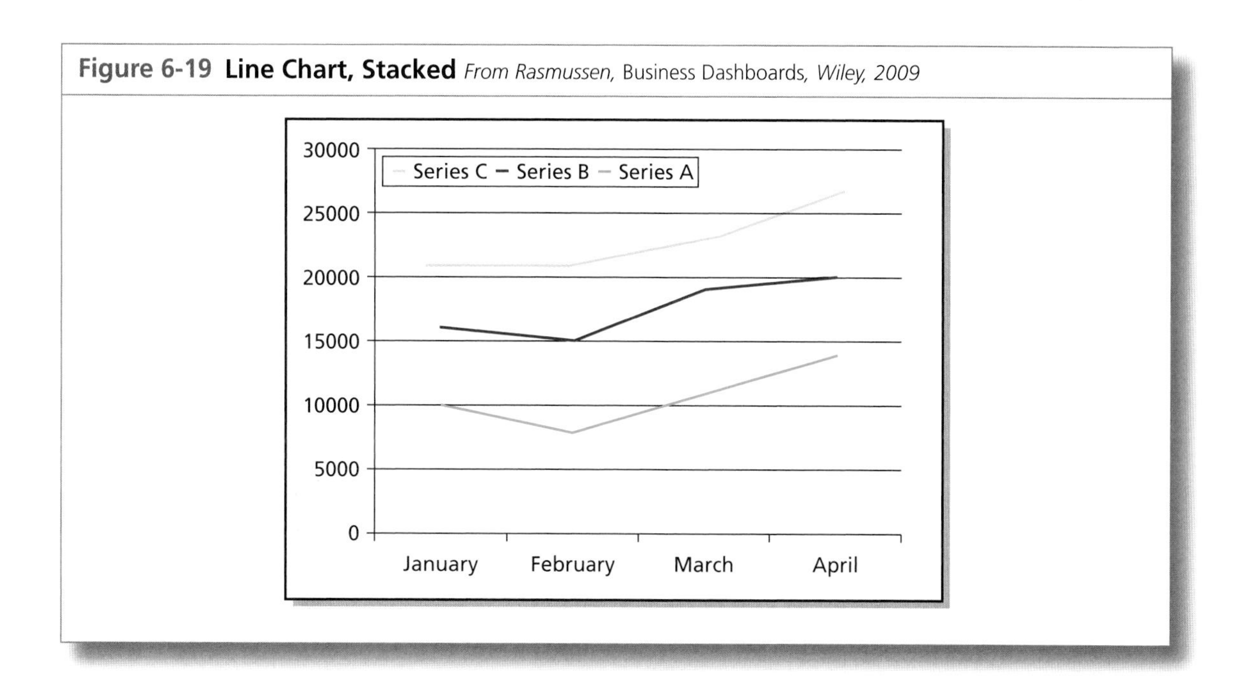

Figure 6-19 Line Chart, Stacked *From Rasmussen,* Business Dashboards, *Wiley, 2009*

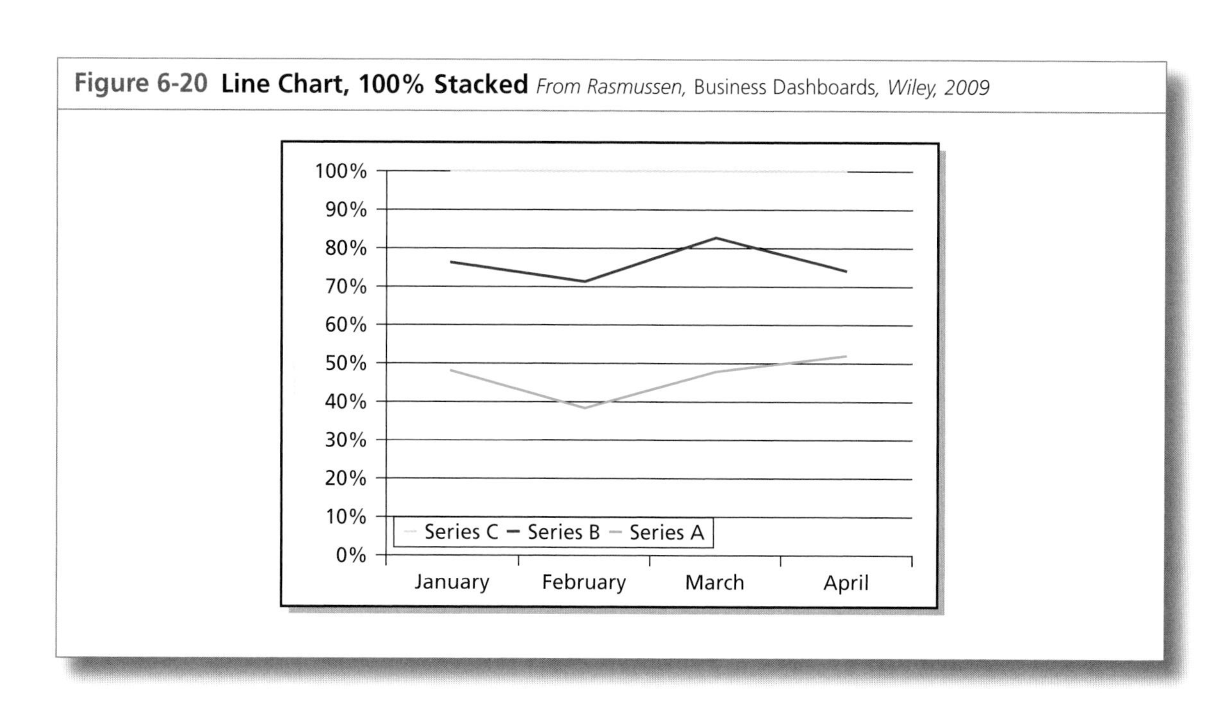

Figure 6-20 Line Chart, 100% Stacked *From Rasmussen,* Business Dashboards, *Wiley, 2009*

Figure 6-21 Tiered Stakeholder Identification in 3-D

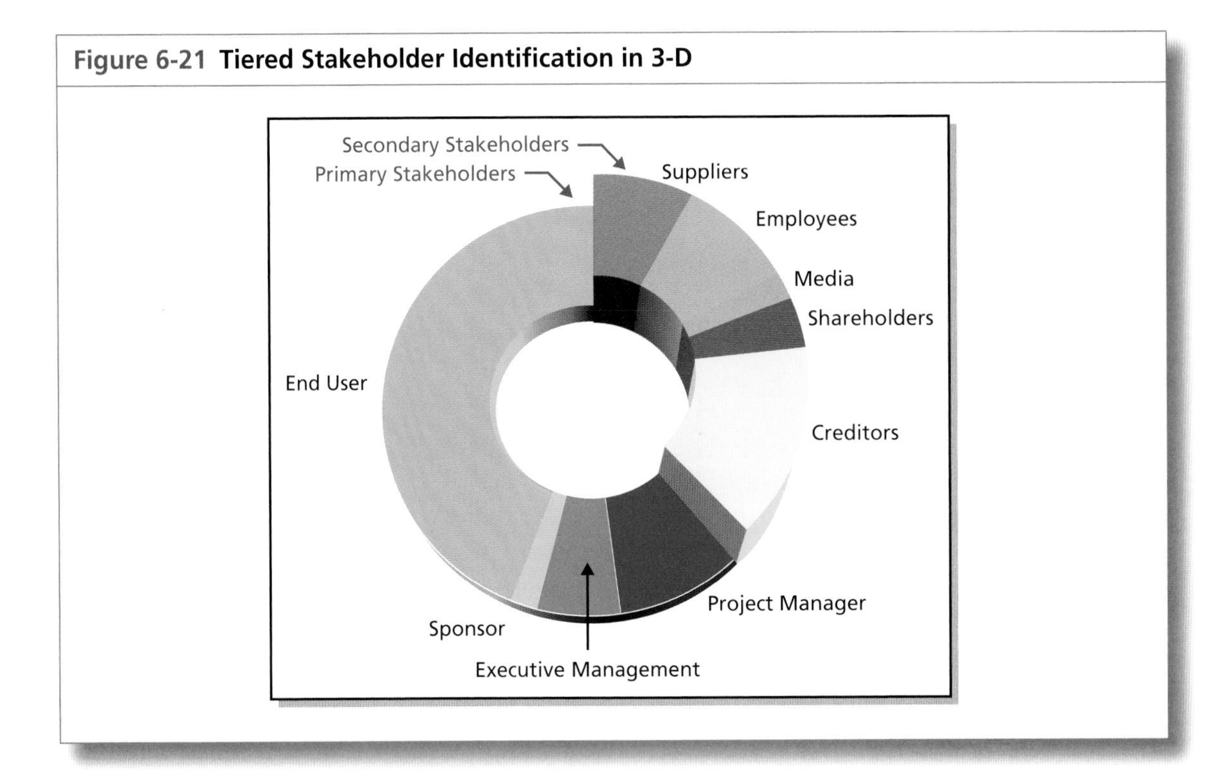

Figure 6–22 is similar to Figure 6–21 but more complex. When you first look at the figure, your eye focuses on the 3-D effect, then you must read the words over and over again to understand what you are looking at even though numbers are provided. Finally, the milestones that were completed within time and cost could have been all of the work packages that did not have a major impact on the project's success, whereas other milestones may have a significant impact. This problem might be overcome by allowing the viewer to drill down to more depth.

Figure 6–23 illustrates the current breakdown of labor hours on a project. Figure 6–23 lacks numerical values for each slice of the pie and would be easier to read as a column chart.

Figure 6–24 represents a 3-D pie chart that would be part of a dashboard for the PMO. The chart illustrates the most common reasons why projects have failed in the past. Once again, even though the image looks impressive in 3-D, the information could be presented more clearly in a line chart and with numbers included. In its current format, all of the slices of the pie look like they are the same size. This may not be the case. As a general rule, any embellishments that are not relevant to the data have no place in the chart.

Figure 6-22 Summarized Milestone Reporting

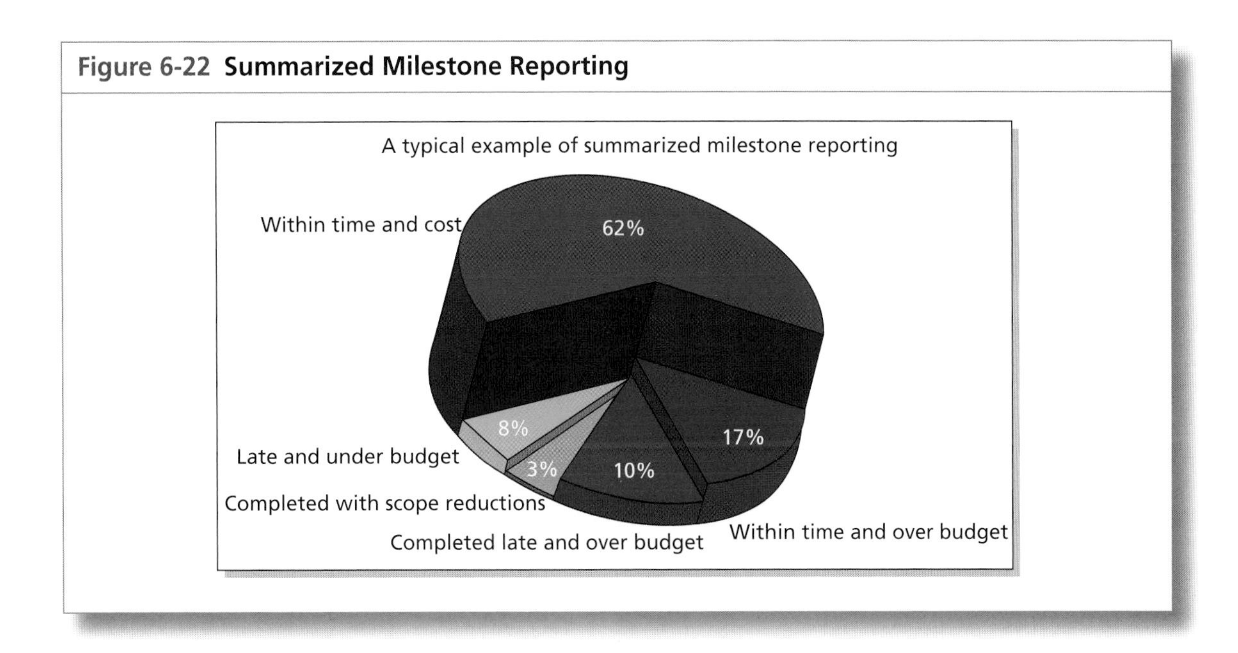

Figure 6-23 Breakdown of Labor Hours

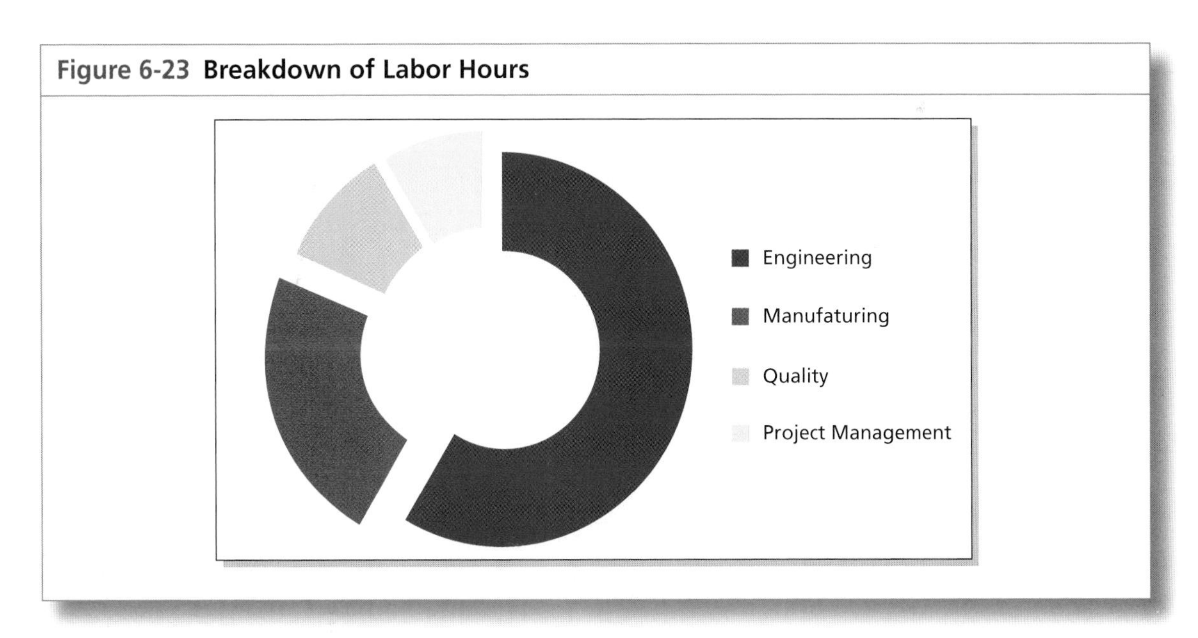

Square pie charts may very well replace the traditional pie chart. Figures 6–25 and 6–26 show the square pie chart in two different rotations. With the square pie chart, the colored-in cells can be added up to get the percentages. Even with the 3-D effect, distortion does not affect the readability of the chart.

Figure 6-24 **Causes of Failure**

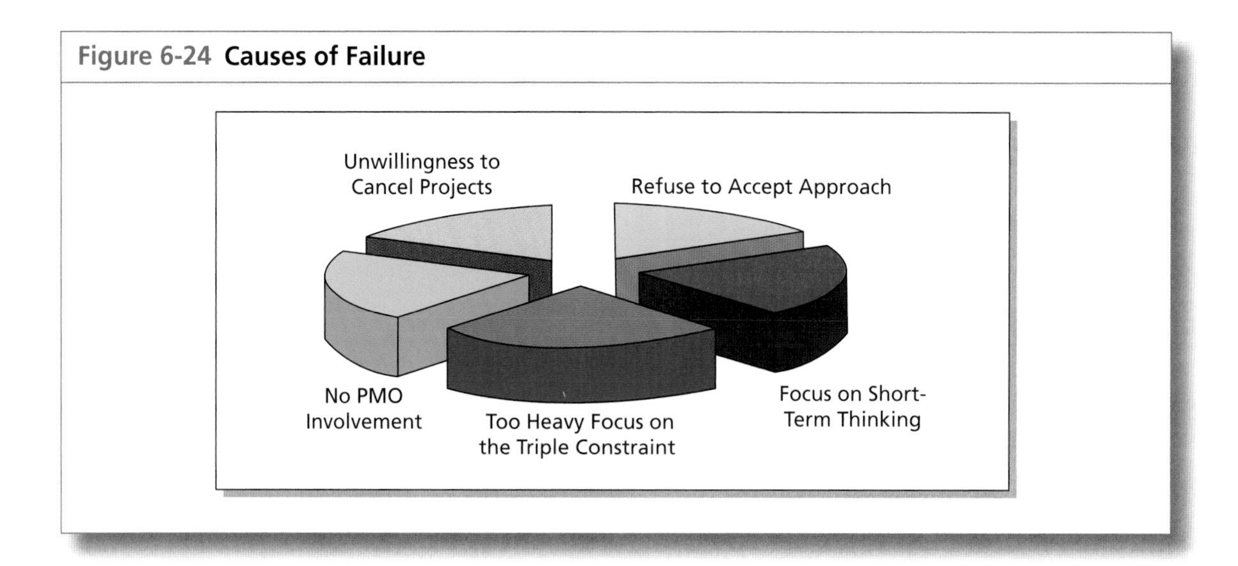

Figure 6-25 **A Square Pie Chart**

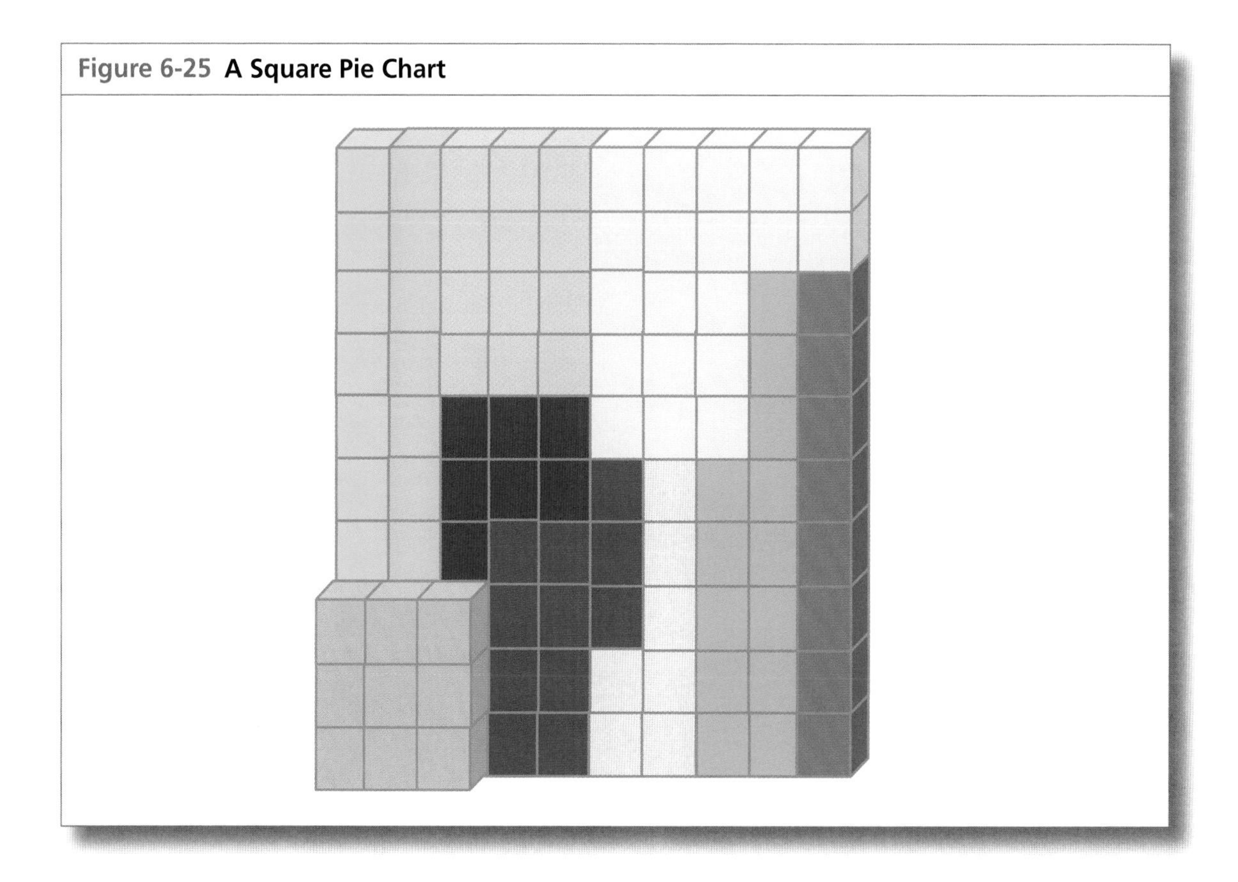

Figure 6-26 A Rotated Square Pie Chart

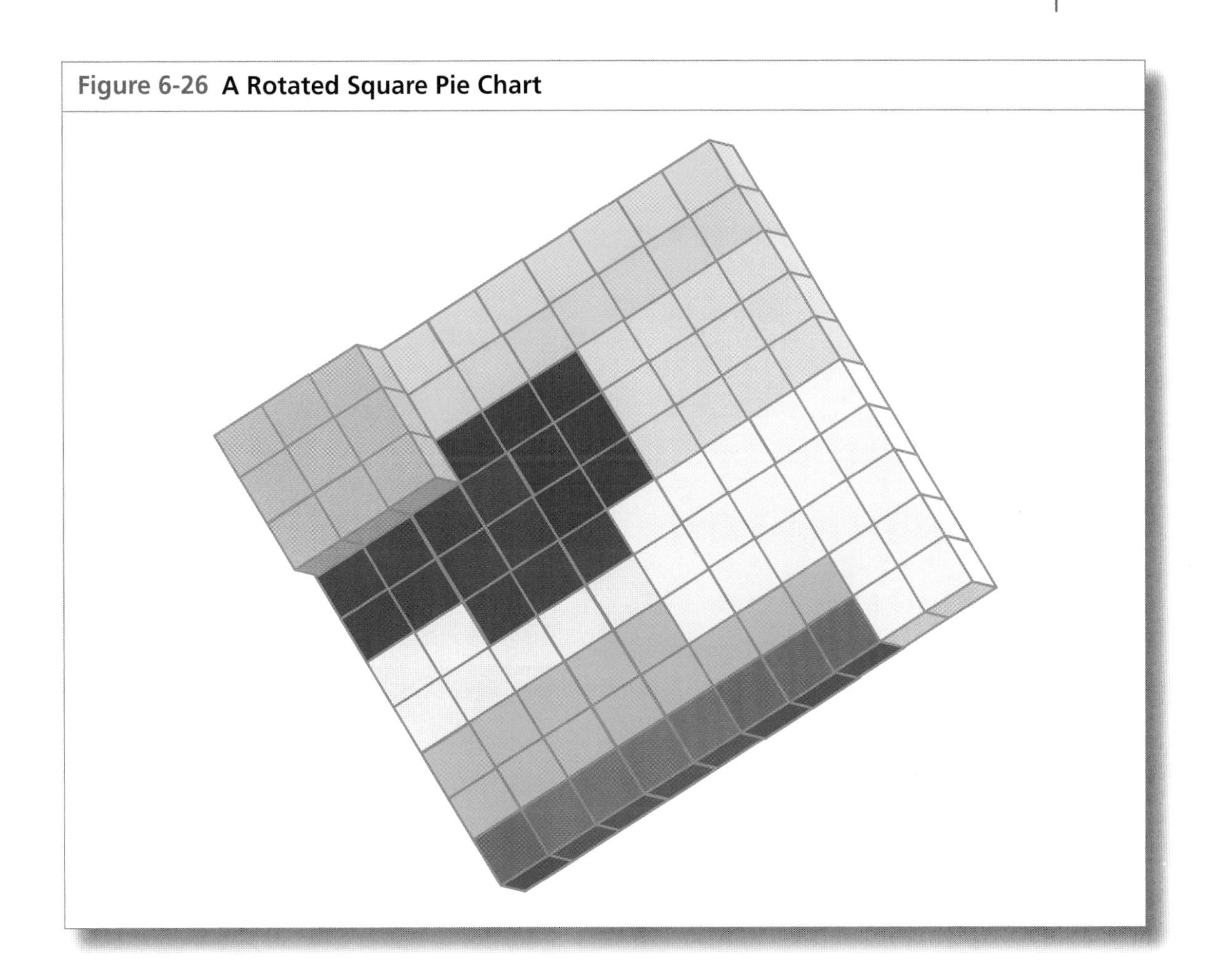

Certain elements can be highlighted by elevating them above the other cells. However, there may be some limitations to the use of square pie charts:

- Each cell generally represents whole number rather than decimals.
- The image may not impress people if one or more of the percentages are very large.
- Cell color selection and placement must be done carefully and more time may be necessary than with the traditional pie chart.
- The image may require more "real estate" on the dashboard screen than the traditional pie chart, thus limiting the number of metrics.
- The 3-D rotation of the chart must be done without sacrificing the aesthetic value of the image and its readability for accuracy.

Figure 6–27 shows the total cost breakdown for four work packages. Although the chart looks impressive, there is no background grid with

Figure 6-27 Total Cost Breakdown per Work Package

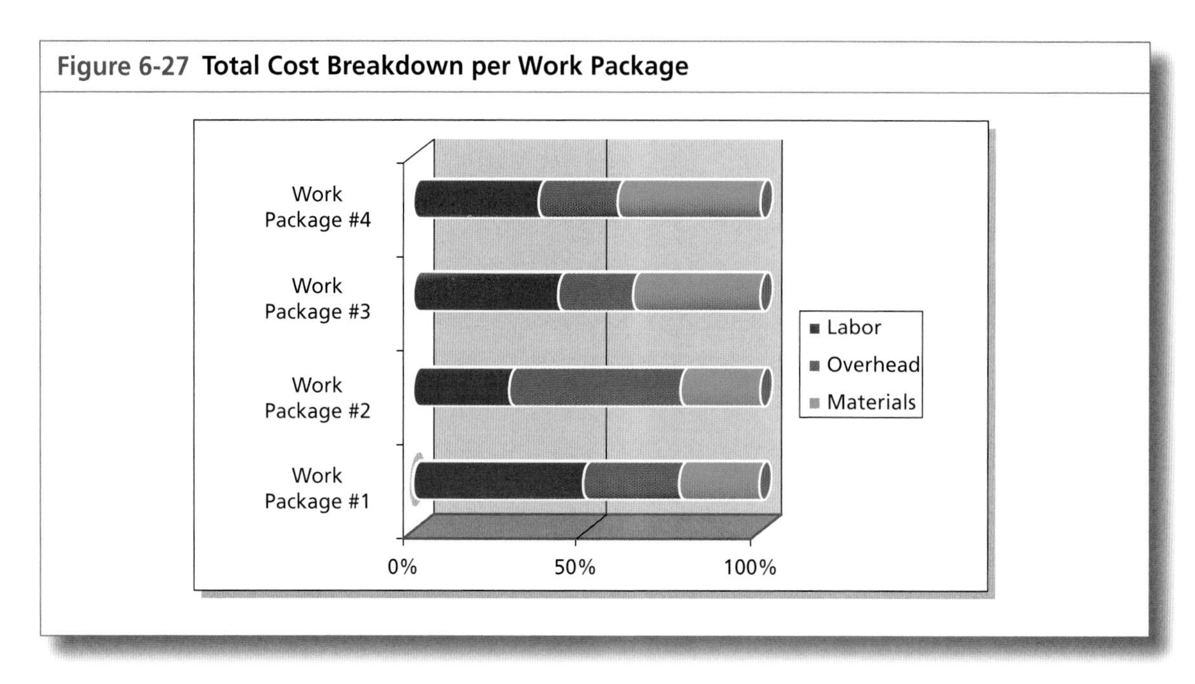

Figure 6-28 Cost Overrun Data

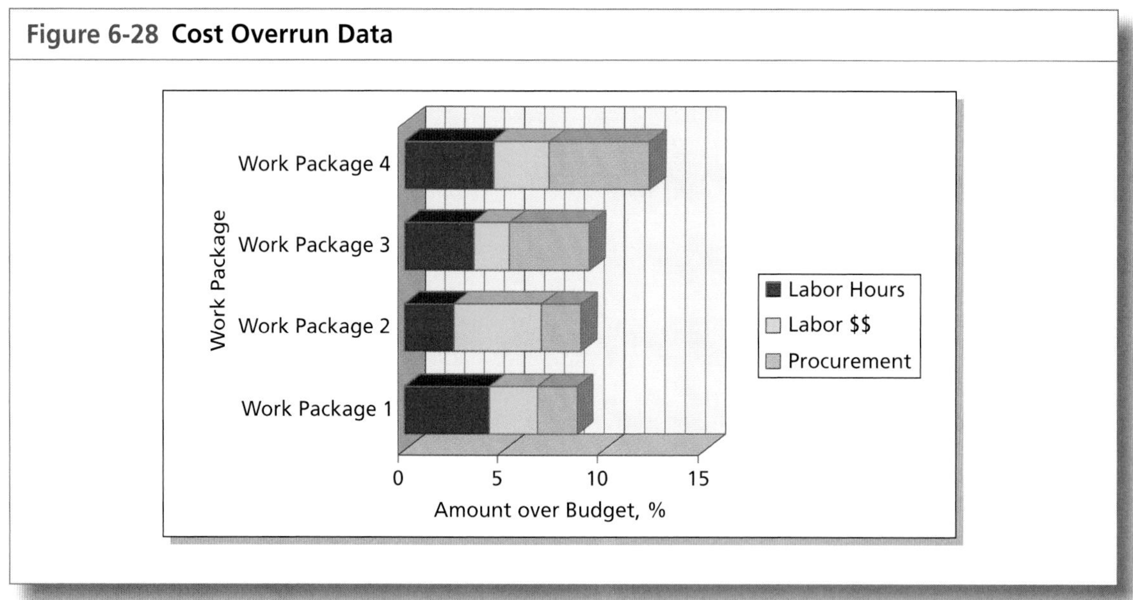

which the viewer can make assessments. Also, the use of red or shades of red might lead the viewer to believe that the labor dollars are excessive or a problem area.

Figure 6–28 shows the cost overrun data for four work packages. In this case, there is a grid, but it difficult to determine the overrun magnitude of labor and procurement. Also, for Work Package 4, should we use the front or back side of the 3-D bar? If we use the front side of the bar, the cost overrun is 11 percent, whereas the back side of the bar illustrates a 12 percent overrun.

Figure 6–29 shows the cumulative month-end CPI and SPI data. On the grid, the parity line at 1.0 should probably be highlighted to show the nearness to the targeted value. Also there should be more grid lines so that meaning numbers can be determined.

There are advantages to using 3-D column charts. However, inserting too much into the charts can make them difficult to use. Figure 6–30 illustrates the complexity in making exact value determinations for Series 1 and Series 2. Also, it might be better to use neutral or standard colors rather than colors designed to emphasize a special situation. Figure 6–31 shows typical neutral colors.

Another common mistake is in the use of textures and gradients, as shown in Figure 6–32. While there are benefits to this in conducting presentations, they may not be appropriate for dashboards.

Figure 6–33 shows a column chart with bright colors. The purpose of bright colors is to emphasize a good or bad situation. If all of the colors are bright, as in Figure 6–33, the viewer may not know what is or is not important.

When using a column chart, standard colors should be used and the shading should go from lightest to darkest for easy comparison, as shown in Figure 6–34. Also, creating shadows or exotic colors behind the columns

Figure 6-29 Cumulative Month end CPI and SPI Data

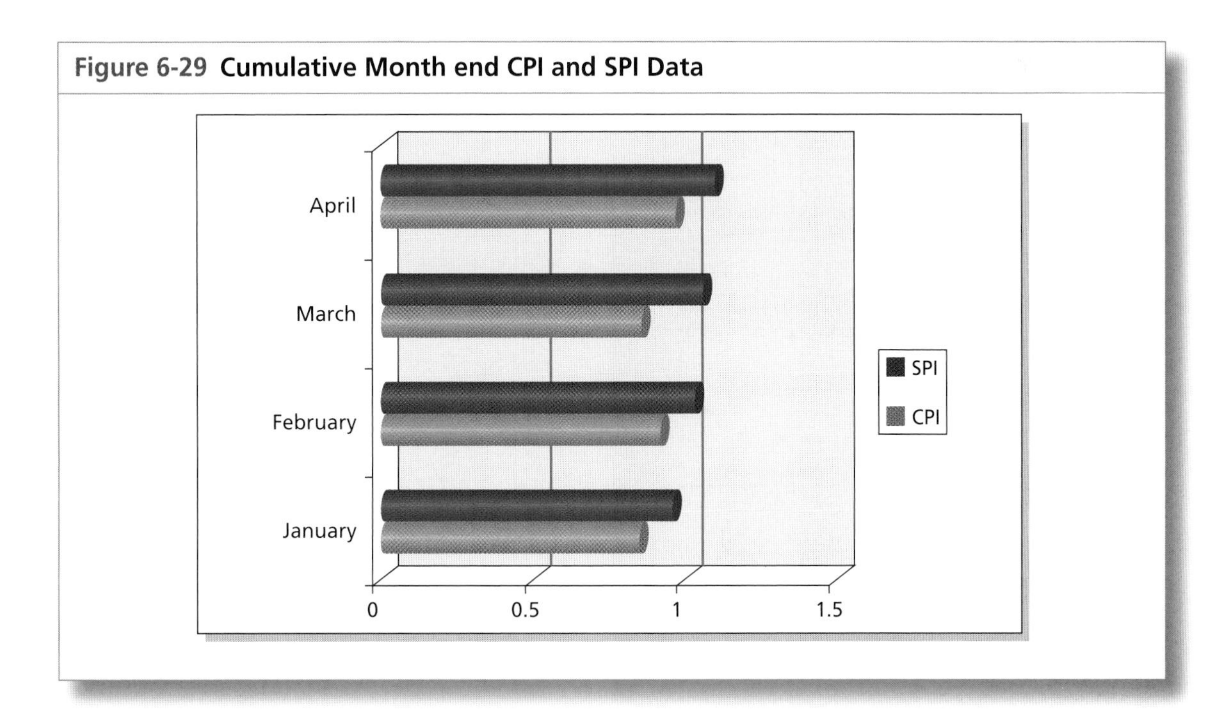

Figure 6-30 **3-D Column Chart**

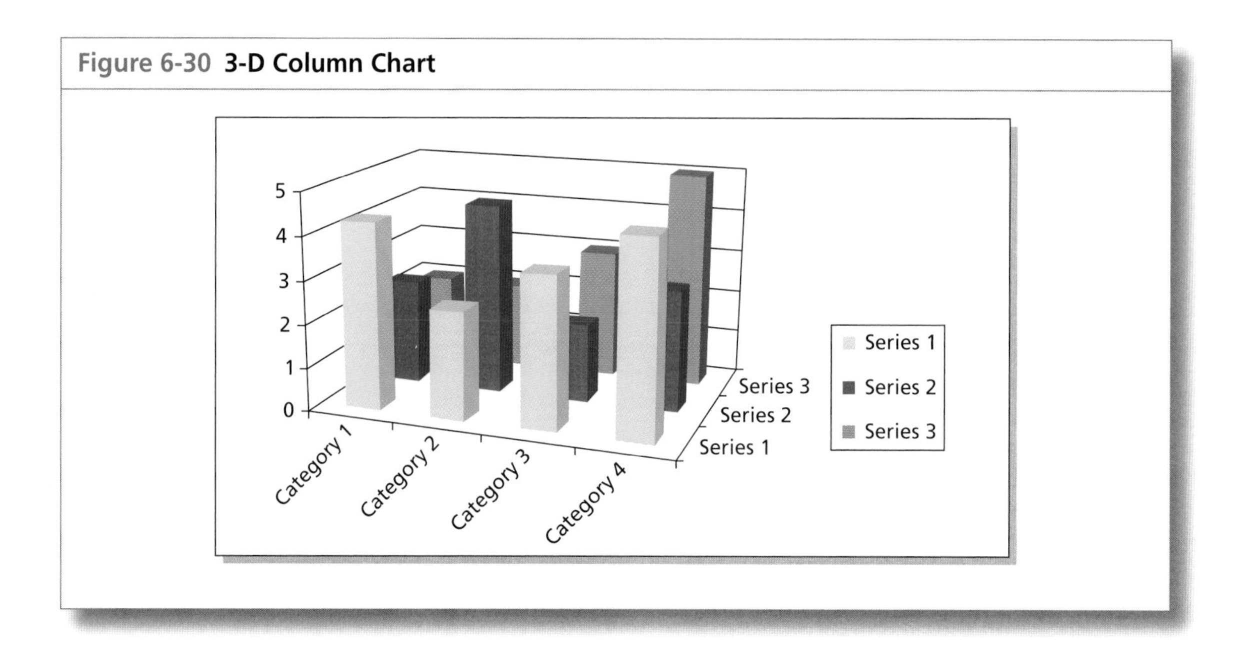

Figure 6-31 **Possible Colors**

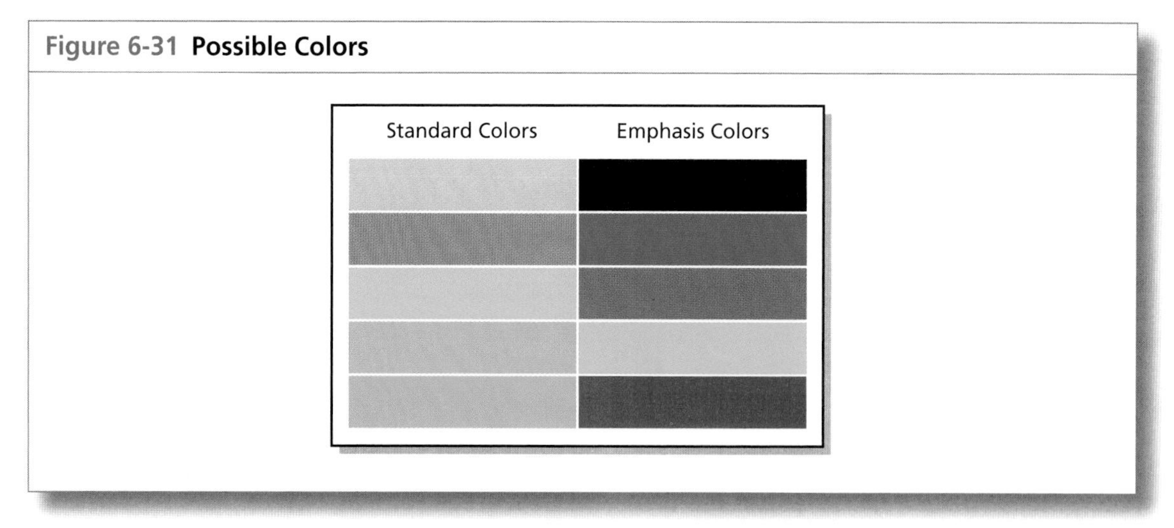

can be distracting and should be avoided because the shadows contain no information or data.

Background colors or shading can play tricks on the eye. For example, in Figure 6–35, the inner squares are all the same size, yet some people perceive the inner square that is on the right side to be larger than the other squares.

Another example appears in Figure 6–36. The outer circles represent the total cost of a work package, in dollars, and the inner circle represents the dollar value of the labor hours that are part of the total cost. Again, the eye

Figure 6-32 **Column Chart with Gradients**

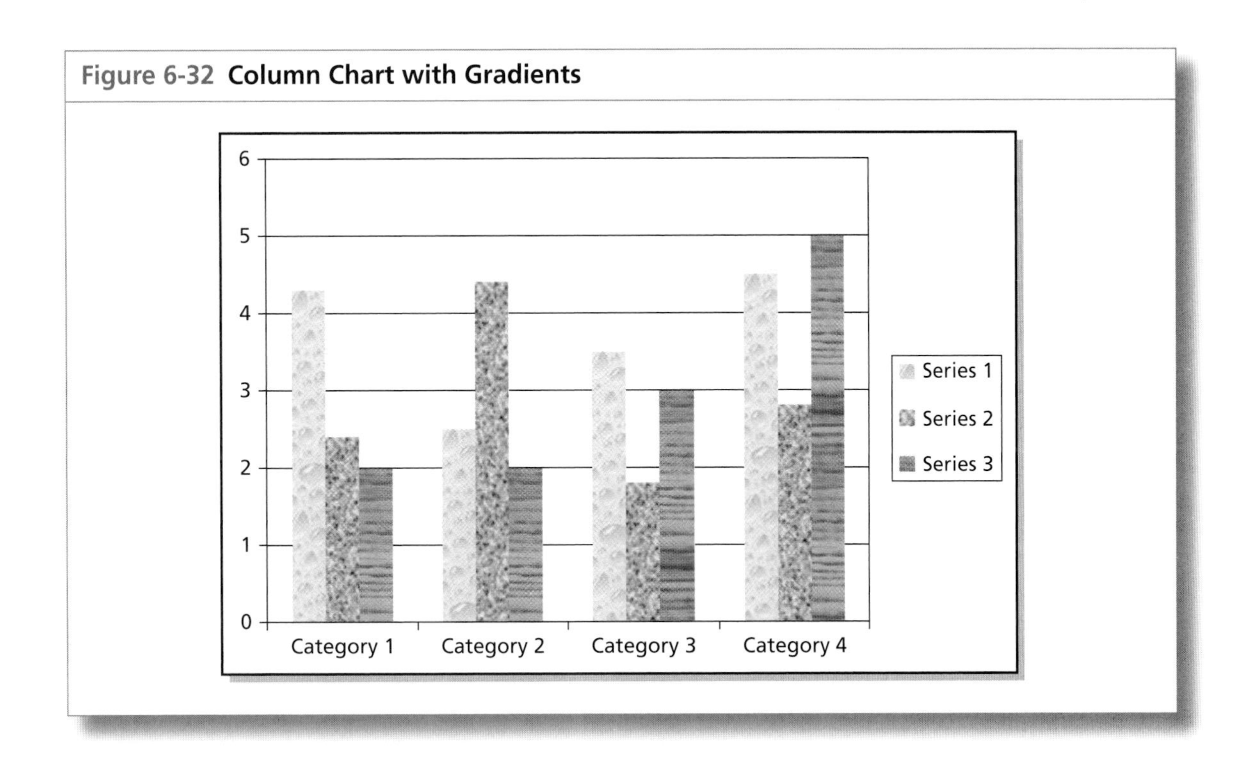

Figure 6-33 **Column Chart Using Bright Colors**

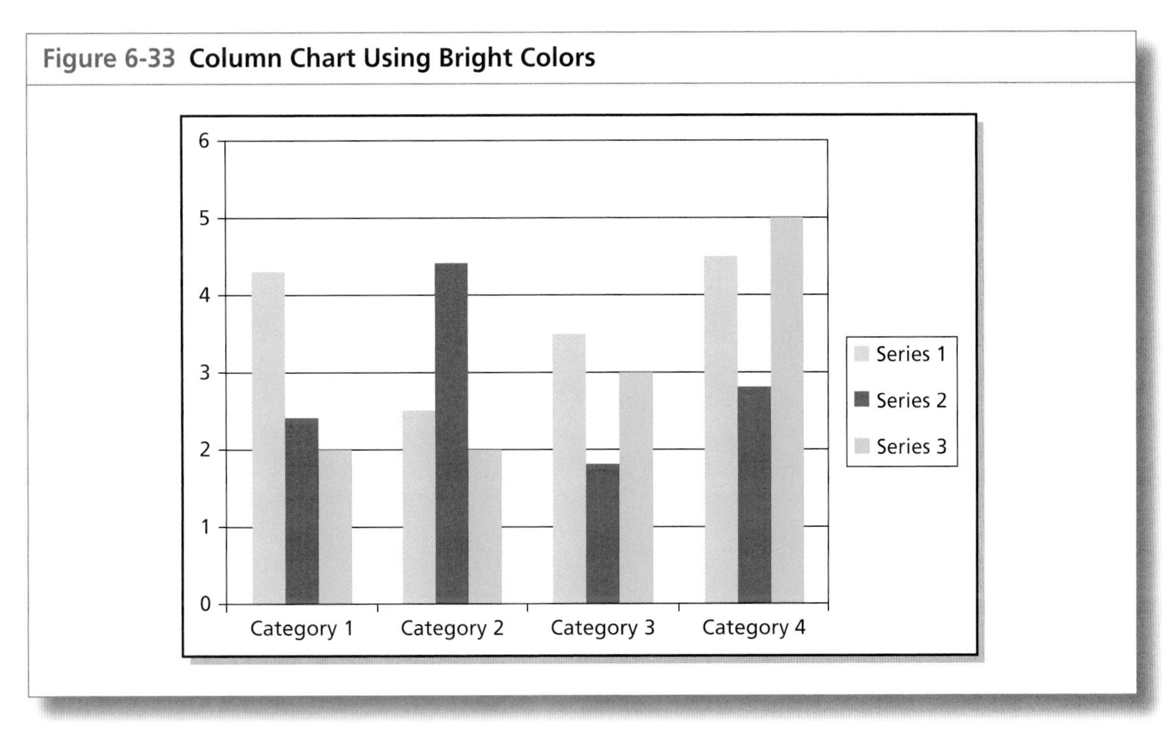

Figure 6-34 Column Chart Using Shading

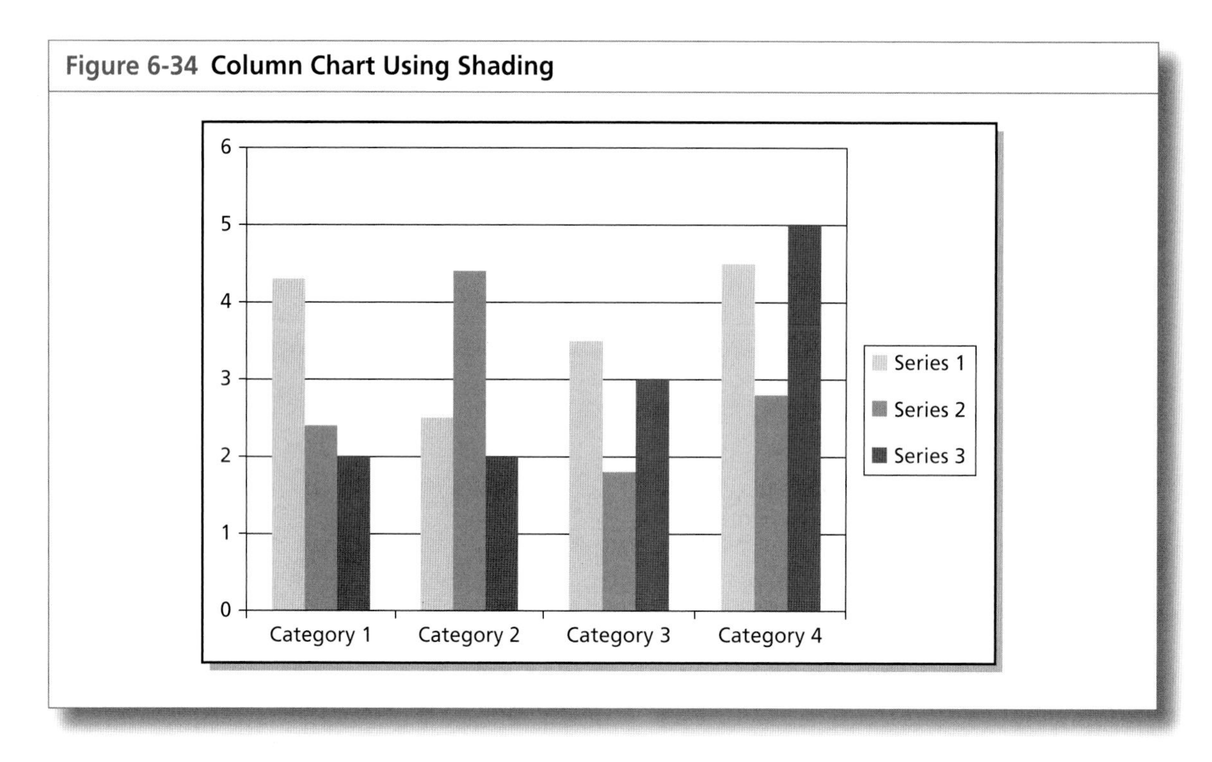

Figure 6-35 Background Colors with Shading

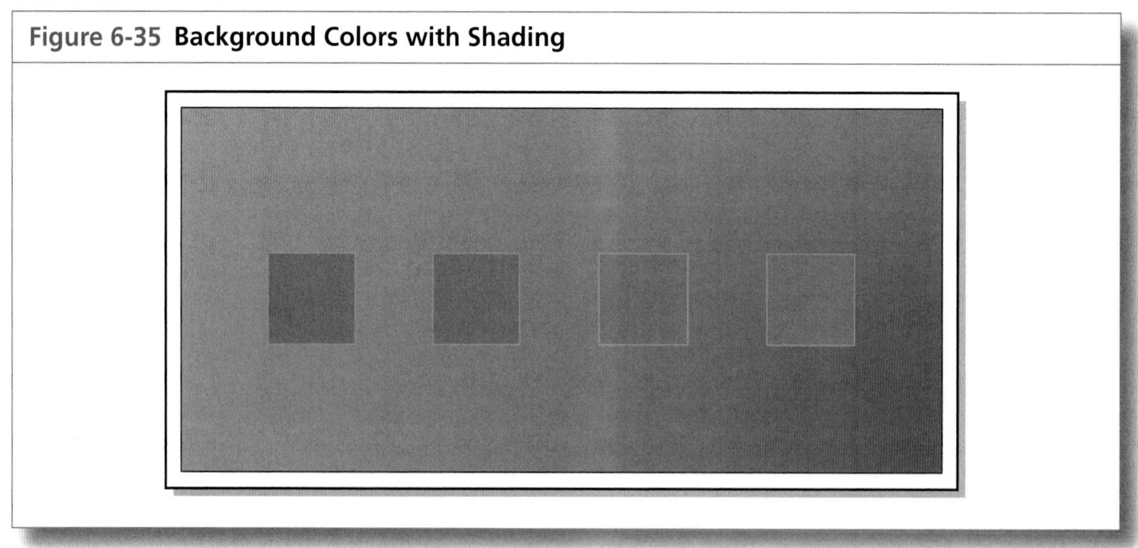

may be deceived because all of the inner circles are the same size. Because some inner circles consume a larger percentage of the outer circles, some inner circles appear larger.

Radar charts, as seen in Figure 6–37 are usually avoided because they are often hard to read, even for people that use them frequently. The

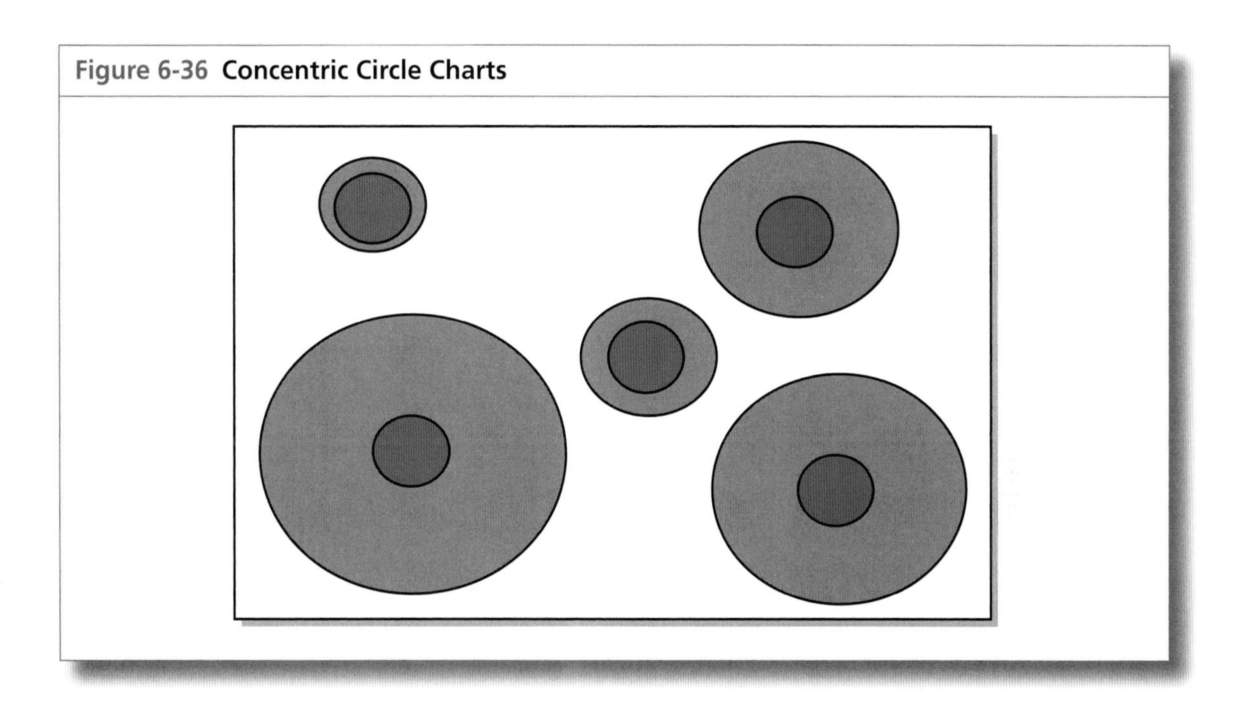
Figure 6-36 **Concentric Circle Charts**

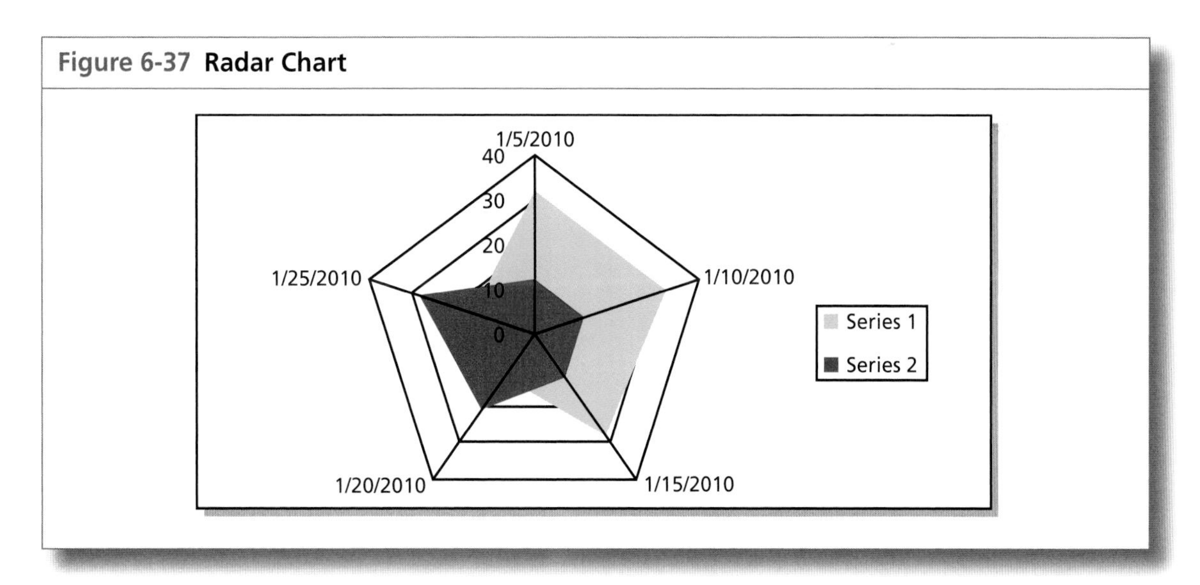

Figure 6-37 **Radar Chart**

information in a radar chart can be displayed in a column or bar chart. However, there are situations where radar charts can be quite effective.

We have emphasized in this chapter that we should place on the dashboard the least amount of metrics that can be used for informed decision making. Unfortunately, this is not always the case. Sometimes, the viewer must have the option to drill down to additional levels for clarification. For example, in Figure 6–38, the column on the left represents buttons. When the button is illuminated in red, the metrics on the screen are for Work Package #1 only. The viewer has the option of depressing any of the buttons.

Based upon the amount of depth in the information needed by the stakeholders, some dashboards must be designed for in-depth levels of detail. This becomes a costly effort if each stakeholder requires a different level of detail.

Figure 6-38 Dashboard with Buttons for Drilling

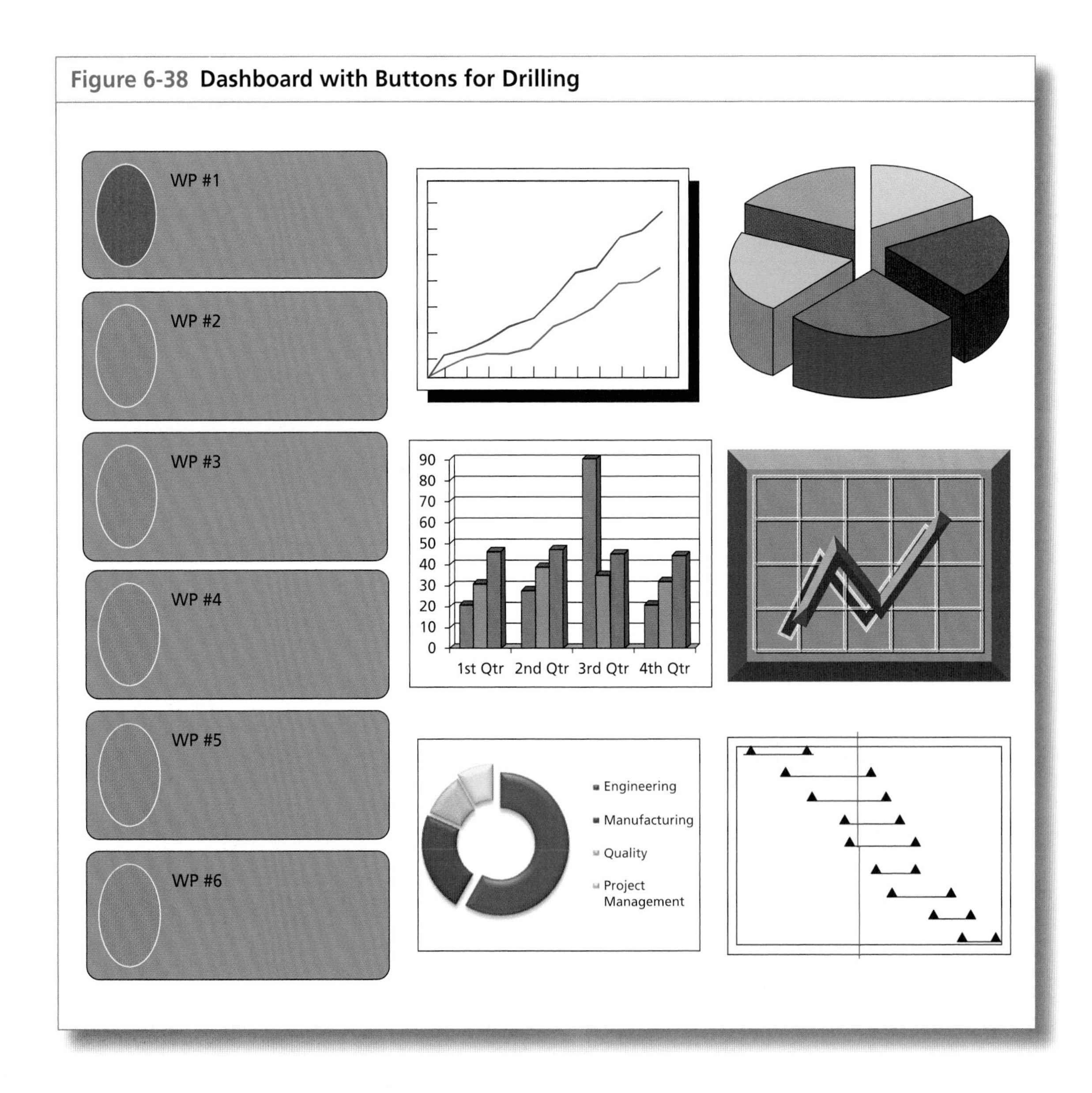

Figure 6–39 is an attempt to show the cost and schedule variances as the project progresses. The chart is good if used just to see the trends in the variances. If actual numbers are required for decision making, however, then the data should be represented in a table.

Some charts are more appropriate when illustrated as a log-log plot or semi-log plot. Figure 6–40 shows a typical learning curve that would be used as part of a manufacturing project. While most project managers may be familiar with this type of chart, it should not be used in dashboards to be presented to stakeholders.

Figure 6-39 EVMS Status Reporting

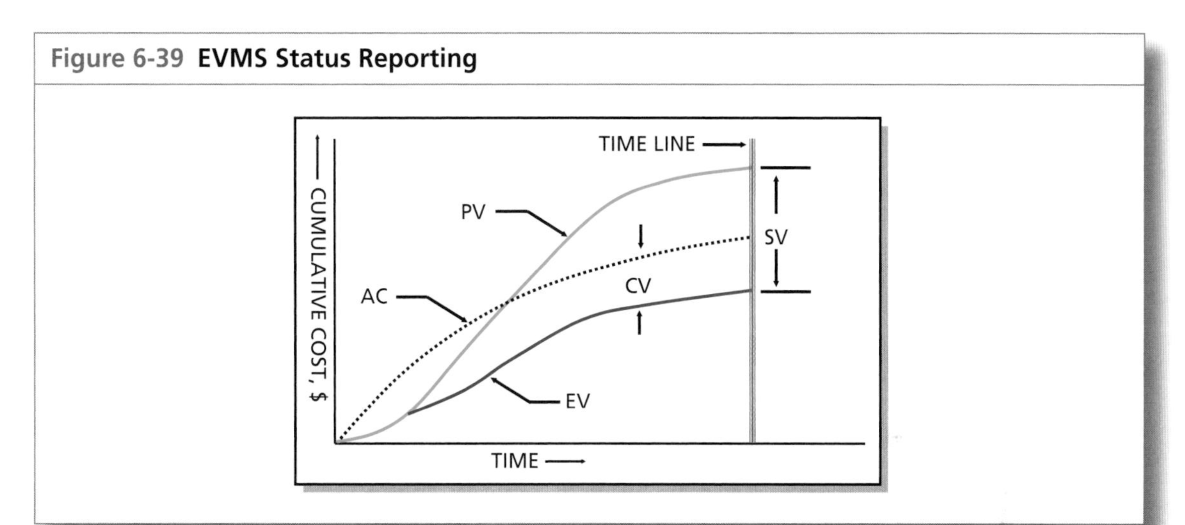

Figure 6-40 Learning Curve on a Log-Log Plot

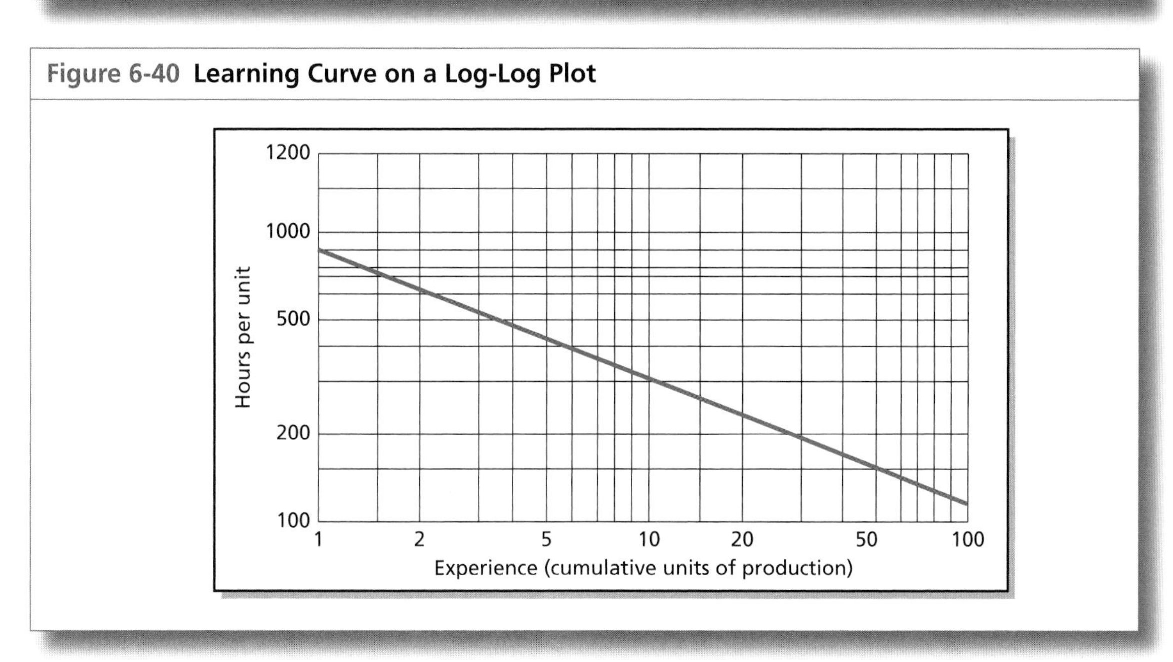

6.8 USING EMOTICONS

Sometimes, we try to use symbols to make the dashboard look elegant. Emoticons can be used in dashboards but the meaning of the emoticons can be misinterpreted. For example, in Figure 6–41:

- Does a happy face mean good or very good?
- Does a sad face mean bad or very bad?
- The color red is normally used to reflect something bad. What does a red smile indicate?
- The color green normally indicates something favorable. What does a green frown indicate?

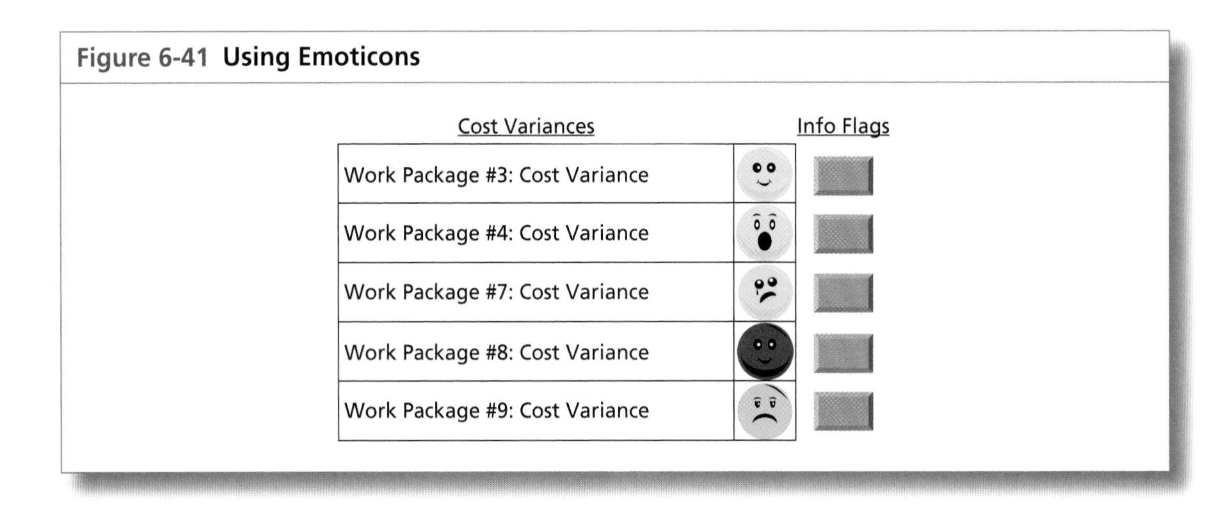

Figure 6-41 Using Emoticons

Figure 6-42 Other Emoticons That Can Be Misinterpreted

Information flag buttons can be used for drill down capability to get clarification on the exact meaning of the emoticon.

Figure 6–42 provides additional examples of other emoticons that can be used. Once again, their intent can lead to misinterpretation. For example, does the police officer emoticon mean the project has stopped or that there is just a small delay? Does the devil emoticon mean that we are in serious trouble or that we are trying to resolve a problem?

6.9 AGILE AND SCRUM METRICS

During the past decade, the introduction of Agile and Scrum techniques have brought with it new metrics. One such metric is the burn down chart. According to Wikipedia, the free encyclopedia,

> A **burn down chart** is a graphical representation of work left to do versus time. The outstanding work (or backlog) is often on the vertical axis, with time along the horizontal. That is, it is a run chart of outstanding work. It is useful for predicting when all of the work will be completed. It is often used in agile software development methodologies such as Scrum. However, burn down charts can be applied to any project containing tasks with time estimates.

A burn down chart for a completed iteration, generated from a Google docs template, is shown below in Figure 6–43 and can be read by knowing the following:

X-Axis

This is the project/iteration timeline.

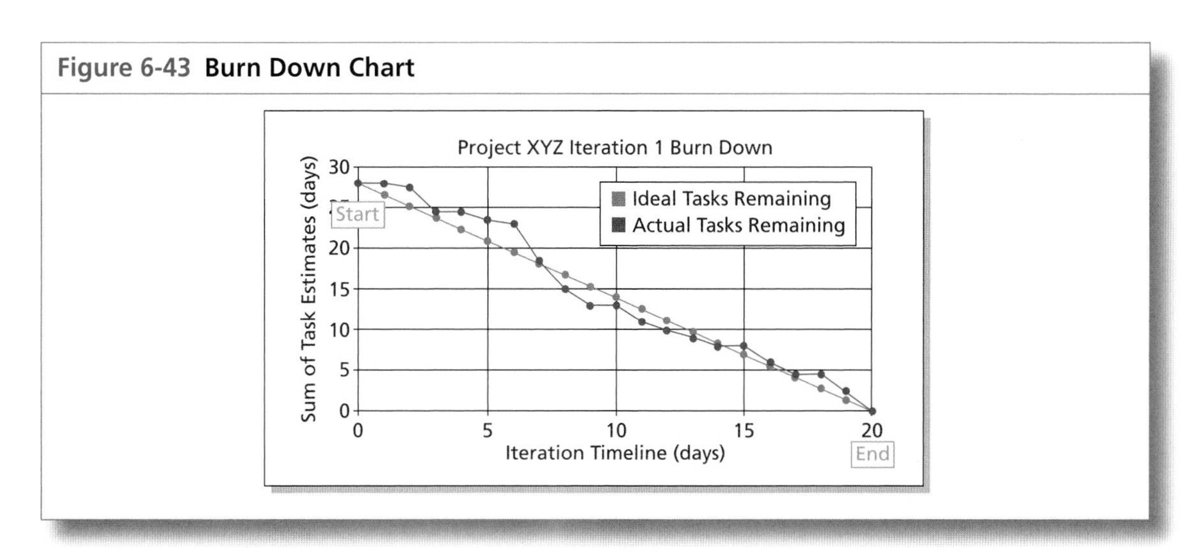

Figure 6-43 Burn Down Chart

Y-Axis

This is the work that needs to be completed for the project. The time estimates for the work remaining will be represented by this axis.

Project Start Point

This is the farthest point to the left of the chart and occurs at day 0 of the project/iteration.

Project End Point

This is the point that is farthest to the right of the chart and occurs on the predicted last day of the project/iteration.

Ideal Work Remaining Line

This is a straight line that connects the start point to the end point. At the start point, the ideal line shows the sum of the estimates for all the tasks (work) that needs to be completed. At the end point, the ideal line intercepts the x-axis showing that there is no work left to be completed.

Actual Work Remaining Line

This shows the actual work remaining. At the start point, the actual work remaining is the same as the ideal work remaining but as time progresses, the actual work line fluctuates above and below the ideal line depending on how effective the team is. In general, a new point is added to this line each day of the project. Each day, the sum of the time estimates for work that was recently completed is subtracted from the last point in the line to determine the next point.

The accuracy of the time estimates is important when using burn down charts. If you continuously overestimate or underestimate the time requirements, burn down charts allow you to include an efficiency factor to remove the variability.

There are always advantages and disadvantages to each metric chosen and the way that the data is displayed. Some people prefer to use burn up charts arguing that burn down charts do not necessarily show how much work was actually completed in a given time period.

6.10 MASHUP DASHBOARDS

During the past 10 years, companies have established project management offices (PMOs), which have become the guardians of the company's intellectual property on project management. Today, project managers are expected to make both project decisions and business decisions. It is, therefore, unrealistic to expect each PMO to have at its disposal all of the business information that will be needed. PMOs may be replaced by information warehouses. On large and/or complex projects, each stakeholder may have an information warehouse and the project team may have to tap into each warehouse to get the appropriate information for the dashboards. Warehouse creation projects are generally performed by the IT departments with little input from perspective user groups.

According to Wikipedia, the free encyclopedia,

> A **mashup dashboard**, as applied to Web developments for example, is a Web page or application that uses and combines data, presentation or functionality from two or more sources [i.e. information warehouses] to create new services. The term implies easy, fast integration, and data sources to produce enriched results that were not necessarily the original reason for producing the raw source data. The main characteristics of the mashup [dashboards] are combination, visualization, and aggregation.
>
> To be able to permanently access the data of other services, mashups are generally client applications or hosted online. Since 2010, two major mashup vendors have added support for hosted deployment based on Cloud computing solutions; that are Internet-based computing, whereby shared resources, software, and information are provided to computers and other devices on demand, like the electricity grid.

Creating real time dashboards that can change over the life of the project, and where information will be extracted from a variety of information warehouses, brings forth the problem with credibility of information. According to Wikipedia,

> Data quality is a general challenge when automatically integrating data from autonomous sources. In an open environment the data aggregator has little to no influence on the data publisher. Data is often erroneous, and combining data often aggravates the problem. Especially when performing reasoning (automatically inferring new data from existing data), erroneous data has potentially devastating impact on the overall quality of the resulting dataset. Hence, a challenge is how data publishers can coordinate in order to fix problems in the data or blacklist sites which do not provide reliable data. Methods and techniques are needed to; check integrity, accuracy, highlight, identify and sanity check, corroborating evidence; assess the probability that a given statement is true, equate weight differences between market sectors or companies; act as clearing houses for raising and settling

> disputes between competing (and possibly conflicting) data providers and interact with messy erroneous web data of potentially dubious provenance and quality. In summary, errors in signage, amounts, labeling, and classification can seriously impede the utility of systems operating over such data.

Another issue is that not all companies will need or require a data warehouse. Also, the project manager must never assume that all of the information needed for the mashup dashboards will be found in the information data warehouse. Finally, good dashboards stimulate interactive discussions between the project team and the stakeholders. With mashup dashboards, this interaction may be difficult to achieve.

6.11 DASHBOARD DESIGN TIPS[10]

Here are some rules of thumb to follow when you design the layout:

- **Colors.:** You have a large number of colors to choose from, and although it is tempting to use a variety of different colors to highlight various areas of importance in a dashboard, most experts agree that too many colors and the "wrong" colors are worse than too few colors.

  Note: remember that some people are color blind, so if you use colors, try to use various shades. A good test is to print out a screenshot of the dashboard on a black-and-white printer and see whether you are able to distinguish what will now be shades of gray from each other.
- **Fonts and font size:** Using the right or wrong fonts and font sizes is like the use of colors; it can make or break the entire look and feel that a user gets from looking at a dashboard. Here are some tips:
  - Do not mix a number of different font types; try to stick with one. Use one of the popular business fonts, such as Arial.
  - Do not mix a number of different font sizes, and do not use too small or too large fonts. Remember that dashboards likely will be use by many middle-aged or older employees who cannot easily read very small fonts. Ideally, you should use a font size of 12 or 14 points and apply boldface in headers (maybe with the exception of a main header that could be in some larger font). Text or numbers should be in fonts of 8 to 12 points, with 10 points being the most often used. When it

10. This section has been taken from Nils Rasmussen, Claire Y. Chen, and Manish Bansal, *Business Dashboards*, Hoboken, NJ: John Wiley & Sons, 2009, pp. 101–102.

comes to font sizes, the challenge for a designer is always the available areas on the screen (also referred to as "screen real estate"). With overly large fonts, titles, legends, descriptions, and so on may get cut off or bleed into another section, and fonts that are too small make the display hard to read.

- **Use screen real estate:** Per definition, most dashboards are designed to fit a single viewable area (e.g., the user's computer screen) so users can easily get a view of all their metrics by quickly glancing over the screen. In other words, the moment a "dashboard" is of a size that requires a lot of scrolling to the sides or up and down for users to find what they looking for, it is not really a dashboard anymore, but a page with graphics. Almost always, users get excited about the possibilities with a dashboard, and they want more charts and tables than what will easily fit on a screen. When this happens, there are various options:
  - Use components that can be expanded, collapsed, or stacked so that the default views after login still fit on a single screen, but users can click a button to expand certain areas where they need to see more.
  - Use many dashboards. If there is simply too much information to fit on a single dashboard (e.g., move sales-related information to a "sales dashboard" and higher-level revenue and expense information to a "financial dashboard"). Many dashboard technologies have buttons or hyperlinks that let us link related dashboards together to make navigation easy and intuitive for the users.
  - Use parameters to filter the data the user wants to see. For example, a time parameter can display a specific quarter in a dashboard instead of showing all four quarters in a year.
- **Component placement:** If you have two tables or grids or scorecards and four charts, how should they be organized on the screen? Here are a few tips:
  - Talk with the key users to find out which information is most important so they can establish a priority. Based on this, place the components in order of importance. Most users read from left to right and start at the top, so that could be the order you place the components. The idea is that if the users have only a few seconds to glance at a dashboard, their eyes first catch what is most important to them.
  - A second consideration for component placement is workflow. In other words, if users typically start by analyzing metrics in a scorecard component and then want to see a graphical trend for a certain metric they click on in a scorecard, that chart component should be placed adjacent to the scorecard to make it easy for users to transfer their view to the chart as they click on the scorecard.

6.12 PURESHARE, INC.[11]

There are several companies that provide metric management and dashboard software solutions to their clients. In the next two sections are three white papers from such companies, and these white papers serve as excellent summaries of the information for dashboard design efforts. The first two white paper presented in this section are from PureShare, Inc.

White Paper #1: Metric Dashboard Design

Designers of metrics management dashboards need to incorporate three areas of knowledge and expertise when building dashboards. They must understand the dashboard users' needs and expectations both for metrics and for the presentation of those metrics; they must understand where and how to get the data for these metrics; and they must apply uniform standards to the design of dashboards and dashboard suites in order to make them "intuitive" for the end users.

This paper outlines dashboard design best practices and design tips, and will help dashboard designers ensure that their projects meet with end-user approval. It concludes with a checklist of design considerations for dashboard usability.

Increasing User Adoption of Metrics Dashboards

Users turn to metrics management solutions to find out what is going on with the business in order to make informed, reasoned decisions.

Good metrics management dashboards show key performance indicators (KPIs) in context so that they are meaningful, and present them in a way that allows users to instantly understand the significance of the information. This presentation lets users quickly evaluate choices and make decisions with full confidence that these decisions are supported by facts.

---

11. The remaining material in this section has been taken from two PureShare, Inc. white papers "Metrics Dashboard Design" and "Pro-Active Metrics Management." Both white papers are reproduced by permission. PureShare, Inc. is a metrics management software vendor that develops proactive, web-based corporate performance monitoring and enterprise reporting applications. PureShare's proactive metrics management applications empower business users to see key performance indicators (KPI) in real time and allow business managers to accurately gauge performance. With PureShare, organizations can harness corporate data into powerful visual metrics that: **Automate** the reporting and monitoring of key performance indicators (KPIs), as well as data transformations; **Enable** discovery of new insights into the business, and to react quickly to those—rather than making after-the-fact corrections; and **Trigger** positive change by focusing on factors that directly impact corporate performance. PureShare's customers include Global 1000, Fortune 500, and mid-size organizations in ITSM, support, financial, insurance, retail, and other industry sectors. PureShare Headquarters: 80 Aberdeen St., Suite 400, Ottawa ON K1S 5R5. Phone: 1-613-236-1644, Toll Free: 1-866-636-6065. Email: info@pureshare.com; www.pureshare.com.

Dashboards are neither detailed reports nor exhaustive views of all data. Good metrics management solutions can offer users the option to "drill down" to as much detail as they require, or even link into reporting systems, but these are only ancillary functions. The primary function of metrics management dashboards is to support—even induce—proactive decision making.

Know the End Users

Users want dashboards that respond to their business requirements.

There is no substitute for understanding end users' needs and getting involved in dashboard development. Even more important than understanding product capabilities is understanding the people who will be using the dashboards, what they need to know to improve the business, and what sort of dashboard organization and displays will work best for them.

Use Context to Make Metrics Meaningful

Users need to understand what the metrics mean before they can make decisions. Data is meaningful only in context.

In order to easily understand metrics, users must see them in context—their context. In fact, context and presentation are integral to any metric; without them the metric is simply meaningless numbers.

Dashboard designers should take time to learn what contextual information users require in order for metrics to be meaningful for them and to facilitate decisions and actions.

Contextual information will differ depending on the specific area being managed. For example, dashboard users in finance may need to track actual expenditures against budget targets, while a support desk may need to track the number of trouble tickets exceeding mean resolution times by more than 15 percent. In an environment where a metrics management solution is being used to help improve processes, users may need to monitor trends, compared to performance during another given period. (See Figure 6–44.)

In the left side of Figure 6–44, the much used pie chart effectively shows proportions, but does not tell anything about performance or progress toward targets. In the right side of Figure 6–44, a simple bar chart effectively shows proportions of allocated budgets and monies spent, targets, and performance: progress toward targets (dotted lines).

Whatever the case, the dashboard designer should spend time understanding not only the data behind the indicator but also the data that creates the context for the indicator (such as revenue targets, time period coverage, maximum capacities, etc.). This rule should guide the dashboard designer through every step of design, from data gathering to detailed composition of displays.

Figure 6-44 User Displays to Show Context and Progress toward Targets *PureShare, Inc. white paper "Metrics Dashboard Design"*

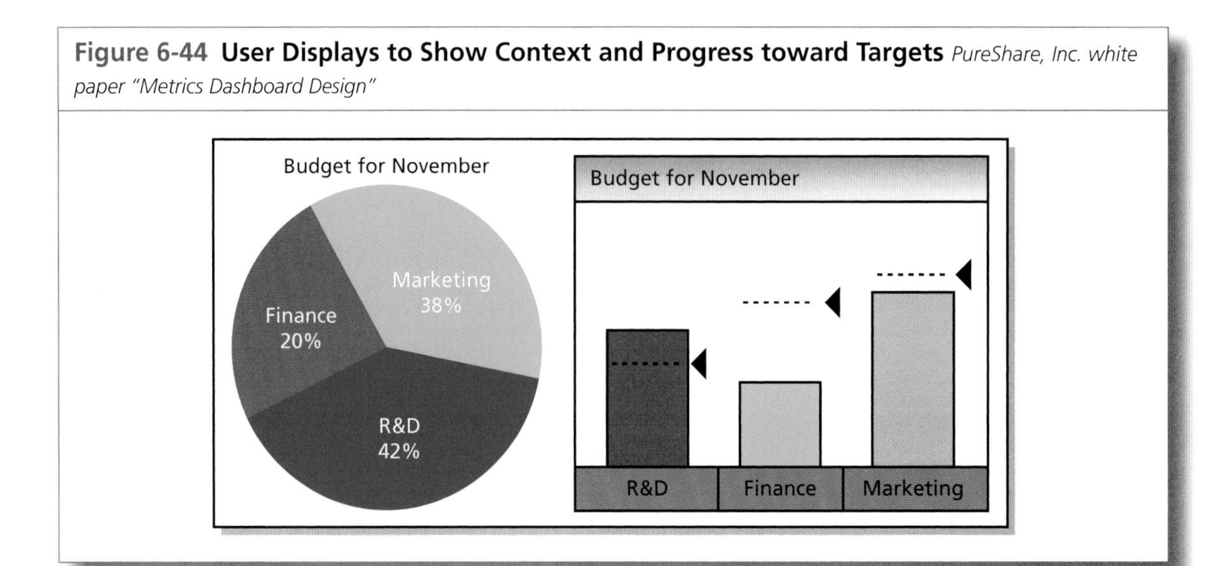

Data Retrieval

Users need to have confidence in the integrity of the metrics they use to manage the business. They need to know that they are acting on facts, not guesses.

There is no easy formula that will guarantee the value of data brought to dashboards. Nonetheless, it is essential to determine the sources, ownership, and quality of the data to be used before starting dashboard design.

The following guidelines will help dashboard designers deliver the metrics dashboard users need and will use.

Identify KRAs and KPIs

Users need to know their metrics so they can make informed decisions for their areas of responsibility.

A common—and effective—approach to understanding what users need to know is to work with end users to identify the key results areas (KRAs) for which they are responsible, then the key performance indicators (KPIs) they need to monitor and manage to improve performance in their areas of responsibility.

Get the Data Whatever Its Format or Location

Users want metrics derived from complete data, no matter where that data is stored, so they do not have to guess. Once the metrics users require have been identified, the data should be retrieved, *no matter where it is located.*

A well-designed metrics management dashboard provides both a current synthesis and details of key metrics. Often the data required to provide this synthesis and these details is spread across a variety of databases technologies and even spreadsheets at different physical locations. It is, nonetheless, essential that all required data be retrieved. A dashboard that provides metrics based on partial information is of little value when a global view is needed.

Refresh the Data According to Users' Needs

Users need metrics that are up to date so they can act on current and probable future situations.

The timeliness of a metric is as important as the metric data itself. Find out how current data must be for it to be valuable to users, and set the polling frequencies for the queries that retrieve the data accordingly.

Metrics used to monitor hourly call levels to a help desk are worse than useless if data is gathered every Sunday at 6 A.M. Similarly, if sales personnel report sales once a week on Thursdays, there is little point in polling the database every hour for updates to this data.

Usability Design Best Practices

Users do not want to be surprised by the dashboard design. They need to be able focus on metrics and decisions.

A metrics management dashboard should function for the user in the same way that an automobile dashboard or traffic signs function for an automobile driver.

Just as a driver knows to stop at a red light, a metrics dashboard user should understand without deliberation that, for example, a red thermometer means that corrective action is required. This requirement means that dashboard design must be consistent with common usage and practices as well as across all dashboards.

Offer Users a Choice of Views

Users need to be able to view metrics differently, so they can see the relationships between the metrics that affect their specific concerns.

Users may need to view metrics differently, and dashboards should allow them to do this. For example, financial metrics dashboards showing actuals versus budgets could offer views by department, and by line item or profit center.

Whatever the view offered, dashboards should be consistent, and the current view should be clearly identified by titles and labels.

Use Commonly Accepted Symbols, Colors, and Organization

Users need to understand dashboards instantly and without specialized training so they can focus on their jobs, not on deciphering data.

Use symbols, colors and organization in ways that are commonly accepted and therefore easily understood. (See Figure 6–45.)

As an example, in the left side of Figure 6–45, the users might be confused by the use of color combined with the expressions on the faces. Red

Figure 6-45 Using Color to Improve Communication of Key Information *PureShare, Inc. white paper "Metrics Dashboard Design"*

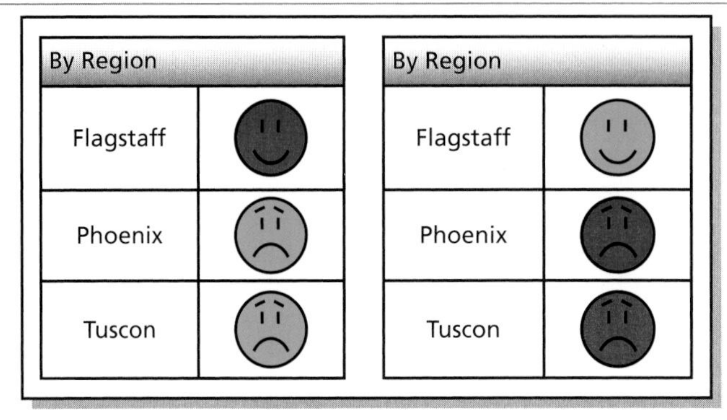

usually means that something is amiss, but the red face has a smile. The user might not know if action is required. In the right side of Figure 6–40, green means that everything is going well. The message is reinforced by the smiley face. Red usually means that action is required. This message is reinforced by the frown.

Red is the color most commonly used to identify something that requires attention. Therefore, unless there is a compelling reason to use another color to identify, for example, sales that are below targets use red to communicate this information. Similarly, where appropriate use commonly accepted symbols, such as stop signs or caution signs, rather than new symbols that users will have to learn.

Organize information using commonly accepted norms. Generally, this means that the most important information is placed at the top of a dashboard and secondary information and details are placed below.

Note, however, that symbols may differ between countries and cultures. Take these differences into account if the dashboards will be used in different parts of the world.

Establish Clear Dashboard Navigation and Hierarchies

Users need to be able to find information—more detail, less detail or a different view—instantly.

Few dashboards are deployed individually. Typically users require a suite of dashboards. Clearly establish an organization for the dashboard suite. Use an easily recognizable hierarchy of dashboards and consistent links for navigating between dashboards.

A good rule is to use the dashboard with top-level information for a user's area of responsibility as that user's entry point, then provide links from individual metrics to dashboards with more details about that area of the business. This is known as "drill-down" capability.

Maintain Consistency of Design

Users do not want to have to learn how to read each dashboard.

Establish and implement a limited set of templates with consistent use of color, symbols, and navigation, and use them throughout the dashboard suite.

Presentation of information on dashboards should be consistent. For example, if dashboards show trends, percentages, and absolute numbers, place each type of information in the same place on every dashboard and use the same display for these types of information across similar dashboards.

If thermometers are used to show progress of revenue against targets in the company financials roll-up dashboard, do not use gas gauges to show revenue against targets in the regional detail dashboards; use thermometers. In the left side of Figure 6–46, the different displays used to the overview (thermometer) and the monthly values (gas gauges) may confuse users who "drill down" to get more detail. In the right side of Figure 6–46, consistent use of thermometers (large for the overview and small for the details) ensures that the users will immediately and intuitively associate the two displays.

Similarly, if display threshold of red-yellow-green are set to 40, 60, and 80 percent for the company roll-up dashboard, use the same thresholds for the details unless there is a compelling reason to do otherwise.

Use Color Judiciously

Users expect color to provide important information they need, not distract them.

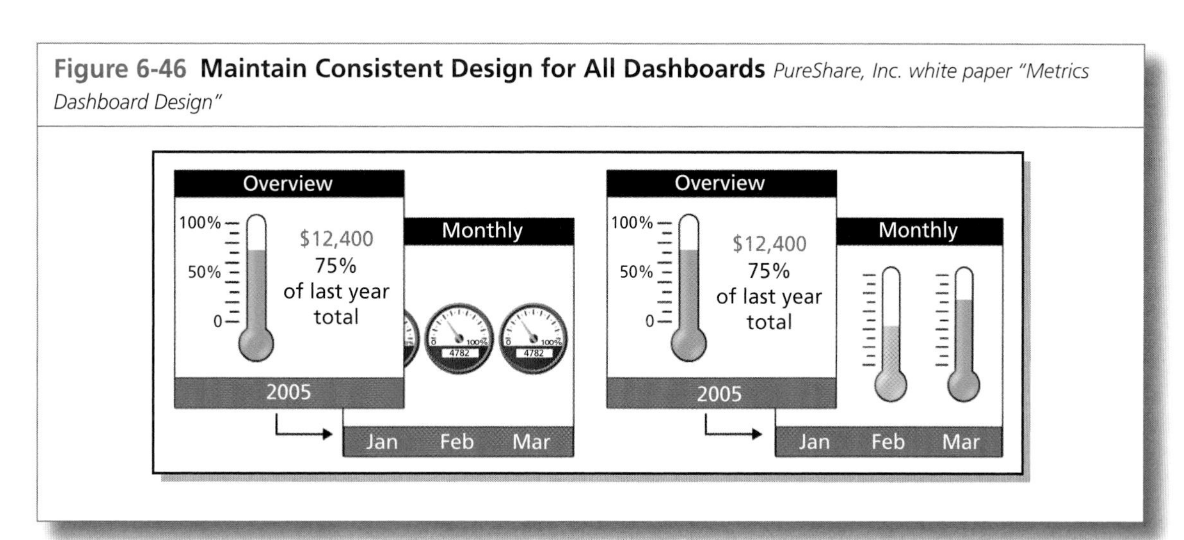

Figure 6-46 Maintain Consistent Design for All Dashboards *PureShare, Inc. white paper "Metrics Dashboard Design"*

Generally, color can be used to create four effects on a dashboard. It can:

- Identify the status of key metrics and areas that require attention. For example, use red to identify expenditures that have increased more than 15 percent over the same period the previous year.
- Identify types of information. Color can be used to help users instantly identify the type of information they are looking at. For example, dark green can be used for monetary values, and dark blue for quantities of items.
- Deemphasize areas or items. Border areas, backgrounds and other supporting dashboard components (the dashboard skins) should use plain, unobtrusive colors that help define dashboard areas without distracting from the information displayed.
- Identify the dashboard type or its level. Different background colors or the color of dashboard titles can help users identify what they are looking at. For example, financial dashboards could use a green skin, while help desk dashboards could use beige as a reassuring color.

Use Dashboard Groups to Improve Organization

Users need to see metric groups and hierarchies so they can understand relationships between different areas of the business.

Group displays together by the type of metric displayed or by functional area. In Figure 6–47, PureShare's Profit Accelerator dashboards group related information together to improve dashboard "legibility."

For example, on a dashboard showing financial roll-ups, put top-level information into one group that shows progress against targets in three ways: absolute numbers against targets; performance compared to the same period in the previous year, or against average performance for the same period during the last five years; and the trend, based on performance over the last 60 days. Alternately, show absolute numbers against targets for each region or department.

Whatever the rationale used to group information on dashboards, be consistent. Use the same rationale when establishing groups on different dashboards.

Set dashboard groups to "open" or "closed" based on the hierarchy of information. Closed groups can be used to provide more detailed metrics or complementary metrics on the same dashboard without distracting the user's eye from the primary information.

Display Selection and Design

Users must be able to understand *what* they are being shown without stopping to analyze *how* it is being shown.

Select the display symbols that are most appropriate for displaying the information the dashboard users need.

Identify a limited set of symbols and use these on all dashboards. Be consistent with location, size, and color of supplementary information

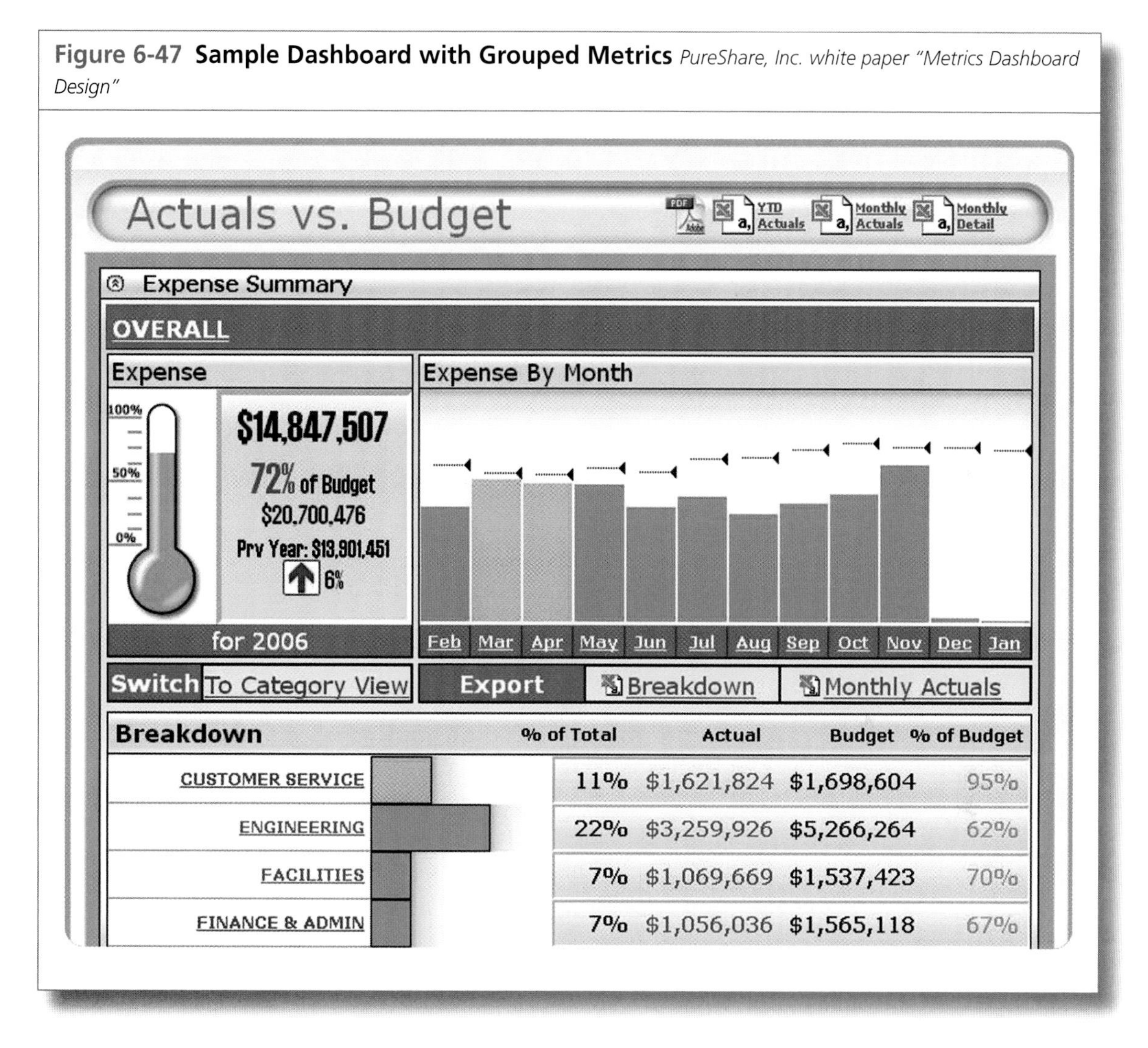

Figure 6-47 Sample Dashboard with Grouped Metrics *PureShare, Inc. white paper "Metrics Dashboard Design"*

associated with the symbols. Consider using different sizes of the same display symbol for different levels of information.

Avoid displays that are overly complex, colorful, or animated. Such qualities are very effective when correctly used. However, the more complex the display, the more difficult it is to process. Overly complex dashboard displays distract users from the information they need and want. Consider offering views that separate the information into grouped displays or even different dashboards.

Do not hesitate to work with the dashboard vendor to design new display symbols that will improve presentation of information.

Actual Values, Percentages, and Trends

Different users need different information in order to make informed decisions and take appropriate action.

Decide with the dashboard users what metrics they need to monitor and manage. Different users may require different information from the same data.

For example, a CEO may want to know trends in budget expenditures, while a CFO and department managers may be more interested in actual numbers (actuals versus budgets). Use data to present the information users need, and make sure that the type of information is clearly identified. A thermometer showing revenue trends that is interpreted as showing actual numbers can lead to costly misinterpretations.

Timestamps

Users need to know when the metric was updated so they know the age of the data on which they base their actions.

Time is an essential part of a metric's context. Ensure that users can know when the data for the metrics was retrieved. In many cases this information is the key to understanding what the metric means. Ensure that this information is available, but that it does not crowd the display. Consider putting the date and time in a mouse-over.

Titles and Labels

Users need to know instantly what they are looking at so they can focus on what it means for the business.

Give all dashboards meaningful, descriptive titles. Descriptive titles are generally more intuitive than cryptic or symbolic ones. Assign labels to dashboard groups and display symbols so that users can clearly identify information. Keep labels in the same location and use the same color standards throughout the dashboard suite.

Do not abuse labels. Overuse of labels can crowd out essential information.

Mouse-Overs

Users often want more information to help them understand the significance of a metric.

Mouse-overs are an effective way to include detailed metric information without crowding the dashboard.

Information such as last and next run time metrics for another corresponding period, etc., can be included in mouse-overs to help users understand the significance of the primary information on the dashboard.

Parameter-Based Views

Users want to see only the metrics that help them do their jobs. Different users give different weight to different information, as seen in Figure 6–48.

Figure 6-48 How Parameters Can Be Used to Simplify Dashboard Design and Implementation and Improve Usablity *PureShare, Inc. white paper "Metrics Dashboard Design"*

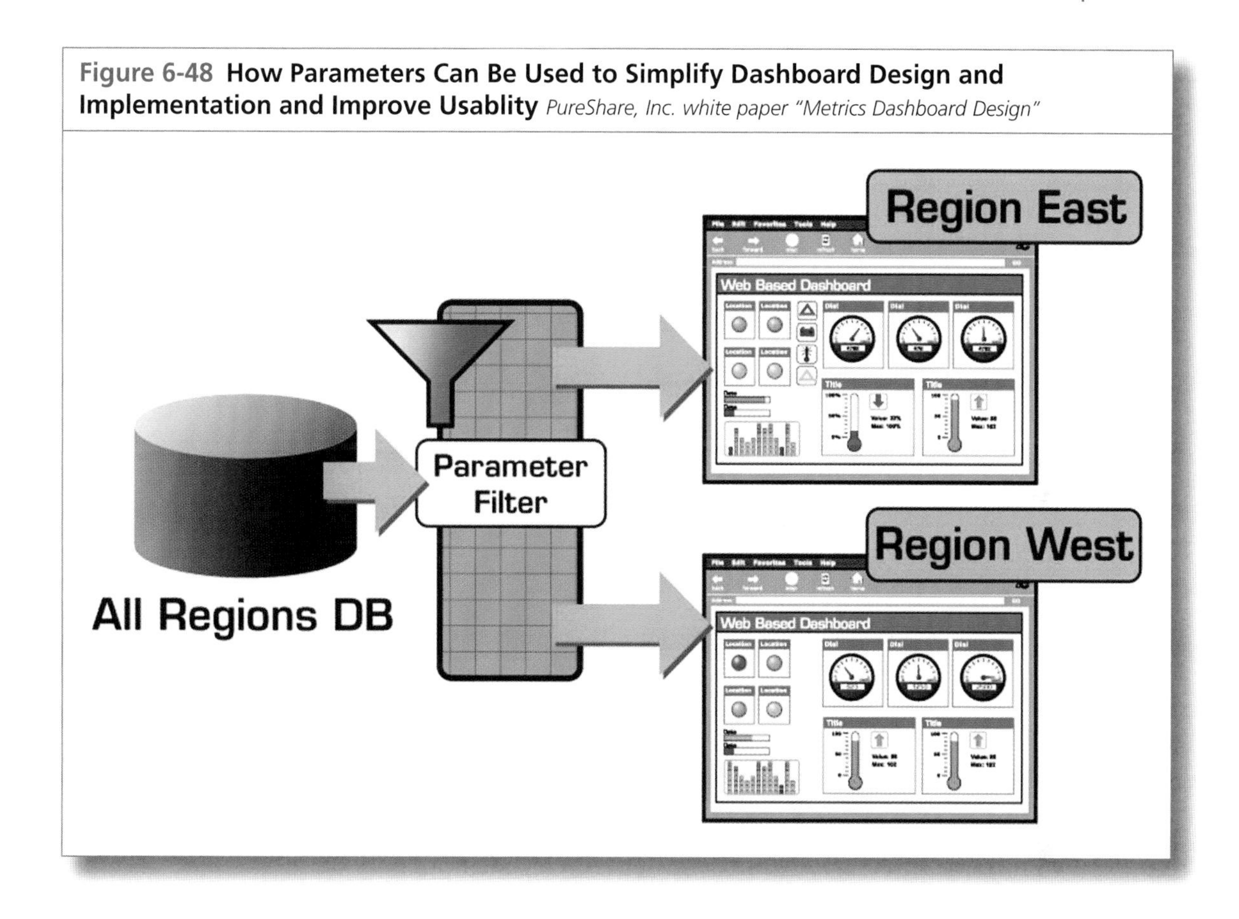

In Figure 6–48, Parameters can be used to filter data so that users see only the information they need. In this example, data is filtered by region. All users see the same dashboard, but users in Region East see information for their region, while users in Region West see information for their region.

Parameters are user-set variables that can be used to filter metric data delivered to the dashboards shown to different users. Consider designing dashboards with parameters, so that users get views of the dashboards based on their needs and permissions.

For example, the same dashboard suite might be used by a help-desk manager and individual employees in the department. The manager may need a consolidated view of all calls, by type, while each employee might need to see the calls for which he or she is responsible, a view created by filtering the information based on the user ID.

If the dashboards use parameters, display these prominently so that users will know what the metrics they are being shown represent.

Use Thresholds and Threshold-Triggered Actions

Users benefit most from dashboards that help them take corrective actions before problems occur.

Thresholds and threshold-triggered actions can be used to transform metrics monitoring into proactive management. Use thresholds to trigger actions that alert users to potential trouble areas, or even initiate corrective action by running scripts. In Figure 6–49, thresholds can be used with alerts to transform dashboards from passive monitoring devices into active vehicles inducing corrective action and, especially, prevent decisions and actions.

For example, set a threshold to send emails to the person responsible for managing a budget if spending reaches more than 70 percent of budget before the middle of the month. Or, in a help-desk environment, launch a script to change phone messages and reschedule nonessential activities if wait times increase beyond SLA requirements.

Roll-Ups and Drill-Downs

Users need to get both the big picture and the details.

A dashboard is most useful when used as part of a suite of complementary dashboards.

Group together on a dashboard the metrics users need to see together, and then use drill-downs and roll-ups to provide details and overviews.

This technique provides users the metrics they need without crowding too much information on any single dashboard.

Figure 6-49 Simple Alert Triggered by a Threshold *PureShare, Inc. white paper "Metrics Dashboard Design"*

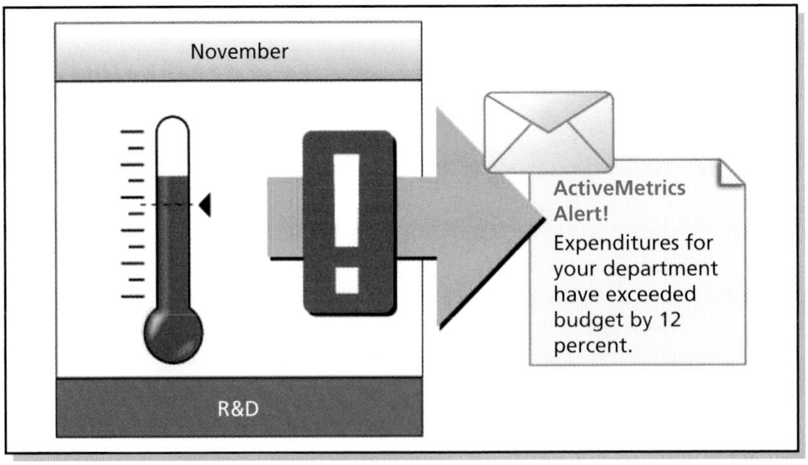

November
ActiveMetrics Alert!
Expenditures for your department have exceeded budget by 12 percent.
R&D

Animation

Users appreciate a bit of fun, but are annoyed by too many gimmicks.

Animation, such as blinking lights, moving figures, and other such displays can add interest to dashboards. Use these sorts of features very judiciously, however, and consider providing non-animated versions of dashboards, or an "animation off" button.

Animation can be fun the first few times but, if not properly used, it can become an annoyance. Consider using animation only for special projects, such as a month-end race to motivate the sales team, or to draw attention to new dashboard features for a limited period of time.

Visual "Noise"

Users turn to dashboards to be informed, not dazzled.

Avoid cluttering dashboards and dashboard displays with unnecessary paraphernalia, such as ornate frames, patterned backgrounds or 3-D effects that add no value to the information displayed.

A minimalist approach is almost always the best approach.

Usability Checklist

Attention given to dashboard design can pay enormous dividends, both in user satisfaction with the dashboards and, especially, in improved business performance founded in proactive metrics management and reasoned, informed decision-making processes.

Dashboard users want their questions about the metrics they are viewing on a dashboard answered even before they can formulate the questions. In fact, by the time a question about business performance is asked, it is often too late to take corrective action. Typical questions are shown in Table 6–4.

Valuable dashboards provide up-to-date information right at the user's fingertips so that problem areas can be addressed as they arise, and opportunities can be taken advantage of immediately.

Ensuring that a dashboard and its components answer the questions in Table 6–5 below should help dashboard designers create more effective dashboards.

White Paper #2 Pro-Active Metrics Management

Choosing a Metrics Management Solution

A business decision is only as good as the metrics that inform it. The quality of these metrics depends on three factors: accuracy, timeliness, and presentation. Understanding the importance of metrics, businesses are investing in defining and monitoring key performance indicators (KPIs). These indicators must not only accurately reflect the state of the business and current trends, but they must also be presented to users in a manner that induces effective, proactive decision making.

TABLE 6-4 User Questions and Design Solutions

USER QUESTION	DESIGN SOLUTION
What am I looking at?	Use clear, descriptive titles and labels.
Does this mean "good" or "bad"?	Use standard, culturally accepted colors and symbols.
Are things getting better or worse?	Employ thresholds, and show meaningful comparisons and trends.
What is being measured, and what are the units of measure?	Clearly identify the units of measure, and provide actual values.
What is the target or norm?	Clearly show targets and norms, and design displays that show progress towards these.
How recent is the data?	Provide a date and time stamp for each metric.

TABLE 6-5 Common User Questions and Design Solutions

USER QUESTION	DESIGN SOLUTION
How can I get more details?	Provide drill-down links to groups with detailed information.
How can I get a broader view?	Provide links to roll-up and overview dashboards.
What do I do with this information: what action should I take?	Always place data in context, and where possible suggest advisable actions based on the metric.
When should I check for an update?	Provide the date and time when the metric will be updated. When business needs warrant, allow ad hoc updates.
How do I get metrics that are not on these dashboards?	Be ready to develop new dashboards. Users will want them!

This paper reviews the capabilities of metrics management software and outlines the most important features and capabilities to look for in a solution.

Metrics Collection and Delivery

One of the key goals of businesses today is the collection and delivery of the metrics decision makers need to make timely, informed, and well-reasoned choices. Even in businesses where the key performance indicators (KPI) have been defined and the data that drives them is available, the collection and delivery of the metrics can fail. Metrics that are timely, accurate, in context and easy to understand are not immediately available to those who need them most: boards, executives, managers, and others at all levels of the organization.

No one is to blame. Until recently the tools needed for effective delivery of these metrics were simply not available. Reporting provides detailed profiles of the state of affairs as they were at a specific moment in the past. Business intelligence (BI) systems provide comprehensive data and deliver metrics, but these systems are prohibitively expensive and the metrics they provide are limited to those derived from the data inside the BI system. In today's business reality, metrics must be gathered from multiple, disparate systems and sources across the organization.

The Metrics Management Investment

Since the last generation of dashboards appeared on the market at the end of the 1990s, "dashboard solutions" have often been promoted as the answer to decision makers' need to know. Unfortunately, however, dashboards alone are not a solution. They are an essential part of the solution; they are effective for delivering metrics and for presenting them and making them accessible. But dashboards are only as good as the metrics they present; their value as decision-making tools is a function of both how well they present metrics and of the quality of the data behind these metrics.

What is needed, then, to transform decision making from educated guesswork based on partial information and hunches to a truly rational and informed process based on timely, accurate metrics is not simply dashboards. This transformation requires a proactive metrics management tool—one that can gather data from across multiple systems and deliver metrics to dashboards in context and with as much or as little detail as decision makers require.

Selecting a Solution

Due to the emergence of metrics management software and to the diversity of dashboard-type products on the market, the search for a solution can be confusing. Metrics management dashboards, as shown in Figure 6–50, should provide users with the information they need to make informed, reasoned decisions. The following is a set of high-level criteria for evaluating metrics management solutions.

Proactive Metrics Management

Look for the ability to trigger effective preventative and corrective action. The crucial differentiator for metrics management solutions is the level of *proactive* metrics management they support. Proactive metrics management is a function of a solution's ability both to trigger appropriate actions in systems with which it interfaces, and to facilitate timely decisions by the persons responsible for key results areas (KRA) in the organization.

It is important to understand here that more than one capability defines a product's level of support for proactive metrics management. Two important capabilities are:

- Thresholds, alerts and notifications. A proactive solution uses thresholds and alerts set from absolute values or trends (or a combination of these) to keep users informed.

Figure 6-50 Sample PureShare metrics Management Dashboard *PureShare, Inc. white papers; "Pro-Active Metrics Management"*

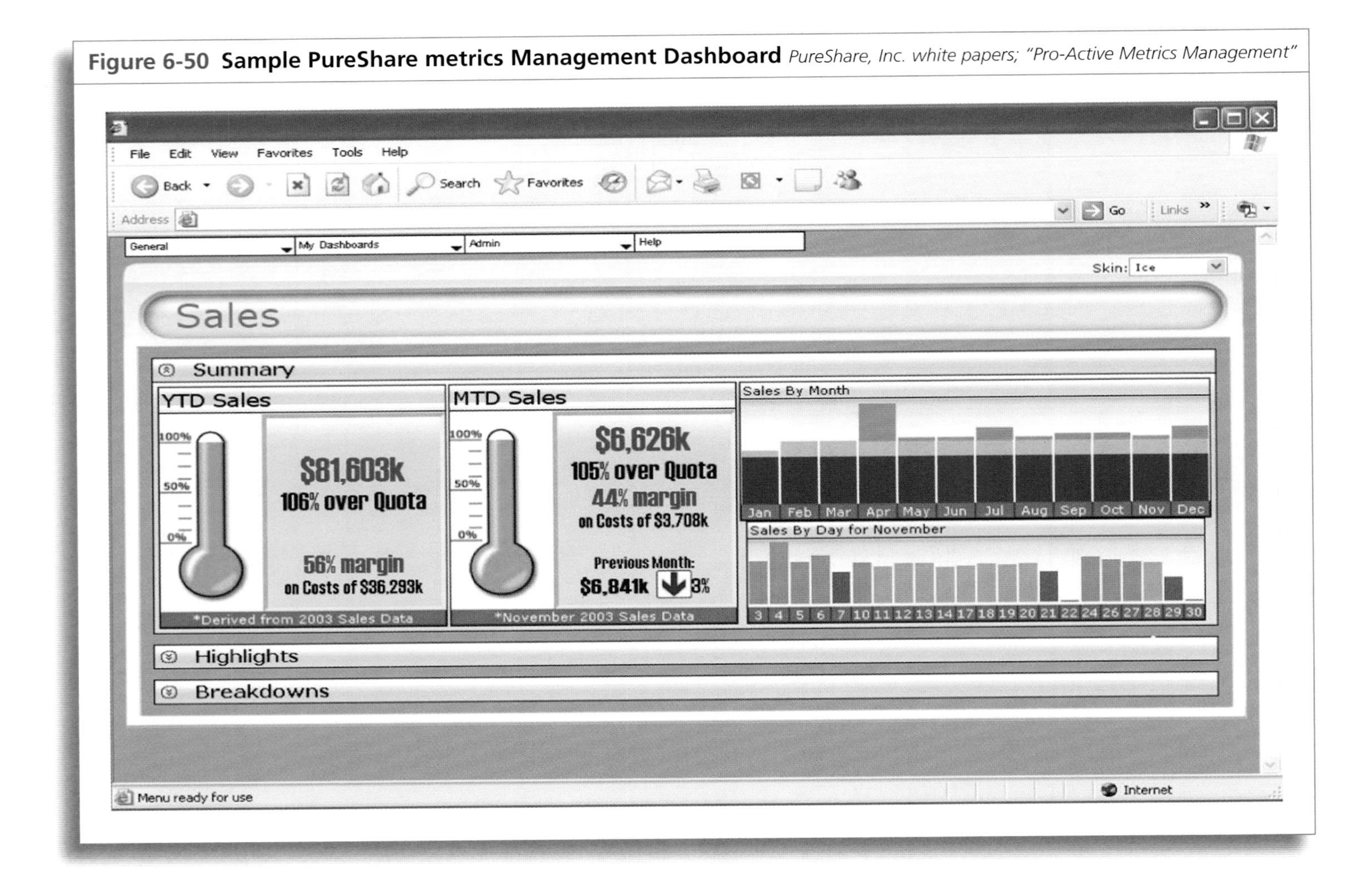

- Triggers and corrective/preventative action. A proactive solution is able to initiate actions when thresholds are compromised. For example, it can send an e-mail alert, or launch a script.

Technology Agnostic

Look for no restrictions on metrics data sources. PureShare ActiveMetrics can retrieve data from virtually any source as shown in Figure 6–51. Very few organizations use a single technology to meet all their data and performance needs.

Typically, different technologies have been implemented because each is the best available for a specific area of the business: financial reporting, trouble-ticket management, etc. A proactive metrics management solution must be able to query seamlessly a wide array of different databases (and even spreadsheets) for key indicators and, if required, use indicators derived from different sources in the same dashboard.

Easy Availability

Look for web-based, easily customizable user interface (UI) design with multi-language support. Organizations today are rarely limited to a single geographic location, and often work around the clock. A metrics

Figure 6-51 Data-Agnostic Metric Dashboard Solution *PureShare, Inc. white papers; "Pro-Active Metrics Management"*

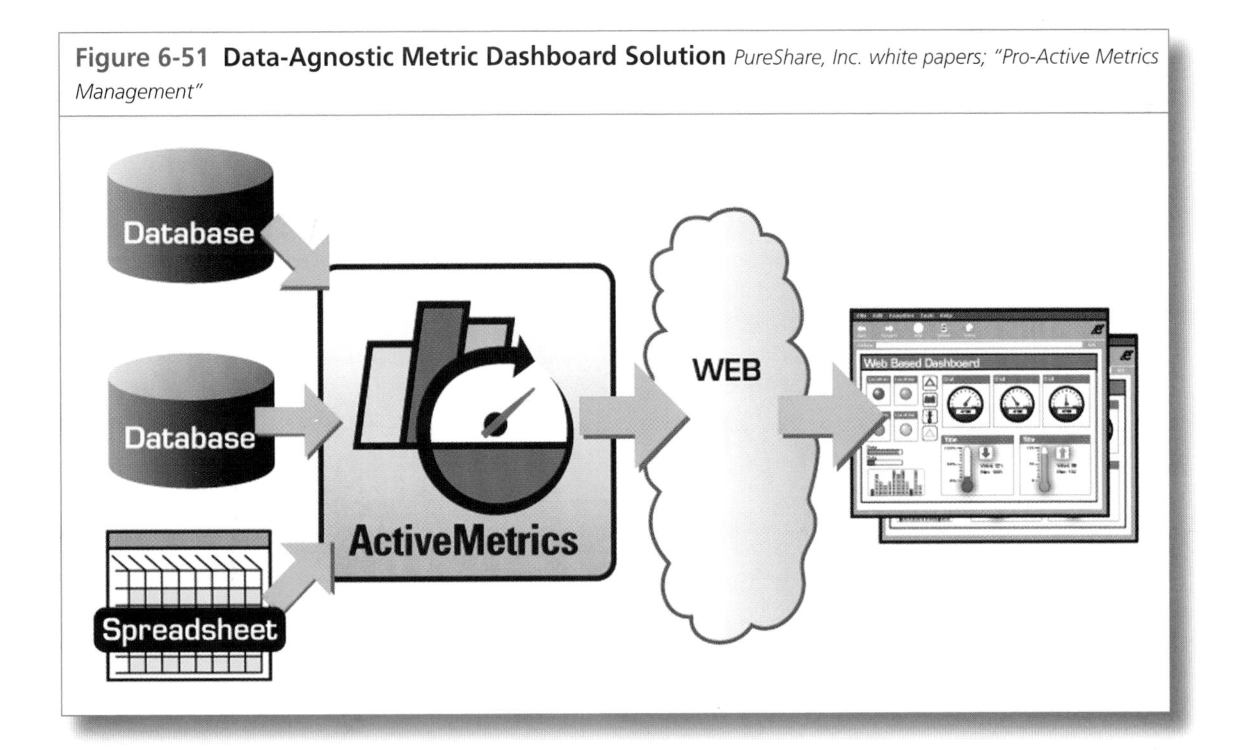

management solution's dashboards must be available at all times and anywhere in the world, and because key metrics must be made available to many people, the solution should not require specialized viewing tools. Users should need nothing more than a web browser to view metrics dashboards over the Internet or an intranet.

Security

Look for configurable, secure user and data source access. Since metrics tell critical information about the business, they usually convey sensitive information. Dashboards and the information they contain should therefore be easily protected without sacrificing their availability. Further, the metrics management solution should be able to access data that is protected behind various security barriers and firewalls without jeopardizing the integrity of these security measures.

Usability

Look for ease of use for both end users and dashboard developers.

As with anything else, a metrics management solution is of little value if it is not widely adopted by users. Users can be classed into two areas: end users and dashboard designers. Any solution must be easily usable by both.

Assuming that the metrics dashboards are easily accessible to those who need them anywhere and at any time, the design of the dashboards themselves and the metrics they present should be evaluated:

- Are the dashboards clear and easy to understand so that users grasp instantly what they are being shown and what it means for their areas of responsibility?
- Can thresholds, trends and warnings be easily and clearly identified by the end user?
- Can legends to explain data be easily integrated into the dashboards?
- Are relationships between metrics easily established by the end users?
- Do the dashboards allow drill-down views for more detailed information, and other view changes requested by users?
- Does the solution offer clean and consistent display of metrics without visual "noise"?

Usability for dashboard designers may be less noticeable day-to-day, but it is critical, for it is the designers who will be responsible for the rapid development and deployment of dashboards to monitor KPIs, and hence for critical business information.

When evaluating solutions consider

- What skill sets do the dashboard designers require?
- Is training necessary, and if so, how much is required? A rule of thumb is that that no more than a few days' training should be needed for staff to become effective dashboard designers.

- Do the solution's dashboard design interfaces permit development of simple dashboards to start, with increasing complexity added as designers increase their expertise and decision makers develop their use of metrics to run the business?
- Does the solution allow users and designers to select dashboard look and feel, including branding the dashboards with your organization's logo and colors?

Deployment

Look for templates that facilitate rapid deployment without restricting future development.

Deployment brings together a host of other criteria that should be examined, and is often the litmus test of these: technology requirements, availability, security, and usability. See Figure 6–52.

Figure 6-52 Sample Dashboard with Grouped Metrics *PureShare, Inc. white papers; "Pro-Active Metrics Management"*

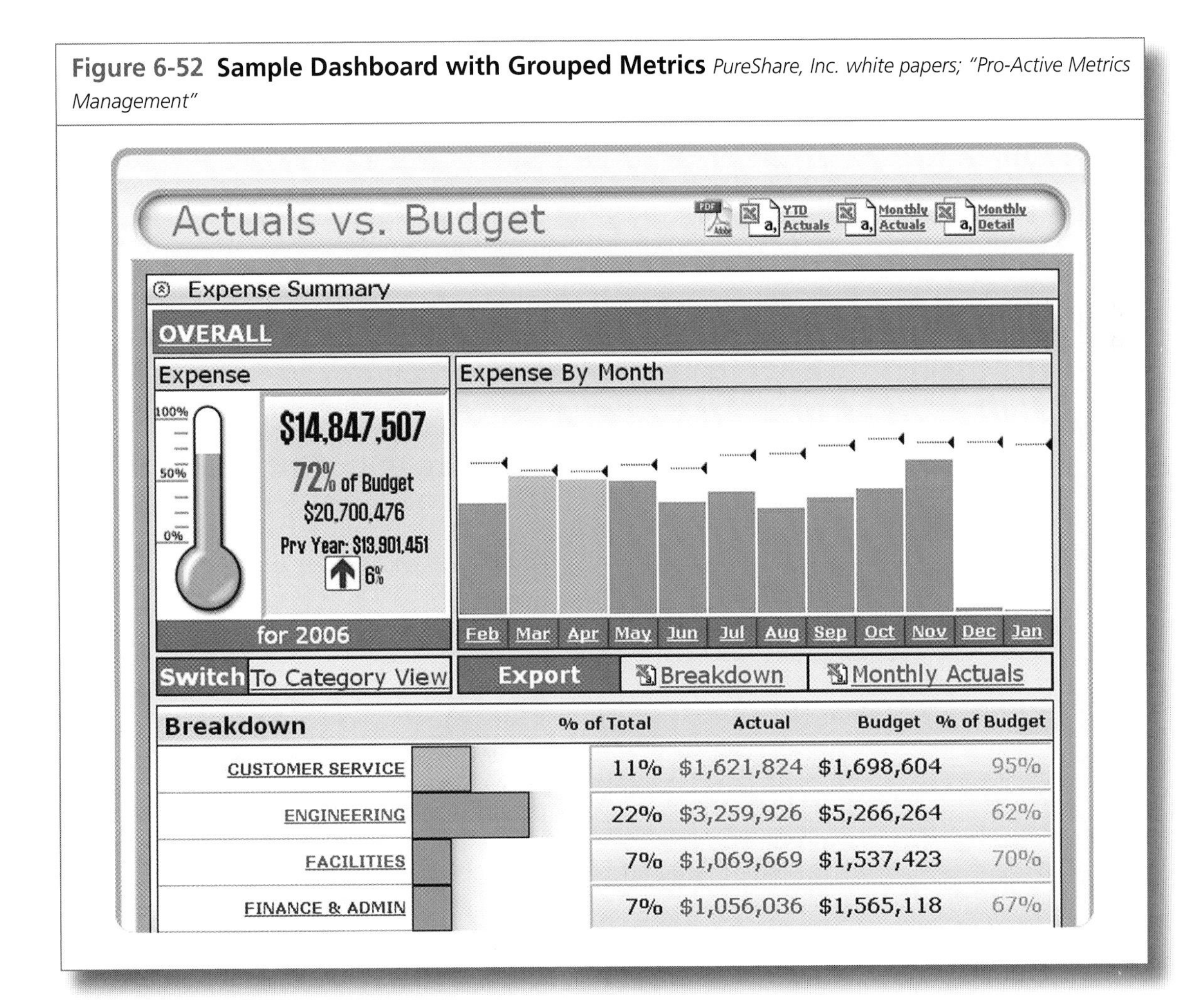

PureShare Accelerators offer rapid deployment of dashboards for specific purposes.

Generally, the more effective the dashboards that can be designed, and the more rapidly that the metrics management solution can be deployed, the better the solution because its effect is immediate. Often, a proactive metrics management solution is brought into a new area of the business to facilitate improved decision making right from the start, or into an area that is struggling in order to effect profound changes. In either case, getting the solution up and running and the metrics dashboards to decision makers quickly is of primary importance. However, rapid deployment in itself is not sufficient reason to select a solution. Beware of instant solutions that limit the ability to grow and improve how metrics are monitored and managed in the future.

Power and Flexibility

Look for the ability to deliver the metrics management capabilities that meet the needs of *your* business environment. The power and flexibility of a metrics management solution encompasses a wide range of capabilities. Questions to ask when evaluating solutions include:

- Can the dashboard pull data from diverse databases and technologies? If the proposed solution cannot do this, if it is limited to one or two proprietary database technologies, go no further.
- Can dashboard designers define the polling periods for queries, and can these be dynamically updated if required?
- Can metrics be based on aggregate queries; that is, can the metrics management solution pull together different types of data from different sources and make sense of it? For example, can it use one type of data for targets, another for progress, and yet another to show trends?
- Can the dashboards maintain histories based on user defined parameters?
- Does the solution offer different dashboard views for different users?
- Can metrics shown on dashboards be filtered, and can this filtering be configured by the dashboard designer?
- Do the dashboards manage thresholds, and what sort of actions can these thresholds trigger: e-mails, scripts, queries, etc.?
- Are metrics presented in context? Does the solution present numbers, or does it present meaningful information: quantities or percentages measured against targets, comparison of quantities for different periods, trends, etc.?

Context includes the date and time the metric was updated. In many cases, a metric has value only if the user knows when the data was retrieved.

Parameters can be used to filter data so that users see only the information they need to see. In Figure 6–53, one data is filtered by region. All users see the same dashboard, but users in Region East see data for their region,

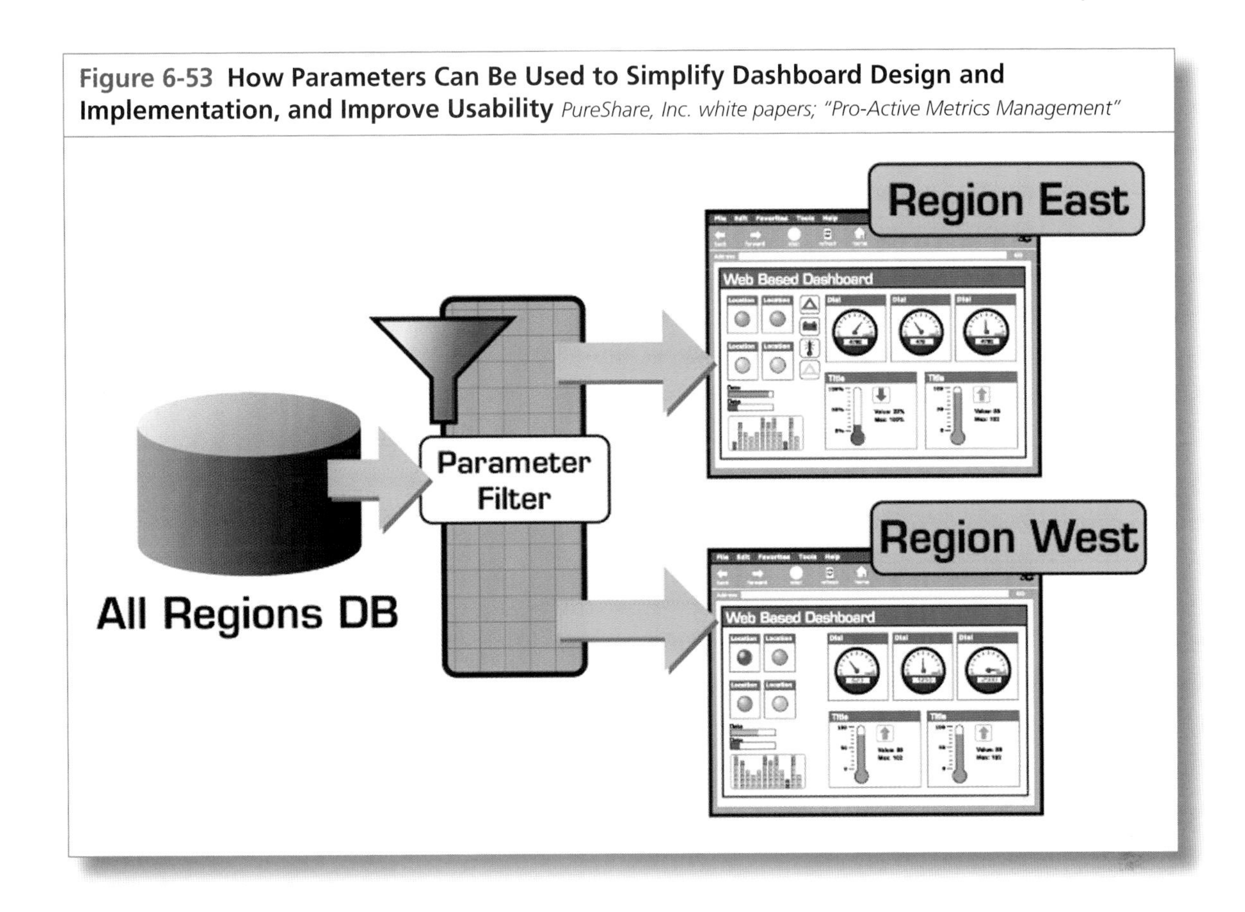

Figure 6-53 How Parameters Can Be Used to Simplify Dashboard Design and Implementation, and Improve Usability *PureShare, Inc. white papers; "Pro-Active Metrics Management"*

while users in Region West see information for their region. Whatever the case, the dashboard designer should spend time understanding not only the data behind the indicator but also the data that creates the context for the indicator (such as revenue targets, time period coverage, maximum capacities, etc.). This rule should guide the dashboard designer through every step of design, from data gathering to detailed composition of displays.

The Future

Look for the capability to initiate and sustain positive transformations of the business.

It is wise to find out if the capabilities offered by a solution will restrict what can be done in the future. Does the solution lend itself easily to additions and improvements, both in the way data is collected for metrics and to the dashboards that users see?

Deployment of a truly proactive metrics management solution will almost immediately provide insights into how business decisions are made and will reveal opportunities for improving the decision-making process, for managing the metrics used in this process—and ultimately for transforming the business.

As shown in Figure 6–54, thresholds can be used with alerts to transform dashboards from passive monitoring devices into active vehicles, inducing corrective and, especially, preventive decisions and actions.

Conclusion

Managers and technical experts looking for a metrics management solution face a growing number of options.

When evaluating solutions for metrics management, prospective buyers should arm themselves with knowledge about what these types of solutions can and cannot do, and with a list of evaluation criteria. And, finally, before buying they should ask two questions that encompass the others:

- Will this solution simply record what happened in the past, or does it reveal what is going on now and, based on current trends, what is likely to happen in the future?
- Is the solution truly proactive? Does it include displays, thresholds, and triggers not just to monitor what is going on, but also and critically to automatically initiate appropriate actions, and to facilitate reasoned, informed decisions?

Figure 6-54 Simple Alert Triggered by a Threshold *PureShare, Inc. white papers; "Pro-Active Metrics Management"*

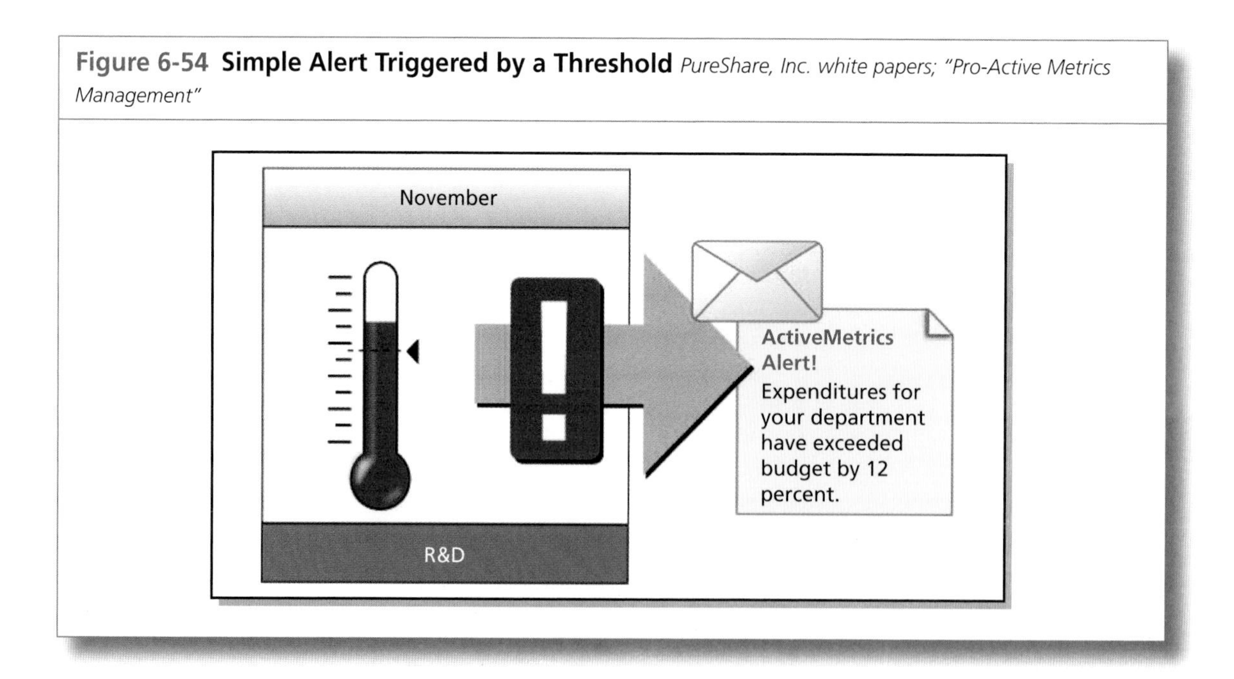

6.13 LOGIXML, INC.: DASHBOARD BEST PRACTICES[12]

Executive Summary

A wise man can see more from the bottom of a well than a fool can from a mountaintop.

—Unknown

Dashboards are becoming the new face of business intelligence (BI), as shown in Figure 6–55. While on the surface, Executive Information Systems (EIS) from the 1980s had a similar look and served a similar purpose, modern dashboards are interactive, easier to set up and update to changing business needs, and much more flexible to use. This, plus their ability to present data and information at both a summary and detailed level, makes them one of the most powerful tools in the business user's kit.

To be useful, however, a dashboard must be implemented around the needs of the business. Its functions should not be dictated by technology or by the whims of the end users. Also, a dashboard should be implemented so that it gets used—and so that the decision makers employing it can act on the information the dashboard presents.

To implement a dashboard that is truly business-driven, IT must take other important factors into consideration. Information included in a dashboard must be carefully selected, useful, and actionable. Too little information will make the feature all but useless; too much will make for a good managers' meeting conversation piece, but will actually render the dashboard cumbersome to use. Also, IT should not overlook adding interfaces that are familiar to end users—such as spreadsheets—and to set up the possibility to print the analysis results. In other words, dashboards must have as much business brain as technological muscle.

The question of how to calculate the return on a dashboard investment can be tricky. There is no one-size-fits-all calculation that can predict this with mathematical certainty. With this and other caveats in mind,

12. The material in this section was taken from a LogiXML white paper entitled "Dashboard Best Practices," written by Gabriel Fuchs, a renowned expert within strategic IT solutions, including business intelligence, performance management, and business analytics. © 2010 by LogiXML. Reproduced by permission of LogiXML. LogiXML, a leader in interactive, Web-based Business Intelligence, which empowers enterprises to turn data into business-critical information with pure Web-based reporting and analysis products. The Company offers a comprehensive platform that addresses all key areas of BI—managed reporting, ad hoc reporting, analysis and data services. Used by thousands of organizations worldwide, LogiXML products are built on standards-based technologies for easy integration, implementation and upgrade. LogiXML's per server pricing model makes its powerful technology the most affordable BI solution on the market. Founded in 2000, LogiXML is privately held and based in McLean, Virginia. LogiXML, Inc., 7900 Westpark Dr., Suite T107, McLean VA 22102. Tel. 1.888.LOGIXML | 703.752.9700. FAX 703.995.4811. Web: http://www.logixml.com. Email: sales@logixml.com.

Figure 6-55 A Modern Dashboard's Ability to Present Data and Information at Both a Summary and Detailed Level Makes It One of the Most Powerful Tools in a Business User's Kit. *© 2010 by LogiXML. Reproduced by permission of LogiXML.*

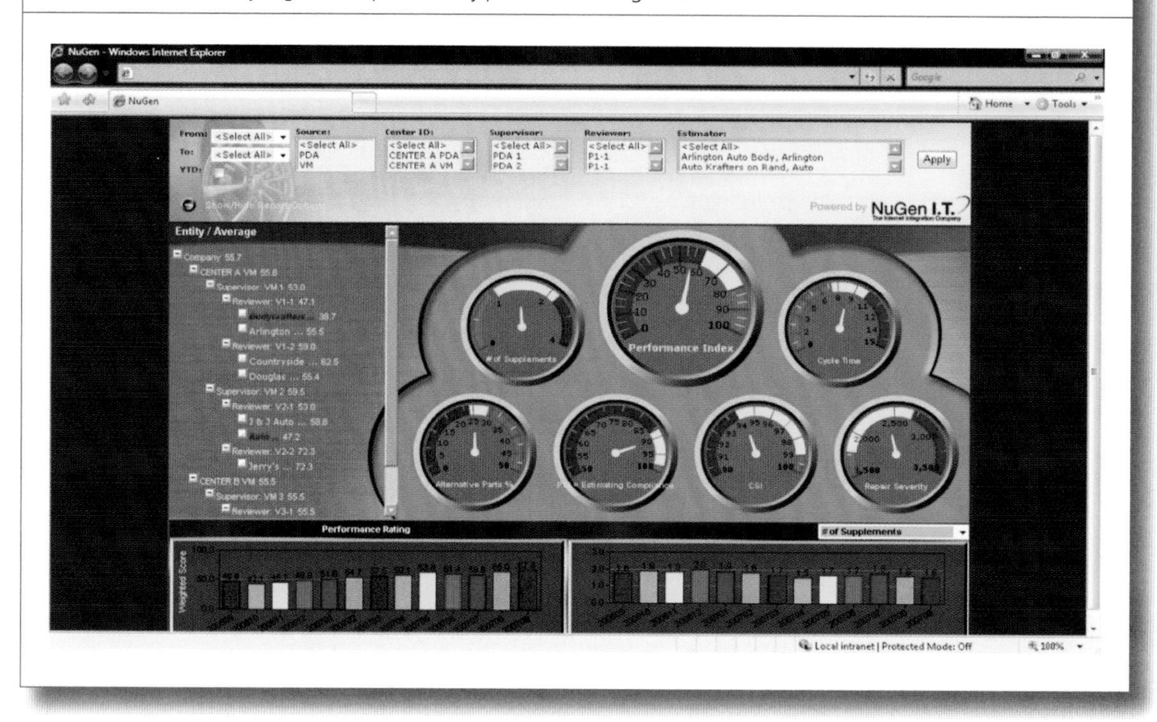

a company can take an approach based on the premises that ROI calculations can be done, but that they will be more a guideline rather than exact science. ROI is something that needs to be recalculated on a regular basis before, during and after the dashboard implementation. And when done regularly, the ROI calculation can actually help ensure that dashboards provide lasting value.

Lastly, today's dashboards can deliver what users and the organization as a whole need to know so as to understand their current business and even conduct informed forecasts.

Introduction—What's New about Dashboards?

If the only tool you have is a hammer, you tend to see every problem as a nail.

—Abraham Maslow (1908–1970)

Knowing what is going on in your business is not merely good; it is a prerequisite to success. Being able to advance this knowledge to make reasonable forecasts about the business is even better: it is what distinguishes the best from the merely good.

Reaching this business "maturity" requires a number of factors. Experience and business acumen are indispensable. A bit of luck will also help—even though it usually seems that the well prepared tend to have more luck than the ill-prepared. Yet, another essential factor is access to hard facts, i.e., timely and accurate business information presented in an intuitive manner.

How Modern Is the Modern Dashboard?

Decision support systems were around as early as the 1960s. With the advent of the client-server architecture in the 1980s, so-called Executive Information Systems (EIS) were presented as the state of the art decision support applications. These EISs from 20 years ago had a look similar to today's dashboards; so why are today's dashboards considered so new and uniquely valuable? As good as EIS systems were as an idea, they still suffered from technological constraints. First of all, they were not very interactive, and they were cumbersome to update to new business demands. Furthermore, they were designed around the approach where the business user's questions had to be predicted in advance—not an easy task. If or when the user had additional questions, the EIS system had to undergo some time-consuming redevelopment.

On the other hand, many of the charts used in older EIS systems were very much like what we see in today's dashboards as shown in Figure 6-56. User needs and wants have not changed much over time. What has changed is the underlying technology on one side, and users' expectations on the other. Modern dashboards meet business needs in a practical and actionable way when they can give quick snapshots of the big picture on one hand while being capable of offering detail on the other.

The Dashboard versus the Spreadsheet

Along with modern dashboards evolving from the old EIS tools, another business intelligence (BI) tool has been with us for a while: the spreadsheet. Most often in the form of Microsoft Excel, the spreadsheet has an intuitive interface and is easy to learn, at least as far as its most basic functions. It provides detailed numbers, which users can analyze adding their own calculations.

However, while the spreadsheet is easy to use and understand, it is often too detailed to give a quick and comprehensive overview of business data. For instance, monthly sales statistics for 50 products sold in 50 different states will in one year generate 30,000 cells of data (50 products × 50 states × 12 months). And this is a relatively simple spreadsheet. Furthermore, users are likely to reformat this business data in other spreadsheets, adding calculations and aggregations. This will create yet more cells of important

Figure 6-56 A Typical Dashboard. *© 2010 by LogiXML. Reproduced by permission of LogiXML.*

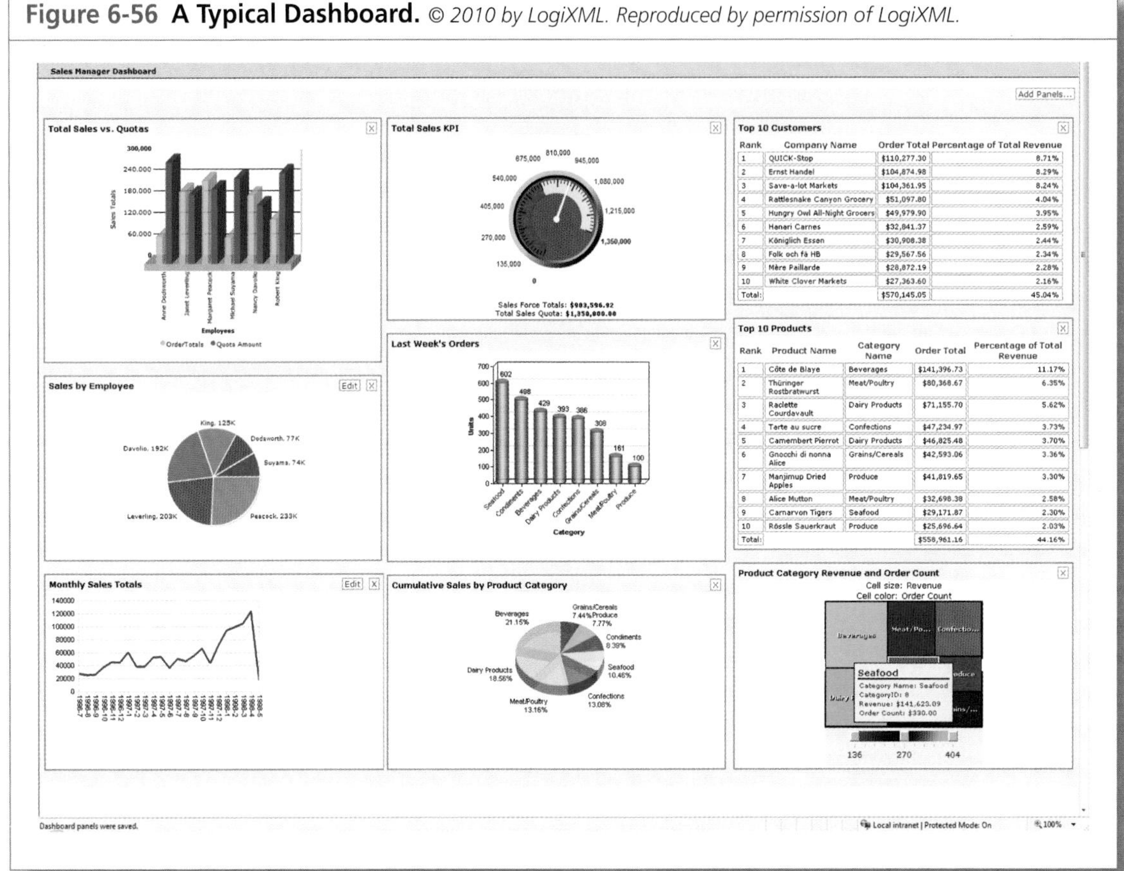

business data. Although it is possible to create complementary charts in most spreadsheets, this is a time-consuming, manual activity that lends itself to easily made mistakes.

Nonetheless, many business users stick with spreadsheets because they feel comfortable with them and are reluctant to change to another model. The reality is that not everything can be done efficiently in the spreadsheet, and one should not get stuck with them simply because that's what they have or have been using. This can lead to a situation where it's the limitations of the program—rather than business needs—that determine the scope of reporting and analysis.

With the right underlying technology, today's dashboards stand out from the spreadsheet, which nevertheless remains the most used BI interface today. Dashboards allow for a quick and easy-to-personalize overview of critical business data in a timely fashion. This added value turns today's dashboards into the new face of BI.

Designing the Dashboard

If you want to seek truth, you must at least at one point in your life doubt all things, as far as possible.

—Rene Descartes (1596–1650)

How does one design dashboards that live up to their purpose as efficient and actionable BI tools instead of merely being fancy playthings? The main answer is that anything presented in a dashboard must have a direct relevance to critical business activities. This means that the business user must be able to act on what is presented as seen in Figure 6–57. If no action is required or expected from the user, the business information presented, ultimately, serves little or no purpose.

Figure 6-57 For Direct Relevance to Business Activities, Business Users Must Be Able to Act on What Is Presented in a Dashboard. *© 2010 by LogiXML. Reproduced by permission of LogiXML.*

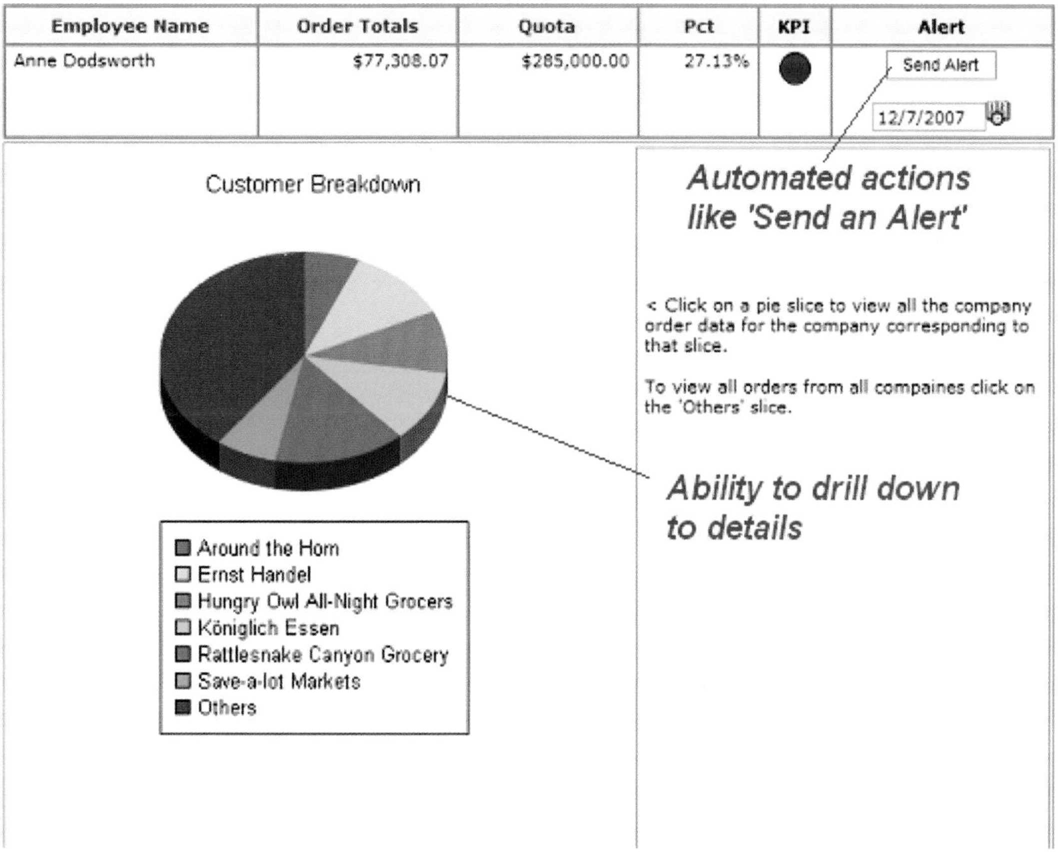

Employee Name	Order Totals	Quota	Pct	KPI	Alert
Anne Dodsworth	$77,308.07	$285,000.00	27.13%		Send Alert 12/7/2007

Customer Breakdown

Around the Horn
Ernst Handel
Hungry Owl All-Night Grocers
Königlich Essen
Rattlesnake Canyon Grocery
Save-a-lot Markets
Others

Automated actions like 'Send an Alert'

< Click on a pie slice to view all the company order data for the company corresponding to that slice.

To view all orders from all compaines click on the 'Others' slice.

Ability to drill down to details

The Business-Driven Dashboard

From the business users' perspective, the efficient use of a dashboard comes with a number of prerequisites. Delivering these requirements allows the dashboard to become an efficient support tool for driving the business.

The following are the main user requirements that need to be taken into account when designing a dashboard:

USER REQUIREMENT	DESCRIPTION
Easy access to information	So that there is little need for the business user to spend time preparing the information.
Standardized format of the information	Which facilitates the understanding of the reports and analyses presented.
Correct and comprehensible information	Meaning, for example, that all business definitions are clear, consistent and unambiguous.
Overview and detailed information	So that any exceptions or outliers can be quickly detected and further investigated. This information can be presented in managed reports that have been set up in advance to answer recurring business questions. Ad hoc reports and OLAP engines can also be used to access and further analyze, sort, group, and calculate this detailed information.
Spreadsheets	Because of their familiarity to most business users, they can certainly be useful for presenting detailed subsets of information.
Paper	I.e., the possibility to properly print reports and analyses. This point is overlooked in some dashboard solutions.
Color, charts, and key performance indicators (KPIs)	These greatly help getting the necessary overview of the state of the business, and drive attention to where performance falls short of expectations. While pie charts and bar charts are common and useful because they are relatively intuitive, heat maps are becoming more and more popular since they provide a better overview for more concurrent data dimensions. GIS Maps are also popular as they are intuitive and reduce the risk of misinterpreting geographical business information.
Ability to share information with colleagues	Which entails exporting information to file formats such as Adobe Acrobat (PDF), Word, PowerPoint, and Excel.
Ability to act	Meaning that the information presented needs to be actionable. If not, what is the user supposed to do with the information?

Overall, the dashboard needs to be easily adaptable to the user's needs. Furthermore, it must present information that allows the user to act where it matters to the business, which means that the dashboard must be business-driven rather than technology-driven. It is true that it is only with today's technology that dashboards can truly help change business behavior, but there must also be business brain and not only technological muscle when designing dashboards.

The Implications for the IT Provider

Given user and business requirements and preferences, the IT provider must design the dashboard according to guidelines that will ensure proper user acceptance. These guiding principles are often different compared to how IT develops applications for the day-to-day business operations.

The following are the design guidelines the IT provider needs to observe:

DESIGN GUIDELINES	DESCRIPTION
Customize the dashboards	Meaning that different interfaces must be adapted for different user needs, groups, etc.
Ensure easy administration	Allowing the dashboard to remain flexible to changing user demands, thereby becoming truly business-driven.
Automate what is regularly analyzed and reported	Do not force the user to spend time on repetitive tasks. This will also help assure user acceptance. Remember the unofficial office maxim: everyone wants to get more done with less work.
Don't discard familiar and widespread solutions	At least initially. This concerns especially spreadsheet programs. Therefore, a dashboard solution should leverage existing solutions and offer the capability to export the results to these widespread solutions.
Ensure plenty of support	So that use of dashboards can be efficient and trouble-free.
Give the possibility to share key information, business actions and their results	Otherwise, it will be easier to repeat mistakes, while good decisions will go largely unnoticed and will remain undeveloped.
Include KPIs that measure the important activities	And not simply the activities that are easily measurable.
Limit the information to what's necessary	Because having too many KPIs will drown the truly important ones.

Taking user preferences and business needs into account will ensure a business-driven dashboard solution. In combination with modern decisions support technologies, dashboards will become an essential part of business decision making.

Implementing the Dashboard

See first that the design is wise and just: that ascertained, pursue it resolutely; do not for one repulse forego the purpose that you resolved to effect.

—William Shakespeare (1564–1616)

Once the dashboard has been designed in accordance with business needs, IT capabilities, and business user demands, its implementation must begin. Implementing a dashboard is not the same as setting up operational IT systems. Dashboards are often new and unfamiliar to many business users. Furthermore, as few organizations have standardized the way they use BI solutions, dashboards must be iteratively customized to fit different individual business needs.

Organizational Challenges

Given the novelty of modern dashboards for many organizations, there tend to be some common implementation challenges. These challenges are not necessarily difficult to overcome, as long as they are properly identified.

Common organizational challenges when implementing a dashboard tend to be the following:

COMMON ORGANIZATIONAL CHALLENGES	DESCRIPTION
Little standardization when working with BI applications	Something that in turn results in a lack of a common business language.
Poor communication between different units	Following the lack of a common business language.
Few power-users	Who feel comfortable leading the work around BI applications.
No clear strategy	On what is expected from the dashboard solution.
Focus on cost	A narrow-minded focus on expenses often obfuscates the vision of the true value of dashboards.
Office politics	Which have a tendency to be reinforced in organizations with relatively strict boundaries between business units. These boundaries will be challenged when implementing dashboards, as business critical information becomes more widely available.

The challenges must be solved with a project management approach that emphasizes change-management activities. When implementing dashboards, change management is absolutely essential to achieve the necessary user acceptance.

Also, let's remember that dashboards may not be the only solution to analyze and act upon key business information. There are other older systems—frequently spreadsheets—with which business users feel comfortable. Consequently, there may not be a feeling among the users that the dashboards are essential, since there are other ways of getting the job done—even if these older methods are actually more time-consuming and inefficient.

Accordingly, flexibility and follow-up both from the IT and the business standpoint are imperative to succeed with the dashboard implementation.

Common Pitfalls

Besides common organizational challenges when implementing dashboards, there are other frequent pitfalls. These pitfalls are independent of the existing organization as they relate more to the behavior and attitudes of the individual business users and the IT provider.

Following are some typical points that will prevent successful dashboard implementations.

COMMON PITFALLS	DESCRIPTION
"Cool" trumps useful.	It is easy to get seduced by dashboards, but their actual business use must always be kept in mind. Looks must coexist with brains.
Users will come automatically.	Simply because the dashboard is such a hip BI tool. Any new solution needs marketing, which is where the change-management aspects are important.
The more advanced it is, the better it has to be.	Which might be true. But is it really worth the extra work to train the users on an advanced solution where they may not use more than a small fraction of all the possible features?
More is better.	Meaning that an abundance of KPIs are better than a few. Well, it is not.
IT-driven implementation	Where many user needs are underestimated or ignored by an IT department that takes the lead in implementing the dashboard.
User-driven implementation	Where business users do not take technology constraints or technology standards into account when pushing for a dashboard.
Little relation between strategy and action	Meaning that many business facts presented cannot be directly acted upon by the end user.
Little understanding of implementing dashboards	As the project approach is different from other more operational IT applications.
Return on investment (ROI) expectations	Which are sometimes greatly exaggerated to overcome resistance when a new project is implemented. Even though dashboards normally have a good ROI when used correctly, ROI will not come automatically.
Data quality	Which is the one thing that is constantly underestimated and because of that, the problem just won't go away. As a general rule, data quality is actually lower than anyone thinks; the result is that if this issue is not confronted, it can and will break the users' confidence in the dashboards.

These wrong perceptions and expectations need to be managed so that the dashboard can be of value. It should also be remembered that, after the dashboard is implemented, lack of further user demands does not mean that everything is fine. On the contrary, it probably indicates that the dashboard is not being used. User demands should be seen as something healthy and a proof that users work actively with the dashboard. A lack of demands typically indicates a lack of interest.

Justifying the Dashboard

> *They are ill discoverers that think there is no land, when they can see nothing but sea.*
>
> —Sir Francis Bacon (1561–1626)

Is it worth implementing a dashboard? Why not just go on with the existing spreadsheet solution? This seems to be cheaper for many cost-concerned managers who tend to see the expenses rather than the return when being asked to do something new.

Return on Investment

The expected return on investment (ROI) for IT applications is often the subject of heated debates. Admittedly, ROI can be difficult to estimate, at least initially, and even more so for dashboards than for other more operational IT applications.

The main reasons for the perceived difficulties of calculating ROI for Dashboards are typically:

DIFFICULTIES IN CALCULATING ROI	DESCRIPTION
No one-size-fits-all ROI model	That can be applied to dashboard solutions. With operational systems, there are usually standardized processes driving business operations. For BI, there are often no implemented standards. Consequently, there are no widely used methods for calculating ROI, even though there are documented approaches that can be used.
Lack of data	That will permit relevant benchmarking activities. This means that it might be seemingly hard to define the present costs of reporting and analyzing information and then compare it with the costs and returns of implementing a dashboard.
No clear business case	Explaining why the dashboard is being implemented. No matter how much a dashboard can help an organization, there must be a clear idea where and how it will be leveraged. Otherwise, it risks falling into the "cool trumps useful" pitfall we have seen.
Investment cost takes the upper hand.	I.e., investment cost becomes a major issue overshadowing eventual returns. This happens because costs are easier to calculate and understand than returns.

With these perceived obstacles in mind, a company can take an approach based on the premises that ROI calculations can actually be done, but that they will be more a guideline rather than exact science. ROI will never be correct the first time. It is something that must be recalculated on a regular basis before, during, and after the dashboard implementation. And when done regularly, the ROI calculation becomes an important tool toward ensuring that Dashboards provide lasting value.

Ensuring Service Level Agreements

There is, however, one area where cost and returns actually become second priority, and this is when ensuring Service Level Agreements (SLAs). SLAs make sure that IT services are provided according to agreed standards. For SLAs to become efficient, modern dashboards are pivotal to measure adherence to the SLAs. This situation is relevant to any organization having outsourced all or parts of its IT.

What is noteworthy is that a dashboard presenting SLA adherence tends to be operational in character, as opposed to the more common tactical or strategic dashboards. Important as operational dashboards may be, one must not lose focus on the goals of the business as a whole. Dashboards linked to the management of SLAs will further increase their value; that is, if the SLAs themselves are defined to truly support the execution of the company's strategy.

Conclusion

You see, but you do not observe.

—Sir Arthur Conan Doyle (1859–1930) (Sherlock Holmes)
A Scandal in Bohemia, 1892

Dashboards must be used. As obvious as this may seem, far too many organizations have spent great resources and much time on implementing something that ends up serving as little more than show-and-tell at monthly management meetings. No one present is really expected to do much once the meeting is over, apart from turning up for the next meeting.

Having a dashboard without using it shows a lack of management direction. It is also a waste of resources, given that today's dashboards can and should be powerful management tools where each and every user becomes empowered. Modern dashboards go further than the common BI solutions that are still to a large extent focused on standardized reports.

Dashboards today can deliver what users and the organization as a whole need to know so as to understand their current business and even conduct informed forecasts. In summary, good dashboards offer insight, explanations, and shared understanding of business critical information, and then allow the business user to act upon the information when and where necessary.

And this will put the organization among the best instead of the merely good.

6.14 A SIMPLE TEMPLATE

Table 6–6 shows a very simple template that can be used in the creation of a metric/KPI library. Rather than call it a KPI library, it may be referred to as a metric/KPI library because not all metrics are KPIs and KPIs on one project may serve as a simple metric on another project. Also, in the template there may be a reference to various parts of the *PMBOK® Guide*.

The real purpose of the template is to record which metrics have been used successfully. We know that humans absorb information in a variety of ways and the metrics must always undergo continuous improvement efforts. Metrics that work well for one company may not work equally well for another company. The images, colors, and shading techniques may also have to be changed.

Organizations must be cautious not to go overboard in the number of metrics that will appear in the library. Too many metrics may be overwhelming to the users. Rad and Levin provide excellent checklists for assessing what may or may not be important as a metric on a given project or a stream of projects.[13] Items considered as critical for a given project or project stream may mandate that measureable metrics be defined.

6.15 SUMMARY OF DASHBOARD DESIGN REQUIREMENTS

We can now summarize the design requirements for dashboards using the rules for design. The remaining information has been graciously provided by Hubert Lee.[14]

The Importance of Design to Information Dashboards

The success of a project management dashboard (or any information dashboard for that matter) lies in the adoption and use of the dashboard by the end user as a truly helpful tool. Successful dashboards are those used

---

13. Parviz Rad and Ginger Levin, *Metrics for Project Management*, 2006, Vienna, VA: Management Concepts, Vienna, Va.

14. Hubert Lee is the founder of Dashboardspy.com. Known in business intelligence circles as "The Dashboard Spy," Hubert Lee is an expert at business dashboard design and user experience. His collection of dashboard examples is among the largest in the world and can be viewed at http://dashboardspy.com.

TABLE 6-6

Description:____

____

____

KPI Advantages:____

____

KPI Limitations:____

____

KPI Sponsor:____

KPI Owner:____

***PMBOK® Guide* Domain Area:**

- ☐ Initiating
- ☐ Planning
- ☐ Executing
- ☐ Monitoring and Controlling
- ☐ Closing
- ☐ Professional Responsibility

***PMBOK® Guide* Area of Knowledge:**

- ☐ Integration Mgt.
- ☐ Scope Mgt.
- ☐ Time Mgt.
- ☐ Cost Mgt.
- ☐ Quality Mgt.
- ☐ Human Resource Mgt.
- ☐ Communication Mgt.
- ☐ Risk Mgt.
- ☐ Quality Mgt.
- ☐ Stakeholder Mgt.

Relationship to CSF:____

Objective/Target Limits:____

KPI Start Date:____

KPI End Date:____________________

KPI Life Span:____________________

Reporting Periodicity:____________________

Graphic Display:

- ☐ Area Chart: Clustered
- ☐ Area Chart: Stacked
- ☐ Area Chart: 100% Stacked
- ☐ Bar Chart: Clustered
- ☐ Bar Chart: Stacked
- ☐ Bar Chart: 100% Stacked
- ☐ Bubble Chart
- ☐ Column Chart: Clustered
- ☐ Column Chart: Stacked
- ☐ Column Chart: 100% Stacked
- ☐ Gauges
- ☐ Grids
- ☐ Icons
- ☐ Line Chart: Clustered
- ☐ Line Chart: Stacked
- ☐ Line Chart: 100% Stacked
- ☐ Pie Charts
- ☐ Radar Chart
- ☐ Table
- ☐ Other:__________

Measurement Method:

- ☐ Calibration
- ☐ Confidence Limits
- ☐ Decision Models
- ☐ Decomposition Techniques
- ☐ Human Judgment
- ☐ Ordinal/Nominal Tables
- ☐ Ranges/Sets of Values
- ☐ Sampling Techniques
- ☐ Simulation__________
- ☐ Statistics__________
- ☐ Other:__________

every day to inform, manage, and optimize business and project processes. Failed dashboards are those that lie unused and forgotten, or provide no meaningful information.

So the big question is this: On your dashboard, what makes the difference between success and failure? How will you ensure successful adoption of your dashboard by your users?

We all know of spreadsheets that have been derided and cast aside. They may contain all the data elements you need to monitor (and then some!), yet these spreadsheets are just not used as much as they should. Why is that? And why is it that if you pull the key metrics out of the same spreadsheets, lay them out on a properly designed dashboard, you suddenly have the attention of the entire user community? We'll examine several factors critical to the design of your dashboard.

The Rules for Color Usage on Your Dashboard

A truly effective dashboard makes good use of color to display information in an easily understood manner. Color theory and the cognitive effects of color are subjects close to the hearts of visual artists but seldom appreciated by creators of dashboards and other user interfaces. We must always be careful as the poor or careless use of color can mangle the real message of the data.

No discussion of color would be complete without a quick word on some fundamental warnings about the use of color in your dashboard projects. It is estimated that up to 8 to 12 percent of the male population suffers from some form of color blindness. Take a look at this series of graphics in Figure 6–58 that show how the colors of the rainbow appear to people with various forms of color blindness:[15]

Note how similar yellow and green can be to some color -blind people and also how similar red and green can be to other color-blind people. Now isn't that eye-opening?

The other classic warning about colors has to do with black and white versus color printers. It is still uncommon at many offices to find a color printer. Many users send their dashboards to the office printer so that they can study the hardcopy later. Most of those copies will be in gray scale.

So, what do we do about these basic color challenges? Do these limitation mean we should not use color on our dashboards? Of course not. It does mean, however, that we must always use color in conjunction with text labels. By that I mean explicitly including the relevant label right on or next to the graphic. For example, if you are showing a red/green/yellow status indicator graphic, put the text value right next to the graphic.

15. Diagram is from Wikipedia: http://en.wikipedia.org/wiki/Color_blindness.

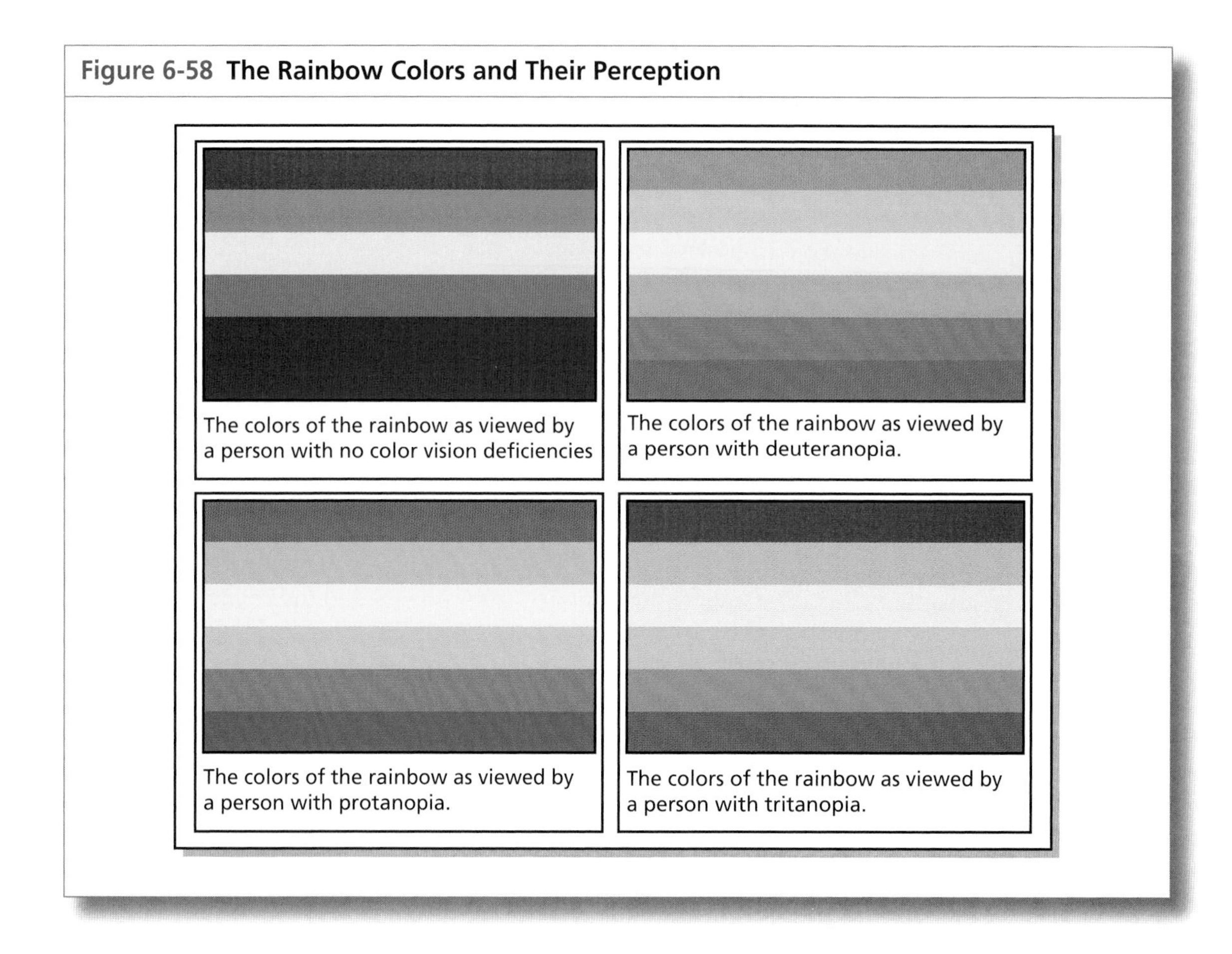

Figure 6-58 The Rainbow Colors and Their Perception

With that basic warning out of the way, let's look at some critical rules to follow concerning the use of color in your dashboards:

- Be aware of the background color of tables and graphs. Use a background color that contrasts sufficiently with the foreground objects. Also, use background color to group and unify different objects.
- Components of graphs and tables that are "non-data," that is, structural elements, should not call attention to themselves. Conversely, the data-centric elements of charts should be highlighted with color.
- When displaying a sequential or related group of metrics, use a small set of related hues and vary the intensity to correlate with the increasing data values if possible.
- Try to make the color usage meaningful. Use different colors to show different meanings.
- Use color consciously—check with yourself that each color (or the particular usage of colors) is meaningful.

Note that the above guidelines refer to the sections of your dashboards that contain data. The rest of the dashboard (page background, navigation elements, branding areas, headers and footers may be designed to accommodate your usual graphic look and feel.

The Rules for Graphic Design of Your Dashboard

The level of design appropriate to a dashboard (or any information visualization) is an age-old debate. Good information visualization practice and business intelligence user interface design calls for effective use of screen real estate with clear and easily understood charts and visually simple charts such as sparklines and bullet graphs. Easy to read and interpret, these charts, combined with a monotone color scheme (so as to focus attention when needed with red/green/yellow alert indicators) have become a "best practice" design for data-heavy business intelligence dashboards.

With its sparseness of color and avoidance of heavy design elements, this approach allows the user to concentrate on the information aspects. However, as dashboard project members often find when embarking on the design phase of the project, such a "sparse" interface often runs into the objections of the "eye-candy people"—often (but not always!) including the project sponsor, who is looking to impress users and peers with visual splash.

Many people want the "sizzle" of gaily colored gauges, complete with 3-D effects, gradients and big splashy presentation. Maybe it "feels" like more of a user experience when you get the "wow" factor in there!

From a development standpoint, libraries of visual components make it very simple to throw all sorts of charts, gauges, dials, and widgets onto a dashboard. It makes it too easy to get carried away by the visual excitement. It becomes the responsibility of the dashboard designer to strike the proper balance between visual excitement and information visualization best practice.

One proven formula is to limit the efforts of the eager graphic designer to clearly defined areas. The following list is usually within the purview of the graphic designer.

Overall layout of the website or software application:

- Header banner
- Logo
- Portlets (widgets)
- Title bars
- Tables
- Sidebars
- Navigation elements
- Footer

Reserve the design of the charts and other data visualization elements for the information visualization expert.

Other tips for the graphic design of your dashboard:

- Avoid the cliché of using a steering wheel—don't take the dashboard metaphor literally! An automobile's steering wheel has no place on your information dashboard.
- Bring a design professional into your team. It is uncommon to find design professionals already on your staff. Hire a proven consultant with a track record of designing dashboards. Find a visual designer with expertise in the layout and design of software applications and pair him/her with an expert information visualization professional. The look and feel of your dashboard will reflect the investment.
- Become aware of current trends in the visual design of software applications. Understand why some looks are considered "modern" and "cutting-edge" and adopt those practices to give your design a longer life.

The Rules for Placing the Dashboard in Front of Your Users—the Key to User Adoption

The most successful dashboards become invaluable tools that users rely on to facilitate their workflows, remind them of what they have to concentrate on, and guide them through metrics and KPI relevant to their changing situations.

To get your dashboard successfully adopted to such a level requires you to carefully think about how the users engage with your system in the first place. From where and exactly how do the users launch your dashboard? Typically, a dashboard is launched by typing the URL of the application into the address bar of a browser. While a common approach, more proactive steps can be taken that can lead to increased adoption by the user community. Here are some ideas:

- Identify launch points within existing company portals. Place branding elements (banner/logo) and provide a "sign in" button right in the sidebar of other applications or web sites.
- Speak with other application owners and insert launch points for your dashboard right into the other application. The idea is to place it at the appropriate portion of the user workflow.
- Consider using the "desktop widget" approach. Have you seen those mini-applications that your operating system displays on your desktop? Some dock on the side of the screen and pop out when you mouse over them. Others reside in full view and allow you to flip them over for more information. Create a widget for your dashboard and have it live on the desktop of your PC. Now all your user has to do is to log onto their PC and they have instant access to their dashboard. No logging in to a separate application or website. This approach is particularly good for delivering messages and alerts to the users.

- Give your dashboard email capability. Send users alerts and updates by email with links that they can click on to go right into their dashboards.

The goal is to make your dashboard, and the information that it delivers, absolutely unavoidable by the user. Don't be afraid to put your dashboard right in the face of your users. After all, that's what is being done by the automobile dashboard and the airplane cockpit. The information and the controls are literally in front of the users' faces.

The Rules for Accuracy of Information on Your Dashboard

This rule is easy to understand and absolutely inviolate: Your dashboard must never be seen as the cause of any problems with the data. The accuracy of the data on your dashboard can be questioned (unfortunately that's sometimes the nature of data)—BUT the dashboard itself must be seen as a trusted reporting mechanism. If there is a problem with the data, the user should blame the team responsible for producing the data, but not the dashboard itself.

If there is a known issue (perhaps a technical problem, or a stale data problem), the dashboard should display a warning message. If the dashboard does not show live, real-time data, but snapshots captured from certain points in time, the tables and charts should be labeled to indicate that.

You must work hard at instilling faith and trust in your dashboard users. Transparency in process, and clear labeling and directions must be used in the dashboard design process for this to happen.

6.16 DASHBOARD LIMITATIONS

Dashboards are designed first and foremost with the viewer in mind. The intent of the project management dashboard is to convert data into a meaningful representation of the project's performance. But how does the viewer intend to use the information presented?

Viewers and stakeholders that are expected to make decisions in support of the project want sufficient data such that they can make "informed" decisions in a timely manner rather than just guesses. Having the correct information is critical. In this case, the dashboard may have to contain detailed information and the viewer may need to examine the information carefully. The amount of information that can be displayed accurately on a dashboard is limited due to space and readability requirements.

However, some stakeholders and viewers that are passively involved in the project and are not expected to make decisions are usually happy with a one-minute or at-a-glace dashboard. This type of dashboard has limitations on both the information presented and its intended use. The one-minute dashboard can be displayed using the three directional icons shown in Figure 6–59. In Figure 6–59, a green arrow always points upward and indicates a favorable condition. For example, if the one-minute dashboard is used to display how well we are managing the project according to the project's constraints, then the green arrow would indicate that we are well within the constraint's limits. A red arrow, which always points downward, indicates that we have an unfavorable condition and have exceeded the limits of the constraint. A yellow arrow might indicated that we are within but close to the threshold limits or tolerances for the constraint. Some people prefer to use the "traffic light" or circular icons rather than the arrows.

In Figure 6–60, we have a project that we assume has only five constraints. For each of the five constraints, we have identified an icon that shows how well we are performing according to the predetermined constraints. This type of one-minute dashboard provides only a cursory picture of the project's health and has limited value for decision making. If the viewers want additional information, they may need to use drill-down buttons for more detail. Dashboards cannot display all of the information needed in most circumstances.

Sometimes, good intentions often go astray. One company created both one-minute dashboards and detailed dashboards for their viewers. In the one-minute dashboard, they decided to go to nine colors, as shown in Figure 6–61, rather than just the three standard colors that they had used previously. This made the dashboard significantly more complex because

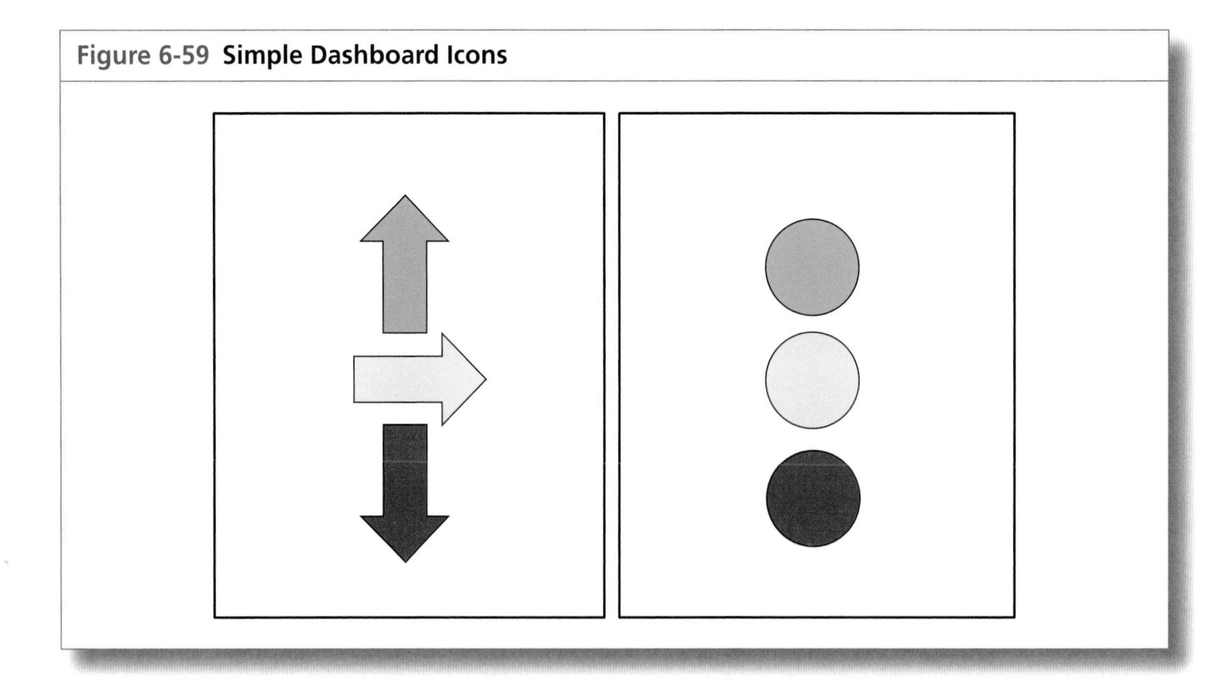

Figure 6-59 Simple Dashboard Icons

Figure 6-60 The At-A-Glance Dashboard for Constraints

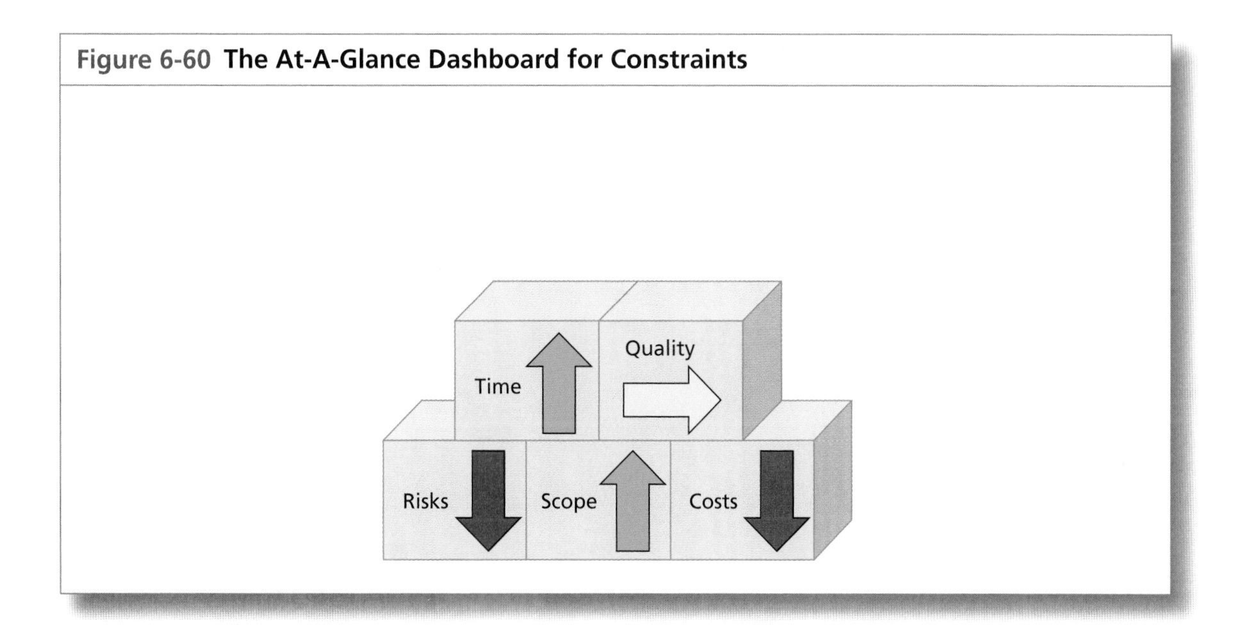

Figure 6-61 Multi-Colored Status Reporting

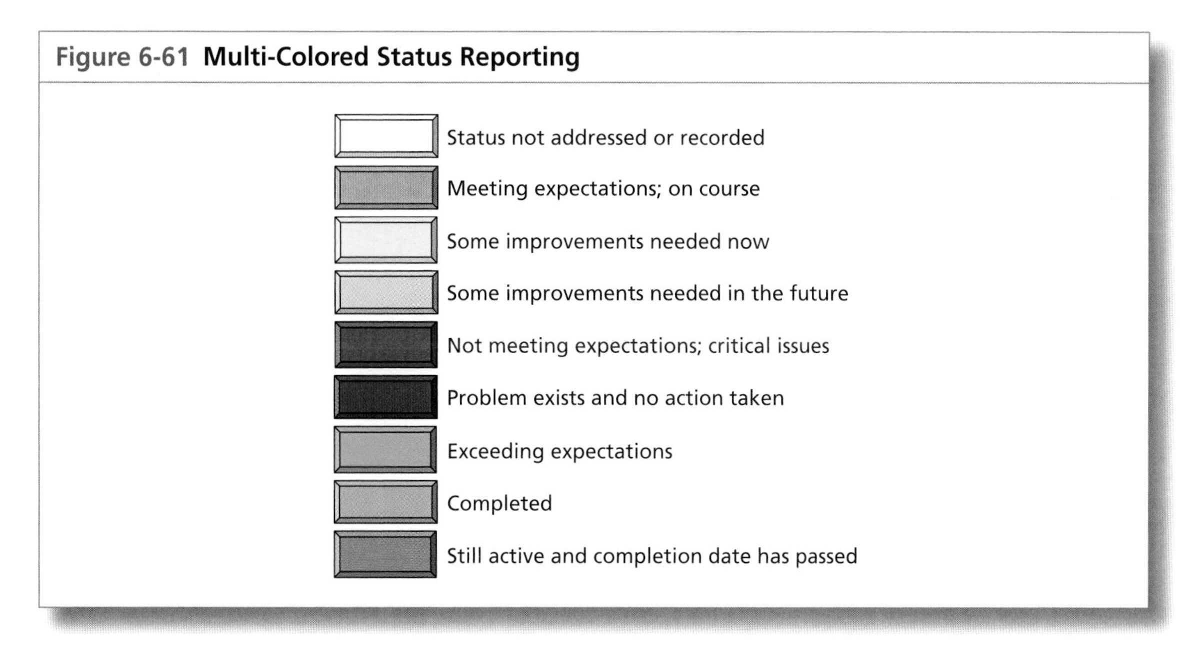

there could be more than one color associated with each activity. For example, an activity could be both red and purple, indicating that the activity has some unresolved critical issues and is still active. It then took longer to read the dashboard because the viewer had to continuously review the meaning of the colors.

Figure 6-62 Color-Coded Variance Reporting

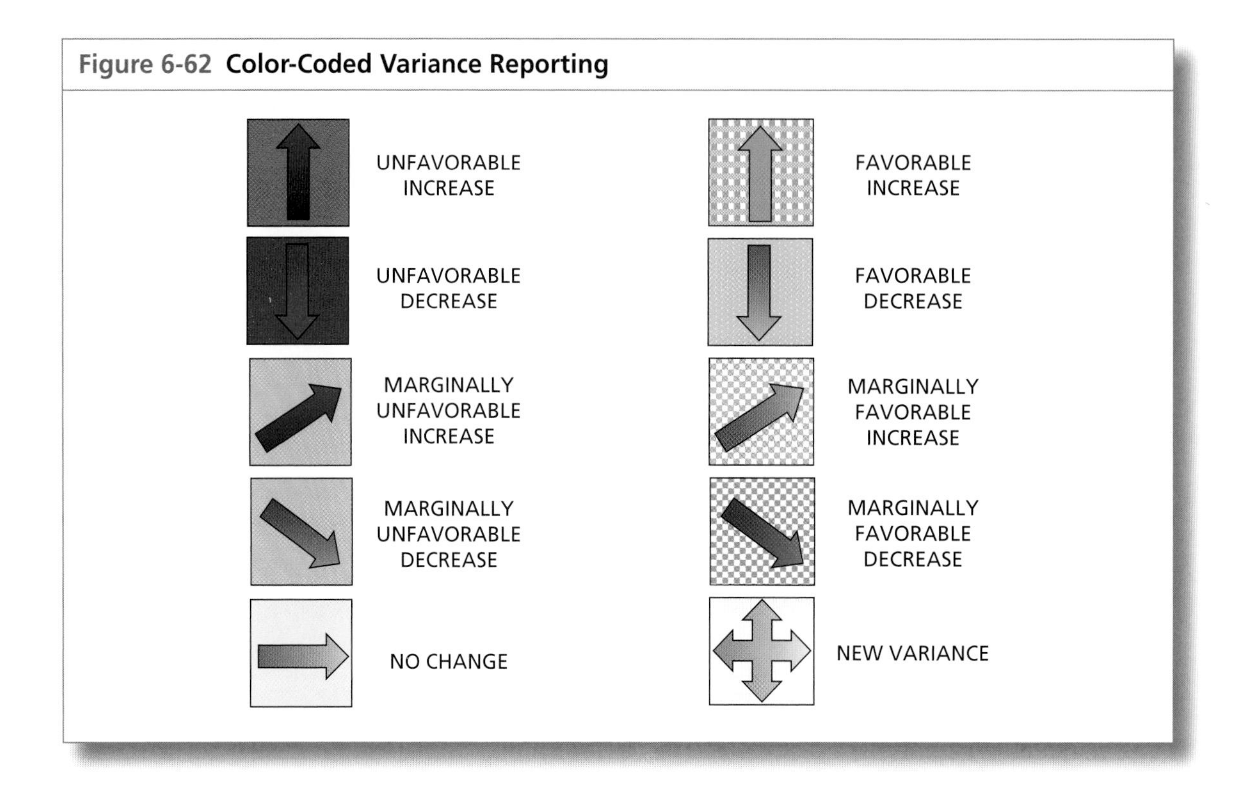

Some of the viewers wanted to see the direction of the variances, specifically time and cost, and the company then added into the dashboard the color scheme shown in Figure 6–62. While there is some merit in using the style shown Figure 6–62, it is inappropriate for the one-minute dashboard. Eventually, the one-minute dashboard was removed and replaced with detailed dashboards that displayed a rainbow of colors and gradients.

6.17 THE DASHBOARD PILOT RUN

While the project manager, the project team, and the dashboard designer have a reasonably good understanding of the information that appears in the dashboard, the same cannot be said for the viewers of the dashboard. If possible given the project's schedule and the size of the project, setting up a dashboard pilot run may be beneficial. It verifies that the client and the stakeholders:

- Understand what they are viewing
- Arrive at the correct conclusions
- Have faith in the dashboard concept
- Are willing to use the information for decision making

It may be necessary for the first few weeks of the project for the project manager and the dashboard's viewers to work together closely to make sure that all of the dashboard information is clearly understood. A great deal of time and money can be wasted if the viewers lose faith in the dashboard concept. Regaining lost faith may be impossible.

6.18 EVALUATING DASHBOARD VENDORS

With numerous companies in the marketplace selling dashboard services, it is important to what how to evaluate both the dashboard features and the accompanying software. Malik has identified 10 categories that must be considered when evaluating dashboard vendors:[16]

- End-user experience
- User management
- Drill-down
- Reporting
- Data connectivity
- Alerts
- Visualization
- Collaboration
- System requirements
- Image capturing and printout

Malik has also identified rules for good dashboard software:[17] It is worth noting that dashboards software must also meet the standards of any good software, which includes the following:

- **Fast response:** Users should not experience an inordinate delay in retrieving their dashboards and associated reports.
- **Intuitive:** End users need not be required to go through a big learning curve or mandatory training.
- **Web based:** Users should be able to access the dashboards through the web, if they have proper access rights.
- **Secure:** System administrators may administer software security easily to reduce and track wrongful access. The software must also provide (when necessary) data encryption to secure sensitive data transmission across the web.
- **Scalable:** A large number of users may access the software without crashing the system or causing it to slow down below an acceptable

16. Shadan Malik, *Enterprise Dashboards*, Hoboken, NJ: John Wiley & Sons, 2005, p.147.

17. Ibid; pp. 9–10.

performance benchmark. The quality assumes a reasonable hardware and network bandwidth.

- **Industry compliant:** The software should integrate with the standard databases of different vendors and work with different server standards (e.g., .Net, J2EE) and various operating systems (e.g., Unix, Windows, Linux).
- **Open technology:** The software should not have proprietary standards that would make it difficult or impossible to extend its reach within a complex IT environment.
- **Supportable:** It should be easy to manage a large deployment within the existing IT staff with limited training on dashboard software. In other words, the software should not be so complex that it requires long-term contracting or the hiring of another expert to support its deployment, assuming that the organization has a reasonably qualified IT staff.
- **Cost effective:** The total cost of ownership should be well below the monetary benefit it provides to justify a strong return on investment (ROI). Therefore, the licensing cost, implementation cost, and support cost should be within a range that provides strong ROI and organizational benefits after deployment.

The main cost factors that should be considered for a successful dashboard solution are:[18]

- Software cost
- Annual support cost
- Additional hardware cost
- Initial deployment cost
- User training cost
- Ongoing support personnel cost

TABLE 6-7 Potential Cost Savings and Business Opportunities

COST SAVINGS	ADDITIONAL BUSINESS OPPORTUNITY
• Reduction or elimination of efforts for consolidating disparate reports • Reduction of time wasted in reviewing overwhelming amount of data and reports • Reduction of time spent coordinating and monitoring complex processes Reduction of effort enforcing regulatory compliances • Elimination of redundancies within the organization for processing of similar data	• Better decision making with more current or live information • Better business insight because of improved data visibility through enhanced visualization • Proactive and timely decision making with exception management and alerts • Greater democratization of information, empowering the front line in the organization • Better customer service and enhanced value delivered to customers and/or vendors

18. Ibid; p.168.

Software purchases require and understanding of the cost savings and business opportunities. This is shown in Table 6–7.[19]

6.19 NEW DASHBOARD APPLICATIONS

Because dashboards can be updated in real time, new dashboard applications have emerged. Companies are using dashboards as part of capacity planning optimization analyses. This is similar to "what if" scenarios used in project scheduling.

With optimization dashboards, a company can look at various ways of assigning resources to either a single project or a multitude of projects. For each change in the way that the resources are allocated, the dashboard will display the impact on the schedules of each project and possibly profitability. These applications are now being used as part of capacity-planning analyses during the portfolio project selection and project prioritization processes. Unfortunately, optimization dashboards focus on the higher levels of the work breakdown structures. As technology advances, these techniques may become commonplace for functional managers as well to perform functional resources optimization.

19. Ibid; p.172

7 DASHBOARD APPLICATIONS

CHAPTER OVERVIEW

Not all dashboards have the same intended use. Some dashboards are for internal use only, whereas others serve as a means of customer communication. You cannot always discern a dashboard's use by looking at it. This chapter contains dashboards from several companies that have recognized the benefits of their usage.

CHAPTER OBJECTIVES

- To understand the various usages of dashboards
- To identify the type of material in each dashboard
- To understand the application of the dashboards

KEY WORDS

- Dashboard design
- Dashboard usage
- Dashboard contents

7.0 INTRODUCTION

Every company has its own use for dashboards. Some of the factors that influence the final design and usage of the dashboards might include:

- Internal versus external usage
- Summary versus detailed information
- Drill-down capability if necessary
- Summary information versus information for decision-making purposes
- Executive levels versus stakeholders' levels versus working levels

In the remaining sections are examples of dashboards that have been graciously provided by various companies.

7.1 DASHBOARDS IN ACTION: VENTYX, AN ABB COMPANY

In recent years, there has been a rapid growth in the number of companies developing expertise in dashboard design techniques and graphical displays of business intelligence. These companies often have the knowledge and capability to construct effective dashboards for clients through a much less costly approach than if the company had to do it by themselves and develop their own expertise. While these companies understand the necessity to custom-design the dashboards to fit the needs of the client, they also try to prevent a "reinvention of the wheel" by first looking at what dashboards worked well for other clients and seeing if any or all of the information in or design of the dashboard can be applied to new clients.

The same holds true for the metrics used in the dashboards. Sometimes, the same metric can be represented differently on two different dashboards, especially if the information is for different clients or a different audience within the same company. There may also be a valid reason to display the same metric differently on a single dashboard. This is shown in the dashboards in Figures 7–1 through 7–5.

According to Paul Bower,

> What we try to provide through our dashboard paradigm is a very rich, multi-dimensional perspective of the metrics being presented. This is intended to provide the end user with a clear picture of where goals and targets are being achieved or not, with the ability to quickly navigate through the data relative to performance. There are many examples where a metric may look fine from one perspective, but display serious problems when looked at from another. This is why you will see multiple charts or other web parts on our dashboards depicting the same metric by different dimensions. These web parts interact with each other allowing the user to interactively filter the data along the pathway of their choice supporting activities such as root cause analysis, performance benchmarking and trending analysis.[1]

7.2 DASHBOARDS IN ACTION: JOHNSON CONTROLS, INC.

Companies invest a great deal of time and money creating dashboards. If the dashboard can be used both internally and for customer presentations, the company can maximize the value and benefits of the dashboard. While this is true, some companies use dashboards for internal control only rather than sharing them with the clients because the dashboards may contain internally sensitive material. In such a case, the same dashboard format may be used on a multitude of projects and, when a best practice is discovered on one project, the change may be made to the dashboards of all similar projects.

1. Paul K. Bower, SVP & GM Advanced BI Solutions, Ventyx, an ABB company. A Leading Business Solutions Provider Offering Software, Data, and Advisory Systems. Visit us on the web at www.ventyx.com.

Figure 7-1 A Sample Ventyx, an ABB Company, Dashboard

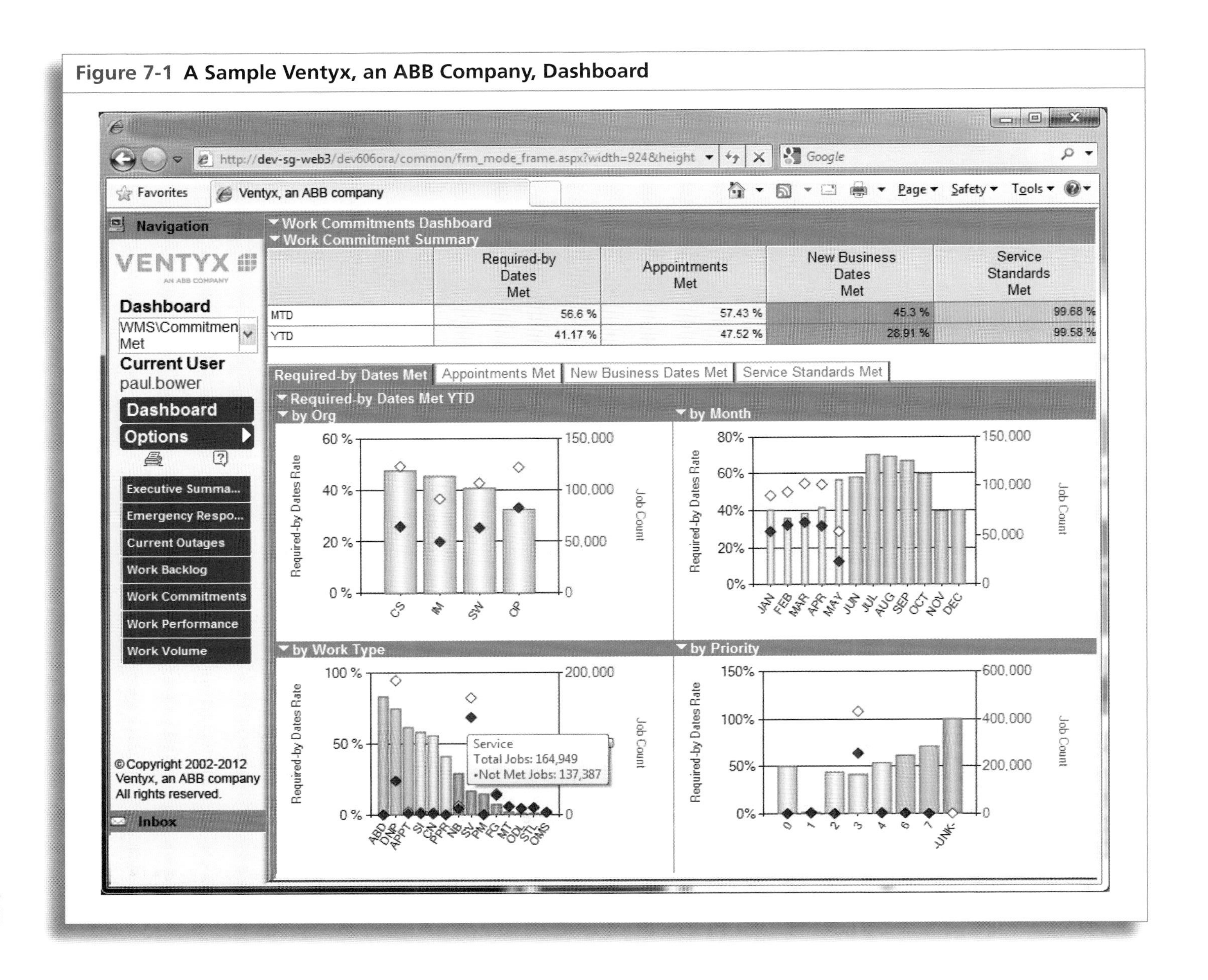

Figure 7-2 A Sample Ventyx, an ABB Company, Dashboard

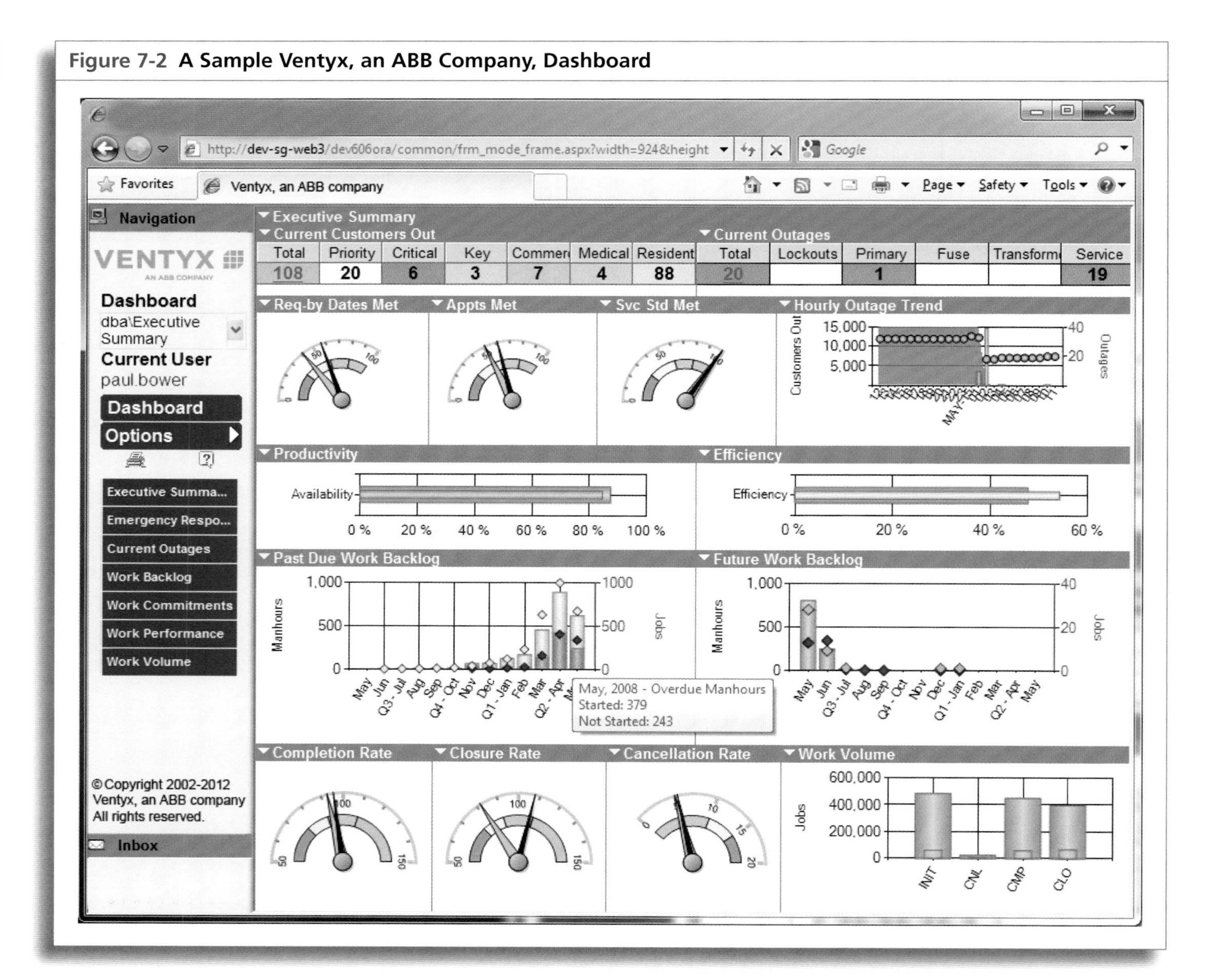

Figure 7-3 A Sample Ventyx, an ABB Company, Dashboard

http://dev-sg-web3/dev606ora/common/frm_mode_frame.aspx?width=924&height

Google

Favorites

Ventyx, an ABB company

Page ▾ Safety ▾ Tools ▾

Navigation

VENTYX

AN ABB COMPANY

Dashboard

VLT\Location Tracking

Current User

paul.bower

Dashboard

Options

Executive Summa...

Emergency Respo...

Current Outages

Work Backlog

Work Commitments

Work Performance

Work Volume

©Copyright 2002-2012 Ventyx, an ABB company All rights reserved.

Inbox

Vehicle Location Tracking

Accuracy | Exceptions | Day View

	Status Exceptions				GPS Exceptions			
	Total Jobs	Exception Jobs	Enroute to Arrival	Arrival to Completion	Arrival Exceptions	Completion Exceptions	Long Duration Stops	Unscheduled Stops
04/30/2009	1,150	321	144	167	126	119	12	0
Last 7 Days	3,131	400	164	206	153	143	40	0
Last 30 Days	3,131	400	164	206	153	143	40	0
Last 365 Days	25,486	6,559	2,814	3,989	1,159	1,048	159	355

Status Exceptions [Yesterday]

Exceptions: 0, 5, 10, 15; Jobs: 0, 20, 40, 60

Butler, Robert L
Arrival Duration Exceptions: 8

Bratton, Linwood K; Butler, Robert L; Chew, Charles E; Collier, Sherry C; Crisp, Robert Edward; Cross, Nathan Aaron; Escoe, Joseph L; Henkel, David C

GPS Exceptions [Yesterday]

Exceptions: 0, 5, 10, 15, 20; Jobs: 0, 20, 40, 60

Bratton, Linwood K; Butler, Robert L; Chew, Charles E; Collier, Sherry C; Crisp, Robert Edward; Cross, Nathan Aaron; Escoe, Joseph L; Henkel, David C

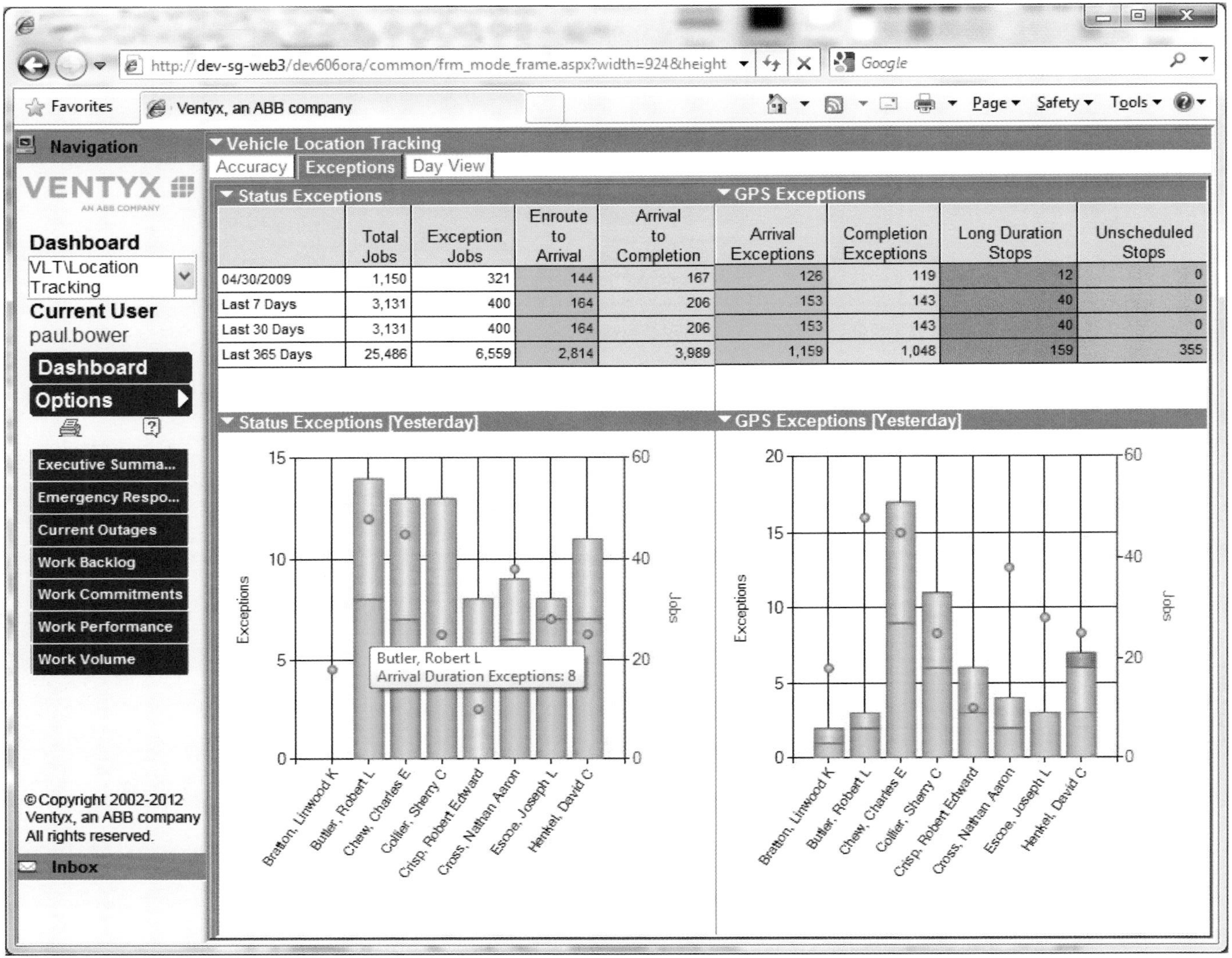

Figure 7-4 **A Sample Ventyx, an ABB Company, Dashboard**

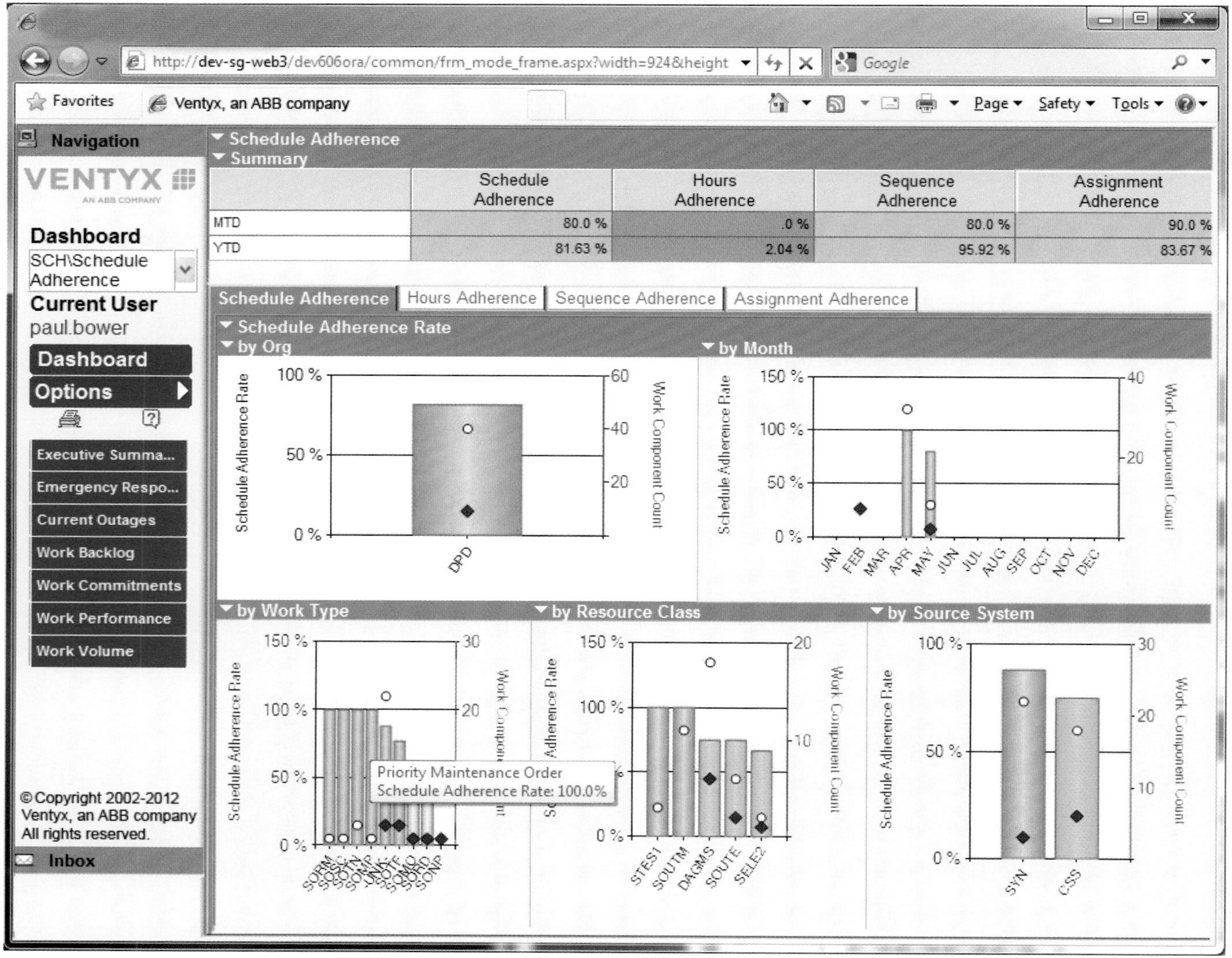

	Schedule Adherence	Hours Adherence	Sequence Adherence	Assignment Adherence
MTD	80.0 %	.0 %	80.0 %	90.0 %
YTD	81.63 %	2.04 %	95.92 %	83.67 %

Figure 7-5 A Sample Ventyx, an ABB Company, Dashboard

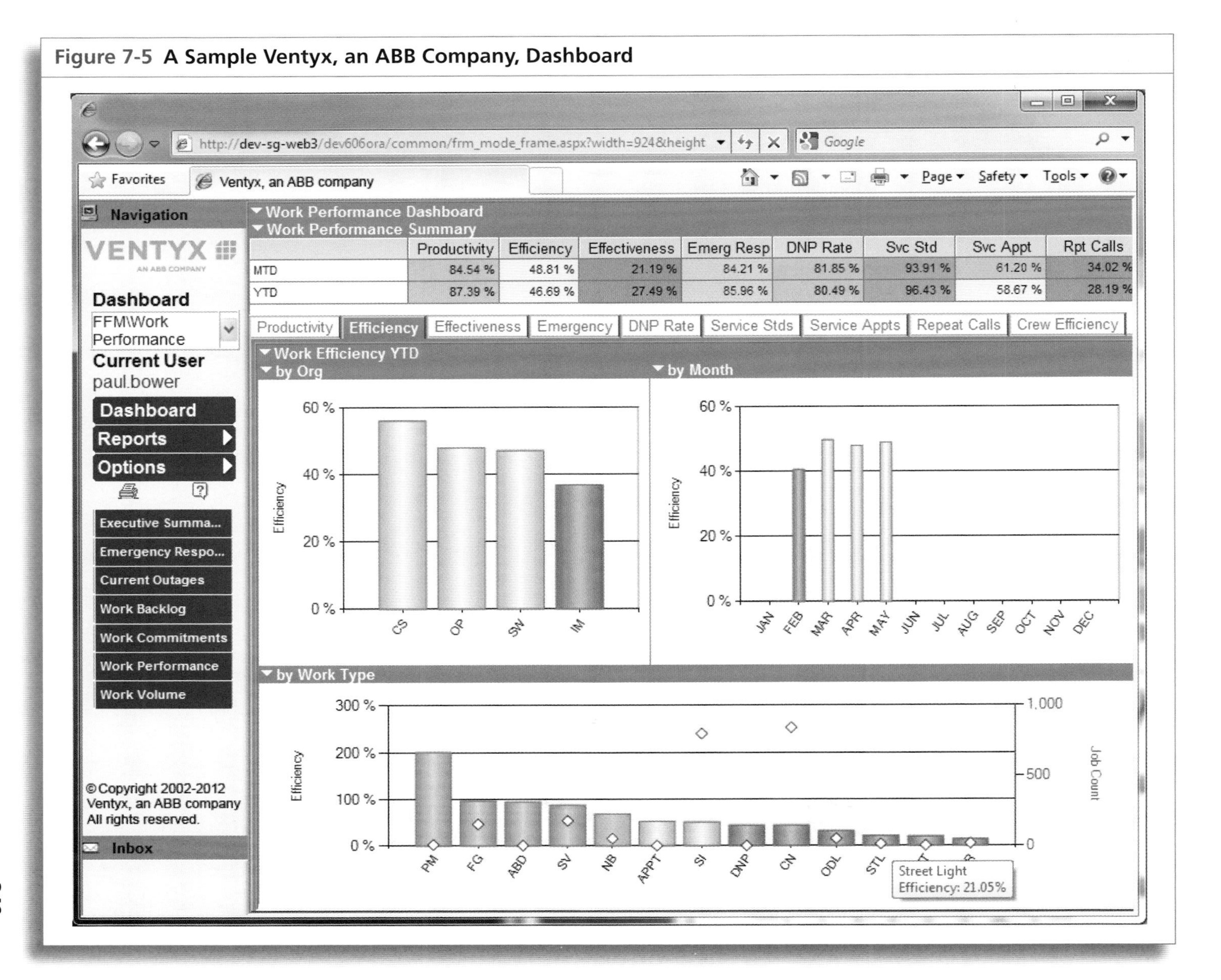

Figure 7–6 show a dashboard used internally at Johnson Controls. Some of the information in the dashboard has been concealed because of its sensitivity. According to Terri Pomfret, D.M., PMP, director of technical training, Automotive Experience Leadership Institute, Johnson Controls, Inc.

> Johnson Controls has found Gate Timeliness to be critical to program success. Gate Timeliness has to do with the completeness of deliverables at various stages of the program. Monitoring timeliness—overall status of deliverables—on a monthly basis highlights potential at-risk programs. Leaders use this information to focus attention and resources on the problem; mitigating and/or eliminating issues before they become significant.
>
> Monitoring Gate Timeliness also helps Johnson Controls continuously improve program management excellence. Timeliness provides trend information that can highlight chronic problems across programs. Such problems are generally addressed through methodology adjustments and/or training.

7.3 DASHBOARDS IN ACTION: COMPUTER ASSOCIATES, INC.[2]

Introduction

CA Technologies utilizes project dashboards to display project and business metrics, view project statuses, and directly access individual project content to view details and/or take action. These configurable dashboards provide convenient status of a group of projects, so that those in need of attention can be readily identified and action taken.

CA Technologies utilizes CA Clarity™ PPM to monitor and manage large numbers of projects and associated resources by rolling up detailed data, which is presented as project metrics. Management and project managers can drill down from rolled up data presented in dashboards into the actual projects to view details and take needed action.

Dashboards are customized displays of real-time project data that support filtering, sorting, and drilling down to underlying data. A number of standard dashboards are utilized to manage all service engagements, but users can also create customized or ad hoc dashboards. Through user-configurable screens, data selection criteria, and thresholds, end users can create dashboards they need regardless of their role within the organization. This means resource managers, project managers, business managers, as well as finance have dashboards that address their unique needs.

2. All of the material in this section has been graciously provided by Mark Elkins, Tom Pengidore, Rob Zuurdeeg and Jon Price, all of CA Technologies.

Figure 7-6 Gate Timeliness Dashboard *Johnson Controls*

Gate Timeliness

Month of May 2010

Johnson Controls

Risk (Lateness Open Gates)

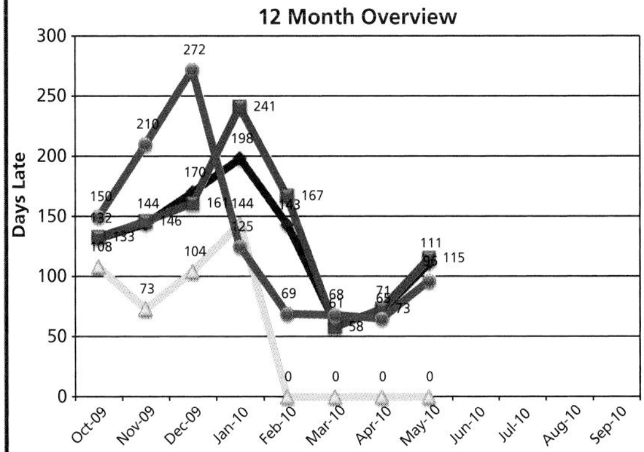

12 Month Overview

Days Late

Oct-09 Nov-09 Dec-09 Jan-10 Feb-10 Mar-10 Apr-10 May-10 Jun-10 Jul-10 Aug-10 Sep-10

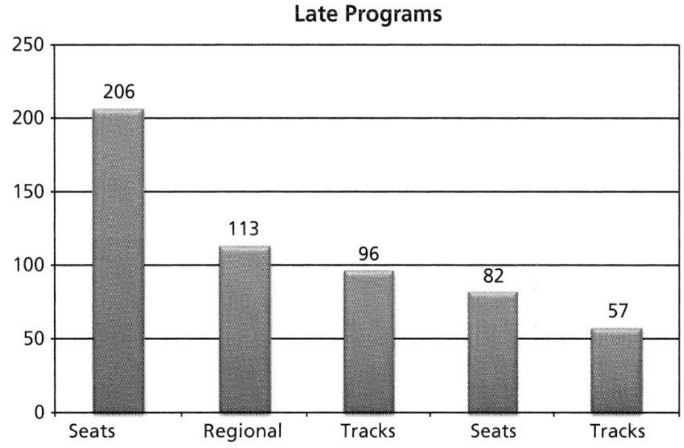

Late Programs

Seats	Regional	Tracks	Seats	Tracks
206	113	96	82	57

Performance (Lateness Closed Gates)

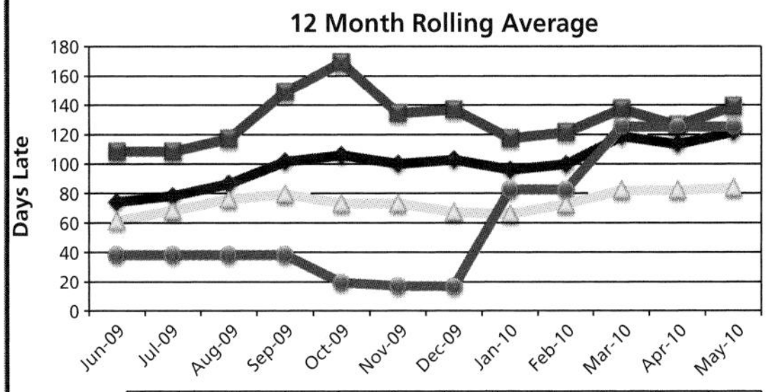

12 Month Rolling Average

Days Late

	Jun-09	Jul-09	Aug-09	Sep-09	Oct-09	Nov-09	Dec-09	Jan-10	Feb-10	Mar-10	Apr-10	May-10
Total	74	78	87	102	106	100	103	96	100	119	113	122
Seating	109	109	117	149	169	135	137	117	122	137	126	139
Interiors	62	68	76	80	73	73	68	66	73	82	82	84
Electronics	39	39	39	39	20	17	17	83	83	126	125	125

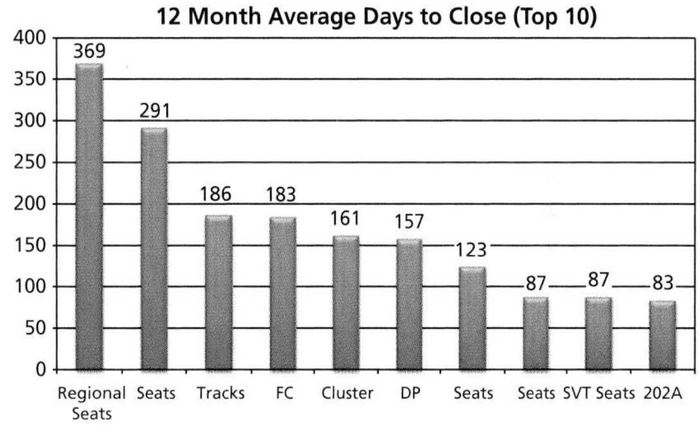

12 Month Average Days to Close (Top 10)

Regional Seats	Seats	Tracks	FC	Cluster	DP	Seats	Seats	SVT Seats	202A
369	291	186	183	161	157	123	87	87	83

PMO Program Portfolio - Launch Program OPEN Gates (SOP Sequential Order)

Status: 1-Jun-10

BU	Program State	EPIC Number	Program / Platform	Zone	Sub Zone	Product	Program Manager	Program Start	Development Start	DV Release	Final Prod Release	Cust Part Approval	SOP	Post Launch
								PLUS BLUE GATES						
rd	Launch	1008622	11	Seating		Seats						03/22/10	04/05/10	05/21/10
rd	Launch	1010300	11	Seating	Mech	Tracks						03/12/10	04/05/10	04/30/10
rd	Launch	1006104	11	Electronics		Cluster							04/12/10	07/29/10
rd	Launch	1010904	11	Interiors		Floor Console						06/18/10	07/12/10	09/01/10
rd	Launch	1008208	11	Interiors		Door Panel						06/30/10	07/12/10	09/01/10
rd	Launch	1009345	11	Interiors		Door Panel						07/01/10	10/04/10	12/01/10
rd	Launch	1009627	11	Seating	Mech	Tracks					04/05/10	08/27/10	10/29/10	01/04/11
rd	Launch	1009786	11	Seating		Seats					11/07/09	09/15/10	11/29/10	12/22/10
rd	Launch	1012855	13	Interiors		Floor Console			06/11/10	N/A	02/11/11	06/27/11	12/05/11	03/05/12
rd	Launch	1012690	12	Electronics		Cluster			02/25/10	06/18/10	03/31/11	12/16/11	01/09/12	05/17/12
rd	NO GWB		11	Seating		Trim								
rd	NO GWB		12	Seating		202A								
rd	NO GWB		11	Interiors	JCIM	Exteriors								
rd	NO GWB		12	Interiors	JCIM	IP Trim								
rd	NO GWB		11	Interiors	JCIM	IP Trim								
rd	NO GWB		12	Interiors	JCIM	Misc								
rd	NO GWB		12	Interiors	JCIM	Misc								
rd	NO GWB		11	Interiors	JCIM	Misc								
rd	NO GWB		11	Interiors	JCIM	IP Trim								
rd	NO GWB		12	Interiors	JCIM	Misc								
rd	NO GWB		11	Interiors	JCIM	Hard Trim								

Pending next 45 days
Late

The following are some of the project-level data available for project list definition or display:

- Project Manager or Business Manager
- Business Classification
- Start/Finish Date
- Budgeted Cost/Revenue
- Planned Cost/Revenue
- Total Effort/Expended Effort
- Contract type
- Contract Amount
- Expenses
- Roles
- Staffing of Resources

CA Technologies has identified key metrics and measures that are critical to understanding how well a project is being executed. Some of the key areas that are tracked and measured include:

- High Reward/Risk Projects (Watch List)
- Underfunded Projects (At Risk)
- Procurement/Purchase Orders
- Financials
- Schedule
- Resource Allocations
- Risks and Issues

The screenshots displayed and described below are production dashboards. Services engagements are managed in CA Clarity™ PPM; these dashboards are just a few in use.

Project Watchlist Dashboard

The Project Watchlist Dashboard, shown in Figure 7–7, provides a user-selectable list of projects with specific evaluations of the project(s) state. The results of the evaluation are compared with PMO- and business-provided thresholds to determine projects that may need attention. The traffic light display, where the icon color provides the state of the threshold that was met, is an effective means of drawing attention to specific areas that need attention.

The list of projects can be sorted by any of the columns. The column icons contain the actual evaluation metric calculation; sorting by the column will order the listing according to the column values worst (or best) to best (or worst).

Additional project attributes can be displayed in the listing, providing additional project description or specific project state.

Figure 7-7 Project Watchlist Dashboard *©2012 by Computer Associates, Inc. Reproduced by permission*

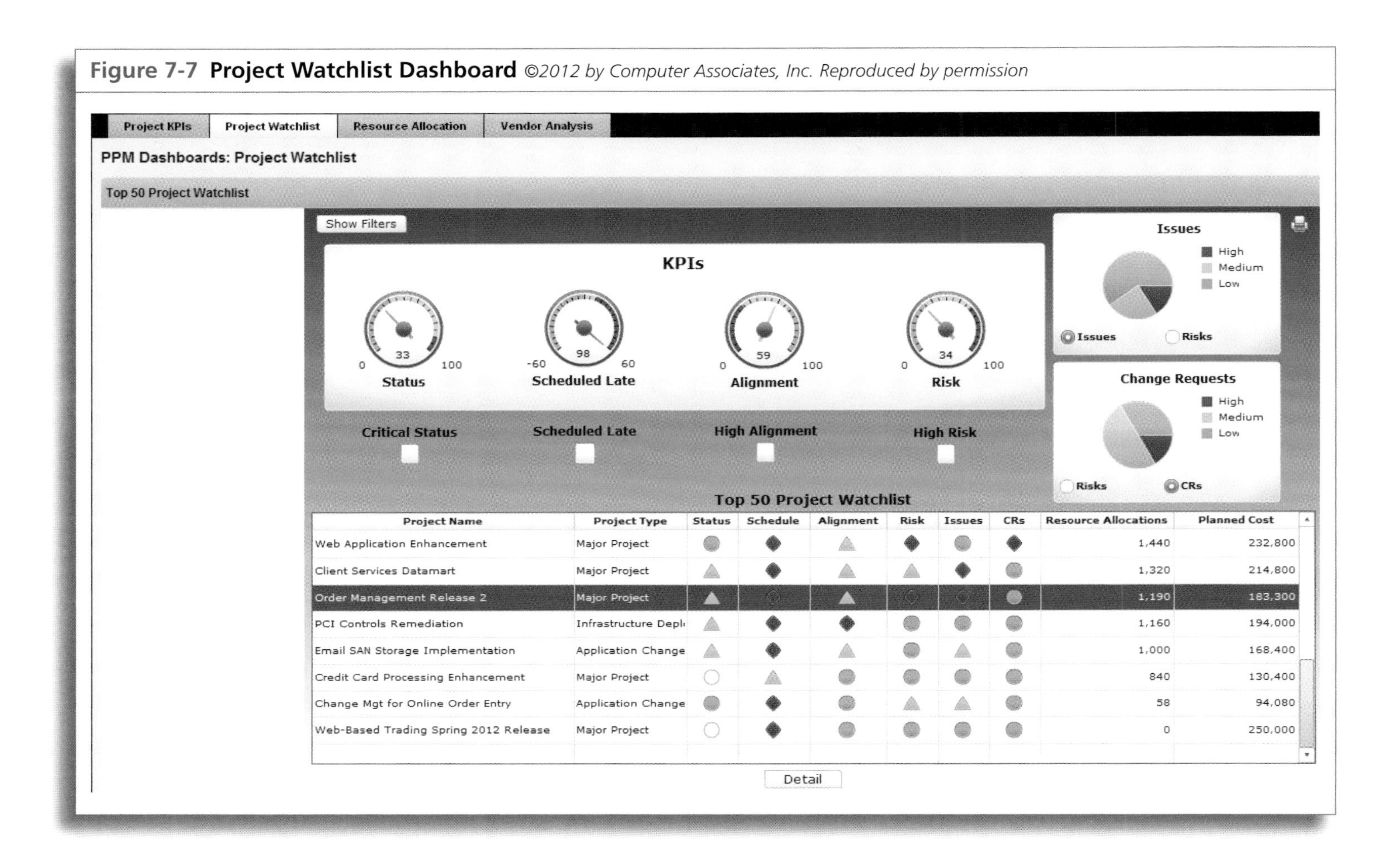

The screenshot of the dashboard in Figure 7–7 is an example of a top 50 project listing. It includes:

- Gauges to show KPIs requiring the most attention (Status, Schedule, Alignment, Risk)
- Filters (Critical Status, Scheduled late, High Alignment, High Risk)
- Breakdown charts for percentages of High, Medium, and Low Risks, Issues, and Change Requests.
- Project listing of key KPI indicators (traffic lights)

Project Watchlist Drill-Down

The user can select a project to go directly into the project or click on the appropriate column icon to go directly to project's specific content used to calculate the metric.

Example:

The Order Management Release project has red indicators for schedule, risk, and issues. This project's metric exceeds the defined threshold and should be investigated. When one clicks on the project, the details are displayed, as shown in the screenshot in Figure 7–8.

PM Alerts Dashboard

The PM Alerts Dashboard, shown in Figure 7–9, directs the project manager to focus on high-priority items, based on due dates and critical alerts.

Project Listing Dashboard

The Project Listing Dashboard, shown in Figure 7–10, provides a user-selectable list of projects with current state information. This dashboard provides the user with specific content that he/she identified to view about the projects. This dashboard is primarily used to find projects with specific attributes.

The following may be used to provide project managers with details on their project state with respect to the schedule, risk, planned and actual hours, and percent complete vs. expended effort. Project managers can also create additional project listing dashboards, if other data and/or thresholds are important for the way they monitor and manage projects.

Users can click on the project's name or the status indicator and immediately drill down to get further detail and/or see the cause of the indicator's coloring.

Figure 7-8 Order Management Release Dashboard *©2012 by Computer Associates, Inc. Reproduced by permission*

Order Management Release 2

Project ID	PR1010	**Status Report Date**	03/16/2012
Project Manager	Reed, Henry	**Previous Report Date**	03/09/2012
Project Type	Major Project	**Status**	Approved
Goal	Cost Reduction	**Progress**	Not Started

Project Objective

Status Report Update

Continued problems integrating with the online order entry system has caused a delay. We are contacting external consultants to help which will also increase the forecasted cost of the project.

Stage	Initiation
Finish Date	03/22/2012
Baseline Finish Date	03/22/2012
Days Late	0
	(Hours)
Baseline	1,190.00
Actuals	0.00
Estimate To Complete	1,190.00
Estimate At Completion	1,190.00
Variance to Baseline	0.00

Status Report Indicators

Overall

Schedule

Scope

Cost and Effort

Project Indicators

Schedule to Baseline

Alignment

Risk

Issue

Change

Project Team

Resource Name	Project Role	Start	Baseline Start	Finish	Baseline Finish	Booking Status	Request Status	Allocation Hours	Actual Hours	ETC Hours
Architect	Architect	03/05/2012	03/05/2012	03/22/2012	03/05/2012	Soft	New	160.00	0.00	160.00
Developer	Developer	03/05/2012	03/05/2012	03/22/2012	03/05/2012	Soft	New	550.00	0.00	550.00
Storage Architect	Storage Architect	03/05/2012	03/05/2012	03/22/2012	03/05/2012	Soft	New	320.00	0.00	320.00
Test Engineer	Test Engineer	03/05/2012	03/05/2012	03/22/2012	03/05/2012	Soft	New	160.00	0.00	160.00

Key Milestones

Name	ID	Status	Finish	Baseline Finish	Days Late	Schedule
Initiating Process Complete	LM.000.100	Not Started	03/05/2012	03/05/2012	0	
Planning Phase Gate Complete	LM.001.100	Not Started	03/05/2012	03/05/2012	0	
Design Phase Gate Complete	LM.002.100	Not Started	03/05/2012	03/05/2012	0	
Construction Phase Gate Complete	LM.003.100	Not Started	03/22/2012	03/22/2012	0	
Deployment Phase Gate Complete	LM.004.100	Not Started	03/22/2012	03/22/2012	0	
Closing Phase Gate Complete	LM.005.020	Not Started	03/22/2012	03/22/2012	0	

Risks

Name	ID	Description	Priority	Probability	Impact	Status	Response Type	Target Resolution
Interfaces to other systems	RS1041	Interfaces to other systems could be beyond the expertise of our staff.	High	Medium	High	Open	Watch	03/09/2012

Issues

Name	ID	Description	Priority	Status	Target Resolution
Funding concerns	IS1056	Since we need external consultants, we need to get additional funding.	Low	Open	03/07/2012
Interfaces to other systems	IS1057	Interfaces to other systems could be beyond the expertise of our staff.	High	Open	03/12/2012

Change Requests

No matching records were found

Figure 7-9 Project Management Alert Dashboard *©2012 by Computer Associates, Inc. Reproduced by permission*

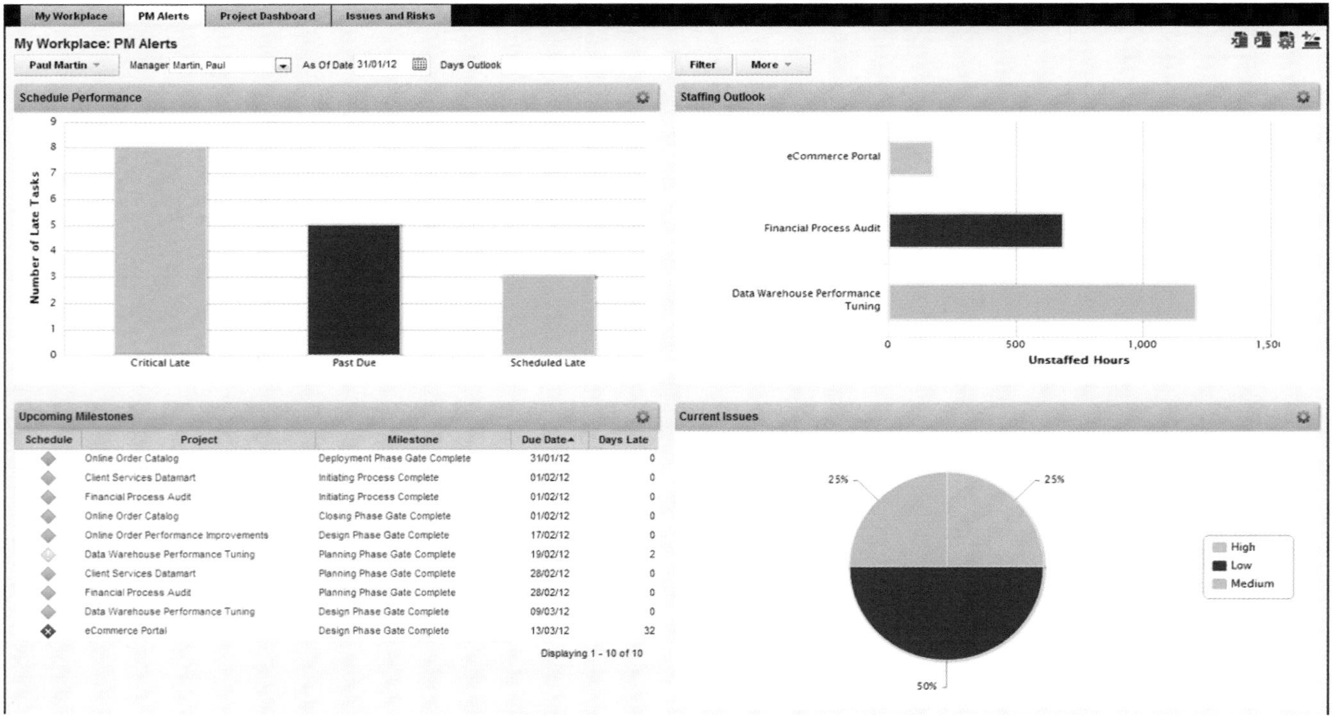

My Workplace | PM Alerts | Project Dashboard | Issues and Risks

My Workplace: PM Alerts

Paul Martin | Manager Martin, Paul | As Of Date 31/01/12 | Days Outlook | Filter | More

Schedule Performance

Number of Late Tasks: 0 1 2 3 4 5 6 7 8 9

Critical Late | Past Due | Scheduled Late

Staffing Outlook

eCommerce Portal

Financial Process Audit

Data Warehouse Performance Tuning

0 500 1,000 1,50

Unstaffed Hours

Upcoming Milestones

Schedule	Project	Milestone	Due Date▲	Days Late
	Online Order Catalog	Deployment Phase Gate Complete	31/01/12	0
	Client Services Datamart	Initiating Process Complete	01/02/12	0
	Financial Process Audit	Initiating Process Complete	01/02/12	0
	Online Order Catalog	Closing Phase Gate Complete	01/02/12	0
	Online Order Performance Improvements	Design Phase Gate Complete	17/02/12	0
	Data Warehouse Performance Tuning	Planning Phase Gate Complete	19/02/12	2
	Client Services Datamart	Planning Phase Gate Complete	28/02/12	0
	Financial Process Audit	Planning Phase Gate Complete	28/02/12	0
	Data Warehouse Performance Tuning	Design Phase Gate Complete	09/03/12	0
	eCommerce Portal	Design Phase Gate Complete	13/03/12	32

Displaying 1 - 10 of 10

Current Issues

25% 25% 50%

High | Low | Medium

Figure 7-10 Project Listing Dashboard *©2012 by Computer Associates, Inc. Reproduced by permission*

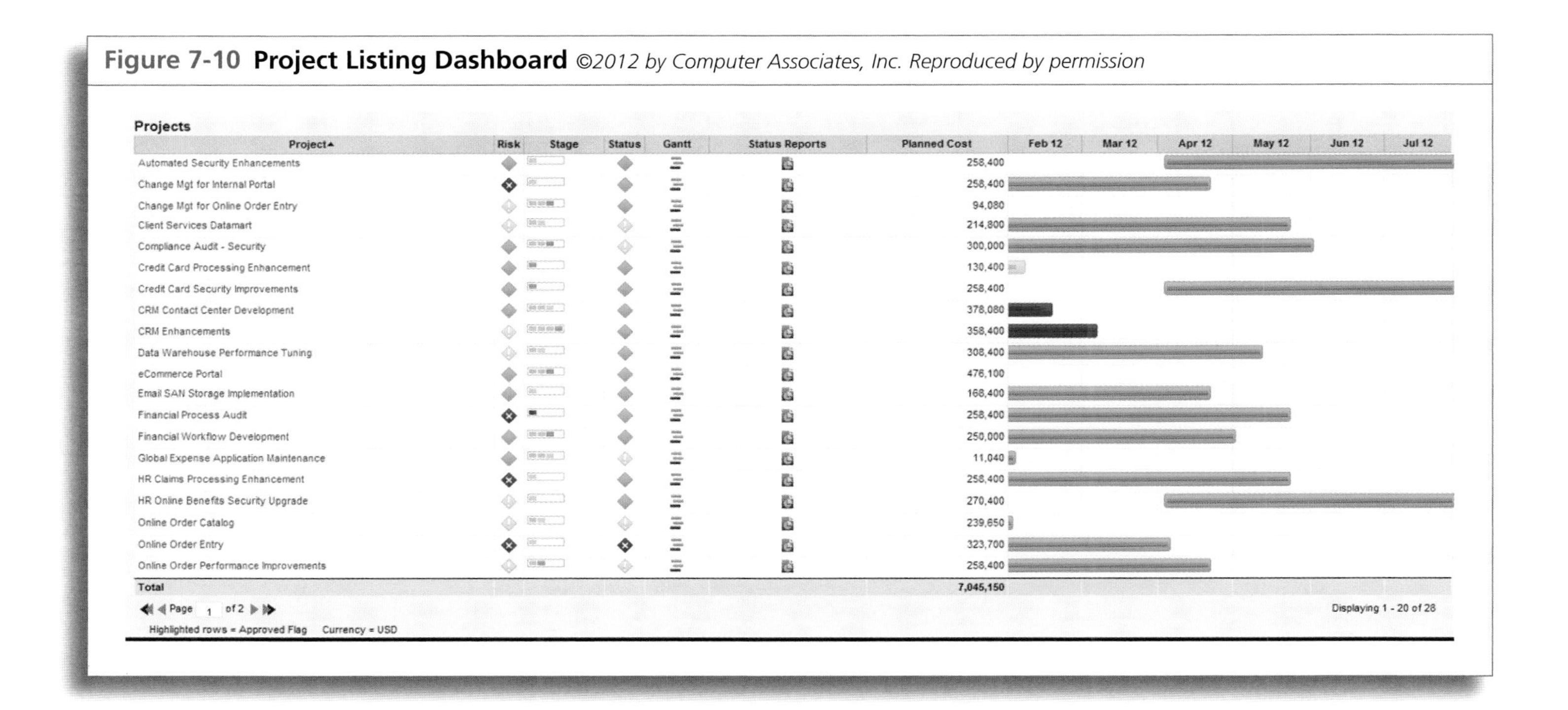

Project Status Reports Dashboard

The Project Status Reports Dashboard, as shown in Figure 7–11, provides an overview of all projects and associated project status report comments, as well as traffic light health indicators. This dashboard is useful for quickly reviewing the latest status on projects, especially troubled projects where management is closely monitoring the details of status changes.

Project names are clickable hyperlinks to the project details.

Resource Planning Dashboard

The Resource Planning Dashboard, as shown in Figure 7–12, provides a user-selectable list of resources with the resource's attributes and relationship to projects. This dashboard has many different purposes depending on the user's objective.

A project manager uses the dashboard to list staffed resources to see when resources are scheduled to participate on other projects when evaluating project schedule adjustments or to find the availability of a specific type of resource for planning. Additional fields, not currently displayed, can be added to show when the resources are planned to conduct the work. This information helps project managers coordinate project effort and resource schedules.

The Resource Management Team uses this dashboard to review assignments and perform whitespace management. This dashboard is used to assist with scheduling adjustments and resolving resource conflicts.

Resource Planning Drill-Down

The user can click on the resource icon to view the details about individual resources.

In the screenshot shown in Figure 7–13, the detailed time allocation for the second resource (Joyce) is displayed. This display has the same content as the Resource Planning Dashboard, but now displays the specific projects that Joyce has worked on. Note that resource managers and project managers dynamically adjust the timescale real-time to create appropriate views.

The Resource Management Team makes resource schedule adjustments from this dashboard. Resources are encouraged to use this dashboard to get short- and long-term views of their schedules.

Figure 7-11 Project Status Reports Dashboard *©2012 by Computer Associates, Inc. Reproduced by permission*

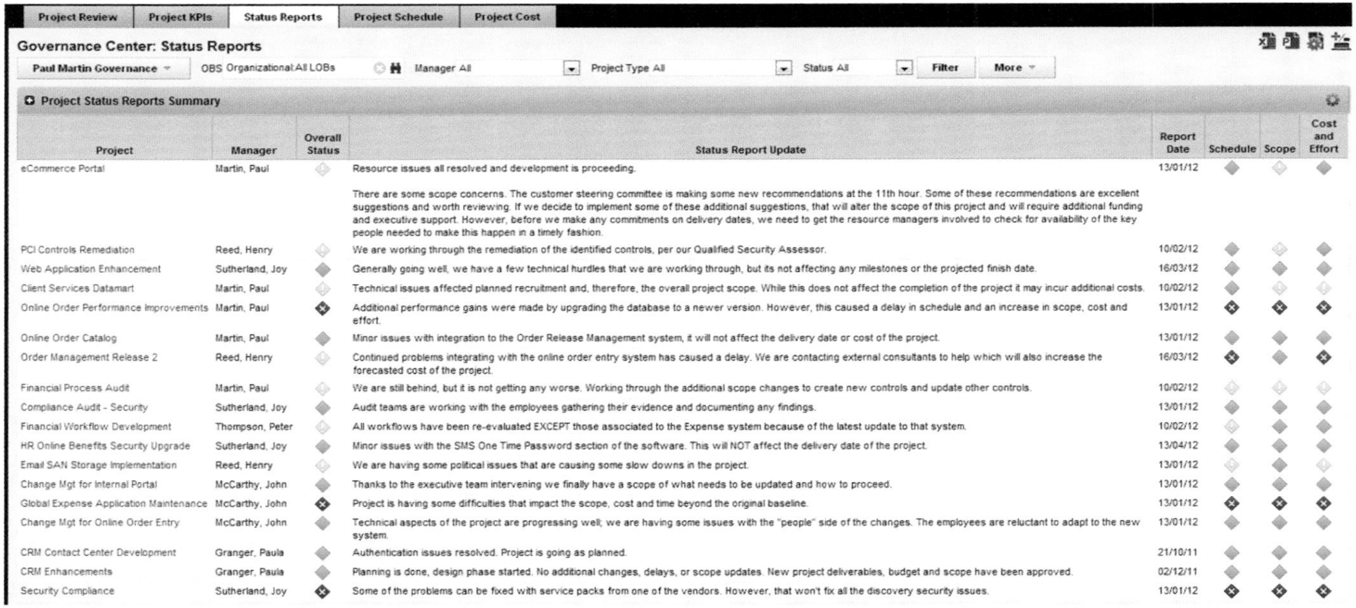

Project Review | Project KPIs | Status Reports | Project Schedule | Project Cost

Governance Center: Status Reports

Paul Martin Governance | OBS Organizational All LOBs | Manager All | Project Type All | Status All | Filter | More

Project Status Reports Summary

Project	Manager	Overall Status	Status Report Update	Report Date	Schedule	Scope	Cost and Effort
eCommerce Portal	Martin, Paul		Resource issues all resolved and development is proceeding.	13/01/12			
			There are some scope concerns. The customer steering committee is making some new recommendations at the 11th hour. Some of these recommendations are excellent suggestions and worth reviewing. If we decide to implement some of these additional suggestions, that will alter the scope of this project and will require additional funding and executive support. However, before we make any commitments on delivery dates, we need to get the resource managers involved to check for availability of the key people needed to make this happen in a timely fashion.				
PCI Controls Remediation	Reed, Henry		We are working through the remediation of the identified controls, per our Qualified Security Assessor.	10/02/12			
Web Application Enhancement	Sutherland, Joy		Generally going well, we have a few technical hurdles that we are working through, but its not affecting any milestones or the projected finish date.	16/03/12			
Client Services Datamart	Martin, Paul		Technical issues affected planned recruitment and, therefore, the overall project scope. While this does not affect the completion of the project it may incur additional costs.	10/02/12			
Online Order Performance Improvements	Martin, Paul		Additional performance gains were made by upgrading the database to a newer version. However, this caused a delay in schedule and an increase in scope, cost and effort.	13/01/12			
Online Order Catalog	Martin, Paul		Minor issues with integration to the Order Release Management system, it will not affect the delivery date or cost of the project.	13/01/12			
Order Management Release 2	Reed, Henry		Continued problems integrating with the online order entry system has caused a delay. We are contacting external consultants to help which will also increase the forecasted cost of the project.	16/03/12			
Financial Process Audit	Martin, Paul		We are still behind, but it is not getting any worse. Working through the additional scope changes to create new controls and update other controls.	10/02/12			
Compliance Audit - Security	Sutherland, Joy		Audit teams are working with the employees gathering their evidence and documenting any findings.	13/01/12			
Financial Workflow Development	Thompson, Peter		All workflows have been re-evaluated EXCEPT those associated to the Expense system because of the latest update to that system.	10/02/12			
HR Online Benefits Security Upgrade	Sutherland, Joy		Minor issues with the SMS One Time Password section of the software. This will NOT affect the delivery date of the project.	13/04/12			
Email SAN Storage Implementation	Reed, Henry		We are having some political issues that are causing some slow downs in the project.	13/01/12			
Change Mgt for Internal Portal	McCarthy, John		Thanks to the executive team intervening we finally have a scope of what needs to be updated and how to proceed.	13/01/12			
Global Expense Application Maintenance	McCarthy, John		Project is having some difficulties that impact the scope, cost and time beyond the original baseline.	13/01/12			
Change Mgt for Online Order Entry	McCarthy, John		Technical aspects of the project are progressing well; we are having some issues with the "people" side of the changes. The employees are reluctant to adapt to the new system.	13/01/12			
CRM Contact Center Development	Granger, Paula		Authentication issues resolved. Project is going as planned.	21/10/11			
CRM Enhancements	Granger, Paula		Planning is done, design phase started. No additional changes, delays, or scope updates. New project deliverables, budget and scope have been approved.	02/12/11			
Security Compliance	Sutherland, Joy		Some of the problems can be fixed with service packs from one of the vendors. However, that won't fix all the discovery security issues.	13/01/12			

Figure 7-12 **Resource Planning Dashboard** *© 2012 by Computer Associates, Inc. Reproduced by permission*

Role Capacity | Organizational Demand | Top Down Planning | Workloads | Allocations | Unfilled Allocations | Task Assignments | Bookings

Resource Planning: Workloads

Scenario: [--Select--]

Resource Workloads

Resource	Primary Role		Allocation									
			27/08/12	03/09/12	10/09/12	17/09/12	24/09/12	01/10/12	08/10/12	15/10/12	22/10/12	29/10/12
Childers, Valerie	Architect											
Coleman, Joyce	Architect											
Hayes, Justin	Architect											
Katect, Art	Architect											
Levert, Ed	Architect											

Figure 7-13 Resource Planning Drill Down Dashboard *©2012 by Computer Associates, Inc. Reproduced by permission*

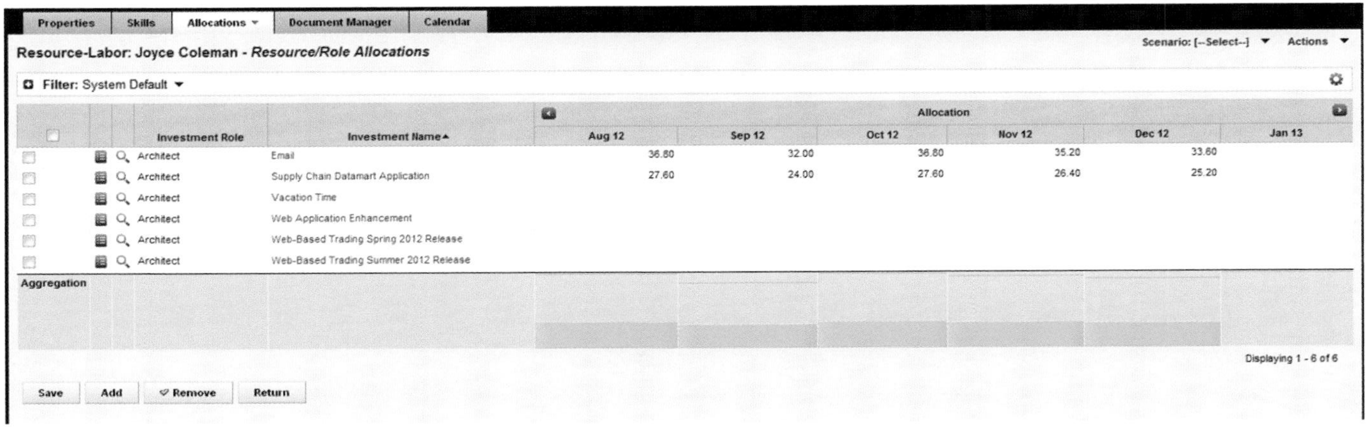

Properties | Skills | Allocations | Document Manager | Calendar

Resource-Labor: Joyce Coleman - *Resource/Role Allocations*

Scenario: [--Select--] Actions

Filter: System Default

Investment Role	Investment Name	Aug 12	Sep 12	Oct 12	Nov 12	Dec 12	Jan 13
		Allocation					
Architect	Email	36.80	32.00	36.80	35.20	33.60	
Architect	Supply Chain Datamart Application	27.60	24.00	27.60	26.40	25.20	
Architect	Vacation Time						
Architect	Web Application Enhancement						
Architect	Web-Based Trading Spring 2012 Release						
Architect	Web-Based Trading Summer 2012 Release						
Aggregation							

Displaying 1 - 6 of 6

Save | Add | Remove | Return

7.4 DASHBOARDS IN ACTION: PIEMATRIX, INC.

Consider the following scenario, which appears to be happening in a multitude of companies. Senior management is actively involved in the selection of projects that will go into the portfolio. Once the projects are selected, however, senior management gets one summary dashboard to look at and cannot easily find any appropriate detailed information that could influence their decisions at the moment. While there is some merit to providing executives with just summary information, there must be a drill-down process in place for easy access to more critical information that may appear on working level dashboards.

At the time of writing this book, there are many project and portfolio management systems but not very many that elegantly tie the executive dashboard down to the front-line team member execution the way PIEmatrix's does. Not only does their software provide this drill-down capability, but it is done with user-friendly software that can be learned in minutes. Customers can turn complex projects into more manageable views that make it easy for executives and front-line people to make informed decisions in a timely manner.

Unlike many dashboard systems that focus only on strategic issues and financial numbers, the PIEmatrix system takes a process focus on project management. This means that the data being displayed in the executive dashboard is not just data about how we do things, but data about how we do things consistently right. Predictability of success becomes more controlled, and we may have more confidence knowing that the execution is being done with truly best practices. The remaining information in this section is devoted to PIEmatrix and their online project management software platform.[3]

7.5 PIEMATRIX OVERVIEW

PIEmatrix Inc. produces a business execution platform called PIEmatrix. It's a software-as-a-service solution. The platform has two key differentiation points. One is its focus on enabling a process framework with best practices to help reduce risk and drive continuous improvement. Two is its visual design made for fast user adoption.

PIEmatrix is made for any business initiative and for any industry. PIEmatrix customers span from federal and state government public sector to private industries like pharmaceuticals, healthcare, and energy. Their customer base includes Fortune 500 firms, midsized organizations, and small "gazelles" (fast-growing companies). The functional use of PIEmatrix

3. The remaining material in this section has been graciously provided by Paul Dandurand, CEO PIEmatrix Inc.; Office 802-318-4891; Mobile 802-578-5653; paul.dandurand@piematrix.com; www.piematrix.com; © 2010 PIEmatrix Inc.; Patent pending #12/258,637. Reproduced by permission of PIEmatrix, Inc. All rights reserved.

covers different departments across the enterprise, such as IT, finance, HR, operations, and marketing. PIEmatrix is a platform, which means it provides all the tools an organization needs to easily setup projects and repeatable best-practice standards for execution and governance.

PIEmatrix is made for project managers, front-line team members, stakeholders, and executives. Although it has many different functions for designing and executing projects, the following pages will focus mainly on the dashboard. The executive dashboard displays quick and important information that shows the progress and compliance with company best practices, standards, or just the right way to get things done. The administrator can setup visibility rights to certain people within the enterprise and even with external partners as needed for transparency and governance.

PIEmatrix Executive Dashboard

The PIEmatrix Portfolio Progress view displays how they came up with their company name (Figure 7–14). The project phases are the slices of the pie. The layering of the stacked projects displays the matrix side. The pie images are circular to represent how projects can be iterative in nature. The labels of the phases (or what are called "slices") can be anything the organization chooses. Using the same phase nomenclature across like projects is helpful for everyone to be on the same page. Imagine two teams building software. One team submits a project report with the phases "Initiate, Plan, Build, Close." The second team submits a report with the phases "Plan, Discover, Design, Construct, Test, Deploy." How will the executive know the difference with words like "Plan"? They could cover the same types of processes or maybe not. The PIEmatrix structure helps with better consistency to help reduce communication issues.

Show me the progress of my preferred project set. Give it to me in real time.

We'll first take a look at the main PIEmatrix dashboard view. Figure 7–14 displays a portfolio of five projects. These are filtered from a larger portfolio set with the Business Units and Public Tags dropdown filter lists. PIEmatrix provides a hierarchical view of the enterprise where a business unit options could show either the entire enterprise or just a business department like IT, HR, or Finance. It could also be used for geographical regions like Asia, North America, EMEA, and so on. The Public Tags can further group within a business unit. For example, a business unit IT could have tags for filtering like Project Proposals, Innovation Projects, System Upgrades, etc.

The simple color codes on the projects show the progress of different states. The dark green bars represent what is completed. The light green is showing in progress. The rest in gray represent the not started state.

The data displayed is automatically updated from the field every ten seconds. PIEmatrix is made not only for executives to govern their portfolio, but also for the teams and stakeholders to execute these projects on a

Figure 7-14 Piematrix Portfolio Progress—Main Page *(Reproduced by permission of PIEmatrix, Inc. All rights reserved.)*

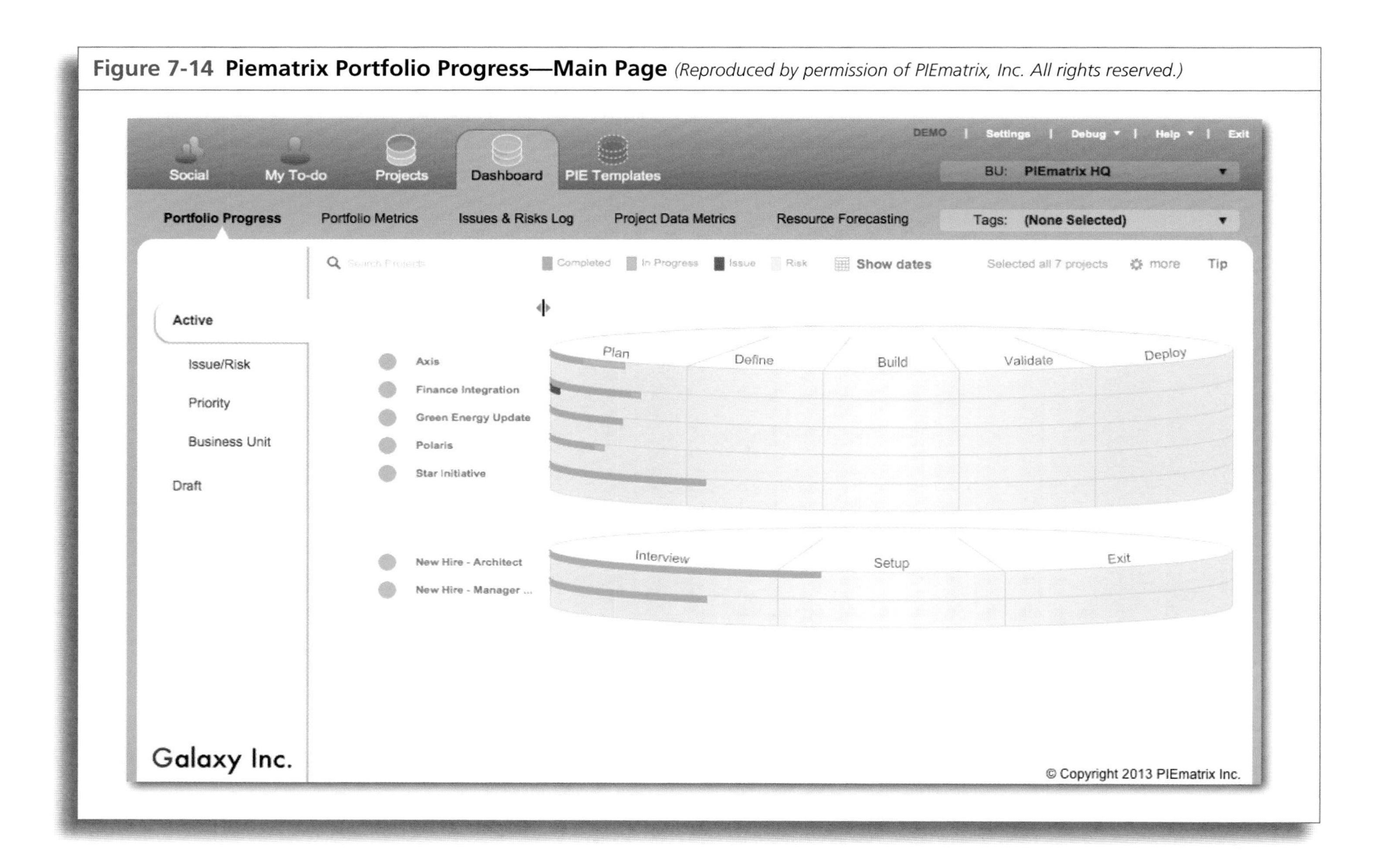

daily basis. This link means the front-line work is automatically sent to the executive dashboard results in real time.

Show me milestones and key status indicators.

PIEmatrix uses a contextual approach to displaying screen information. Their approach is to keep the views simple and clean and then allow the user to expand upon the view with a click. The visuals are very simple and accurate. With a click, the executive can choose to look at more information. Figure 7–15 shows the same dashboard with phase milestones turned on. Red dates represent milestones that are behind schedule. Black dates are on schedule.

Show me only what's important.

An executive can select only the states that are important for him or her to review. Figure 7–16 displays a portfolio view with the status set to show only projects in risk and issue states. This is helpful for executives who don't want the noise of all activity and would rather see what requires their action.

A good feature in PIEmatrix is that the application remembers the user's last view, so when he or she logs back into PIEmatrix, it will take that person automatically to their last page including their last filtered states. This is important for setting our quick viewing states when we just want to keep our important view at our disposal.

Show me all projects grouped by priority.

What if we want to look only at projects in the high-priority state? PIEmatrix has a feature where the executive or director can set relative priority states based on certain criteria. Figure 7–17 displays a complete portfolio of projects within a business unit that are grouped by priority. The high-priority projects are in the top grouping. The executive can change a project's priority by clicking the priority icon to the left of a project's name. In doing so, the system records the person's name, date, and action for accountability and governance.

Show me project data metrics.

The PIEmatrix platform allows the user to create project specific data fields for capturing and displaying project-level data, such as budget information. Figure 7–18 displays a set of columns that list each data element. These are easily defined within minutes with the add data field tool. The user can click on a data cell to pop up the data entry window. The data capture fields can also be set up within any action step inside the project. As team members answer the step question with the data, the dashboard displays the answers to all users within 10 seconds.

Show me project details.

Clicking on a project pie in the main portfolio progress view brings up the project's detail view as shown in Figure 7–19. What is interesting with PIEmatrix is how the layering of the processes being executed is displayed. In this sample project called Axis, the project has three process flows: Project Management, Development, and Governance. These are similar to work streams or swim lanes. They represent a set of process steps across

Figure 7-15 Piematrix Portfolio Progress—One Click to Display Milestone Dates and Project Status Indicators

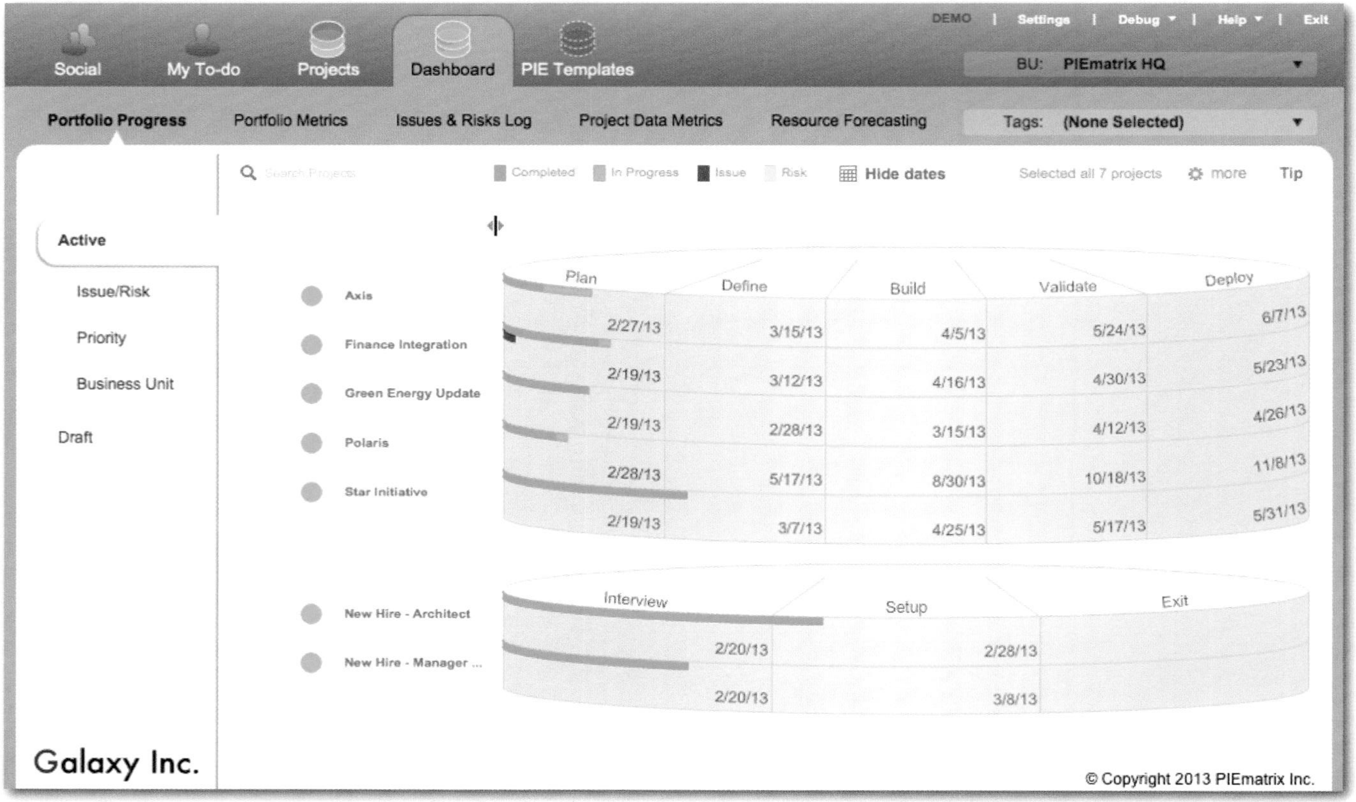

Figure 7-16 Piematrix Portfolio Progress—Filtered to Only Show Risk and Issue State Projects *(Reproduced by permission of PIEmatrix, Inc. All rights reserved.)*

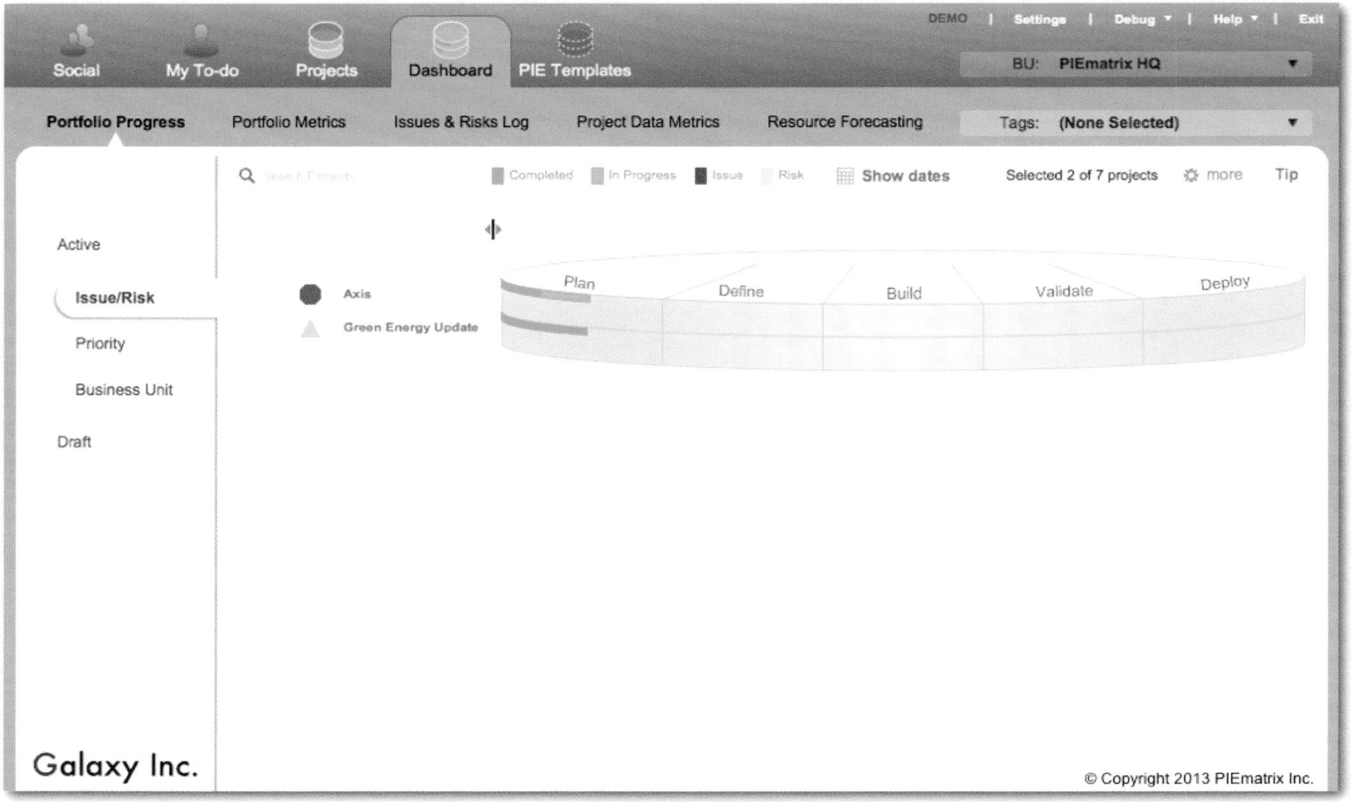

Figure 7-17 Piematrix Portfolio Progress—Full Portfolio Grouped by Priority *(Reproduced by permission of PIEmatrix, Inc. All rights reserved.)*

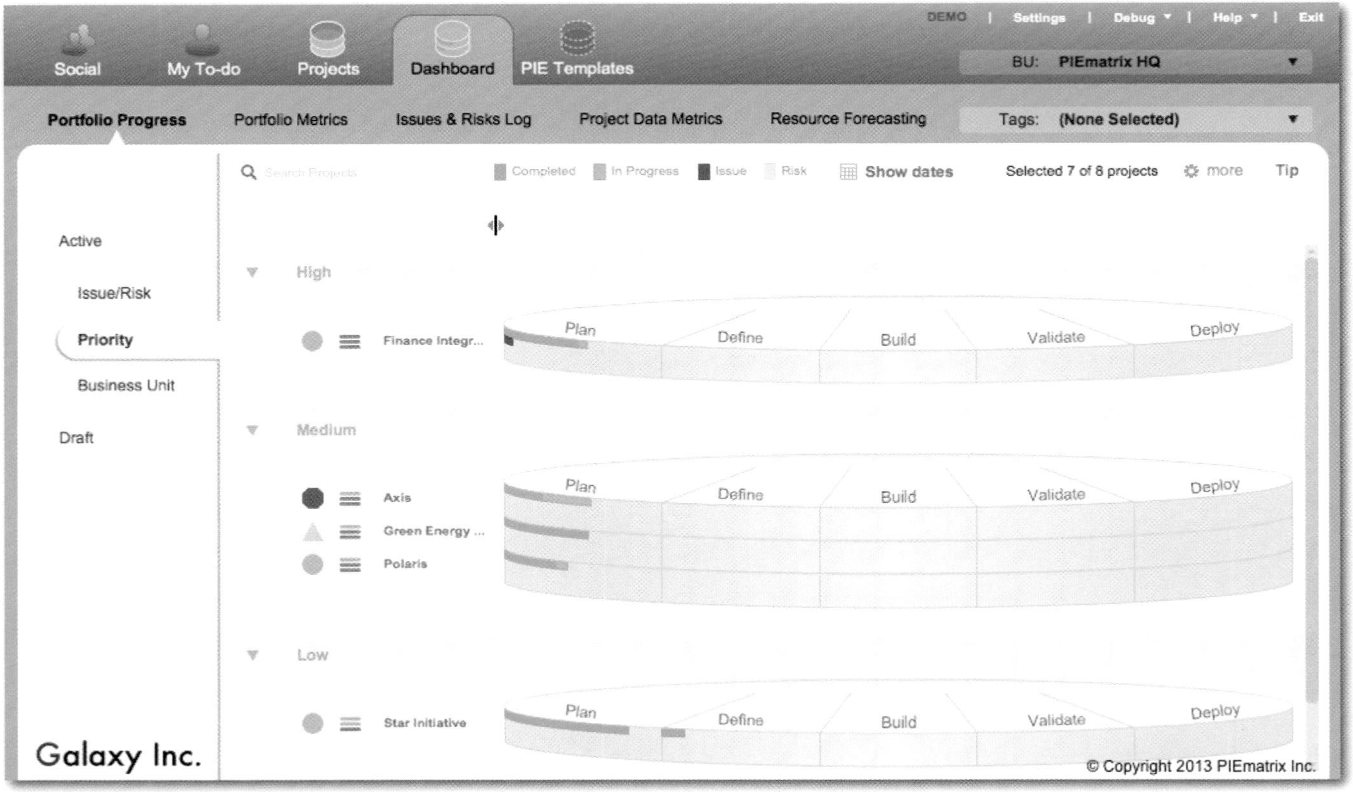

Figure 7-18 Project Data Metrics—Custom Project Data Fields for Tracking Budgets and Other Portfolio Information *(Reproduced by permission of PIEmatrix, Inc. All rights reserved.)*

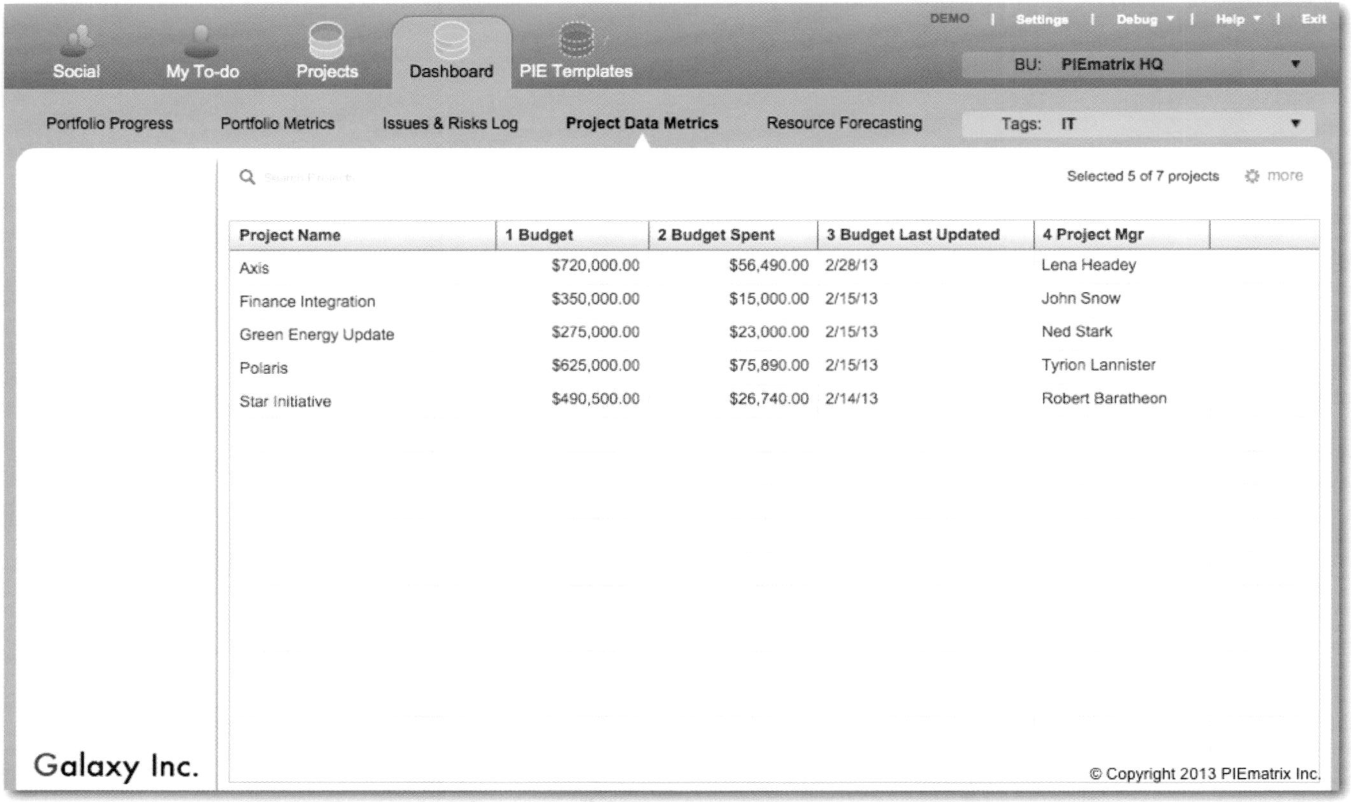

DEMO | Settings | Debug | Help | Exit
Social
My To-do
Projects
Dashboard
PIE Templates
BU: PIEmatrix HQ
Portfolio Progress
Portfolio Metrics
Issues & Risks Log
Project Data Metrics
Resource Forecasting
Tags: IT
Selected 5 of 7 projects
more
Project Name | 1 Budget | 2 Budget Spent | 3 Budget Last Updated | 4 Project Mgr
Axis | $720,000.00 | $56,490.00 | 2/28/13 | Lena Headey
Finance Integration | $350,000.00 | $15,000.00 | 2/15/13 | John Snow
Green Energy Update | $275,000.00 | $23,000.00 | 2/15/13 | Ned Stark
Polaris | $625,000.00 | $75,890.00 | 2/15/13 | Tyrion Lannister
Star Initiative | $490,500.00 | $26,740.00 | 2/14/13 | Robert Baratheon
Galaxy Inc.
© Copyright 2013 PIEmatrix Inc.

Figure 7-19 Piematrix Portfolio Progress—Project Drill-down View Showing Process Layers *(Reproduced by permission of PIEmatrix, Inc. All rights reserved.)*

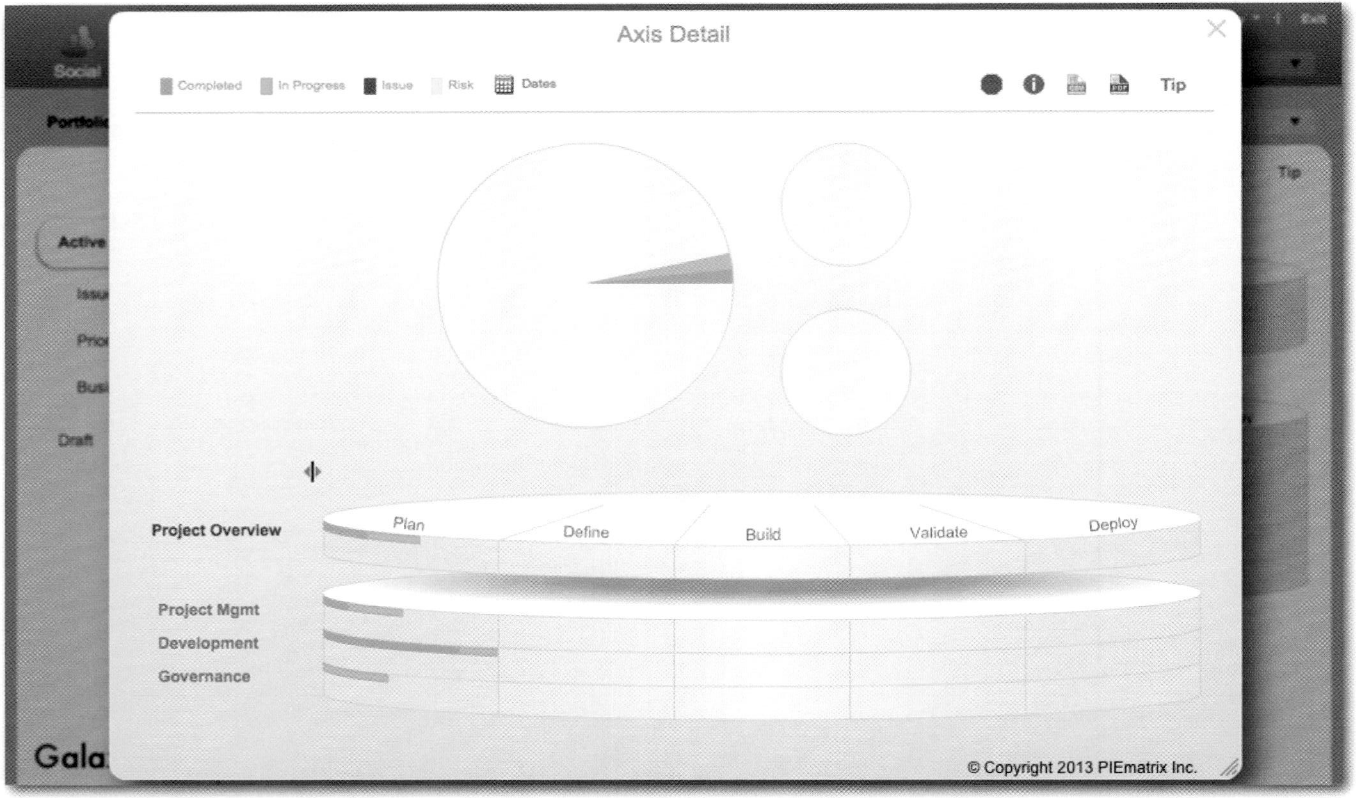

phases that are managed and executed by different groups of people getting things done. The Project Management layer could contain process steps for the project manager to execute such as defining the plan, getting the budget approved, and managing the scope. The Development layer could be for getting the project's deliverables built and delivered. This could contain steps for engineering leads, architects, business analysts, business users, etc. The Governance layer could be for risk planning and control along with sponsor management. This could be managed and executed by a risk team.

PIEmatrix makes it easy for the organization to define and execute steps for any process needs (details shown later). For example, other process layers could be for quality alignment, budget management, or ongoing project value analysis. This approach is a great way to break down silos between groups on complex projects. And coming back to the executive, this dashboard view at the project level provides a great view for governing the execution of the process flows.

Show me the issues and let me solve them in real-time.

In the previous figure (Figure 7–19), we notice the Governance layer has a red indicator in the Plan phase. The executive can click on the red line to view the issue at hand. Figure 7–20 displays an issue popup window. This is a collaboration view of comments made to date. PIEmatrix makes it easy to not only view the issue details, but to also respond with comments on the fly. The executive can enter his or her comments as needed to help resolve the issue. Entering a reply will automatically send an email and a message post to the project's manager and those responsible for the issue. PIEmatrix also has a separate issue list page for overall issues management.

Show me detailed metrics.

Sometimes, the executive wants to view metrics in more detail or different formats. Figure 7–21 displays the PIEmatrix Portfolio Metrics page. As an extension of the visual Dashboard tab, this page presents the portfolio data in a table format. The top section of this page shows the summary for the business unit and the filtered set of projects. The executive can easily click on any of the blue metric numbers to obtain more detail about that area. In this example, we clicked on the number 6 under the Total Projects column. This expanded the view to show the Project List section. Hovering over the small pie icons on the right will give a quick, semitransparent snapshot of the progress bars going across the project phases like we saw in the earlier main dashboard progress page.

Executive Dashboard and To-Do List—Where Does All This Data Come From?

For a dashboard to be at all useful, its data needs to come from someplace. The PIEmatrix dashboard data is automatically derived from people getting their work done, in real-time, day to day. The organization executives, managers, team members, stakeholders, partners, and anyone else who has

Figure 7-20 Piematrix Portfolio Progress—Collaboration Window Displays Conversations Regarding a Project Issue *(Reproduced by permission of PIEmatrix, Inc. All rights reserved.)*

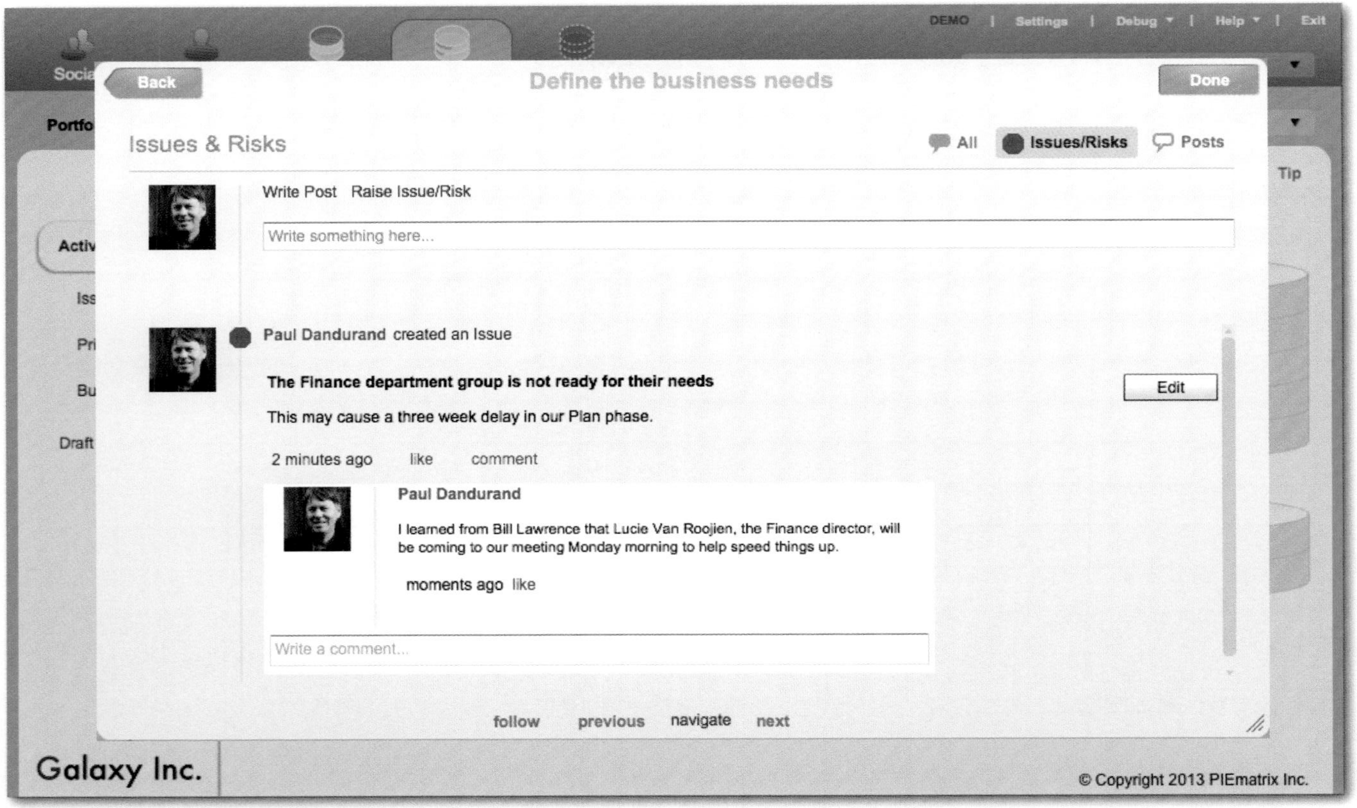
DEMO | Settings | Debug | Help | Exit
Back
Define the business needs
Done
Issues & Risks
All
Issues/Risks
Posts
Write Post Raise Issue/Risk
Write something here...
Paul Dandurand created an Issue
The Finance department group is not ready for their needs
This may cause a three week delay in our Plan phase.
Edit
2 minutes ago like comment
Paul Dandurand
I learned from Bill Lawrence that Lucie Van Roojien, the Finance director, will be coming to our meeting Monday morning to help speed things up.
moments ago like
Write a comment...
follow previous navigate next
Tip
Galaxy Inc.
© Copyright 2013 PIEmatrix Inc.

Figure 7-21 Piematrix Portfolio Metrics—View Multiple Drill-down Options with One Click *(Reproduced by permission of PIEmatrix, Inc. All rights reserved.)*

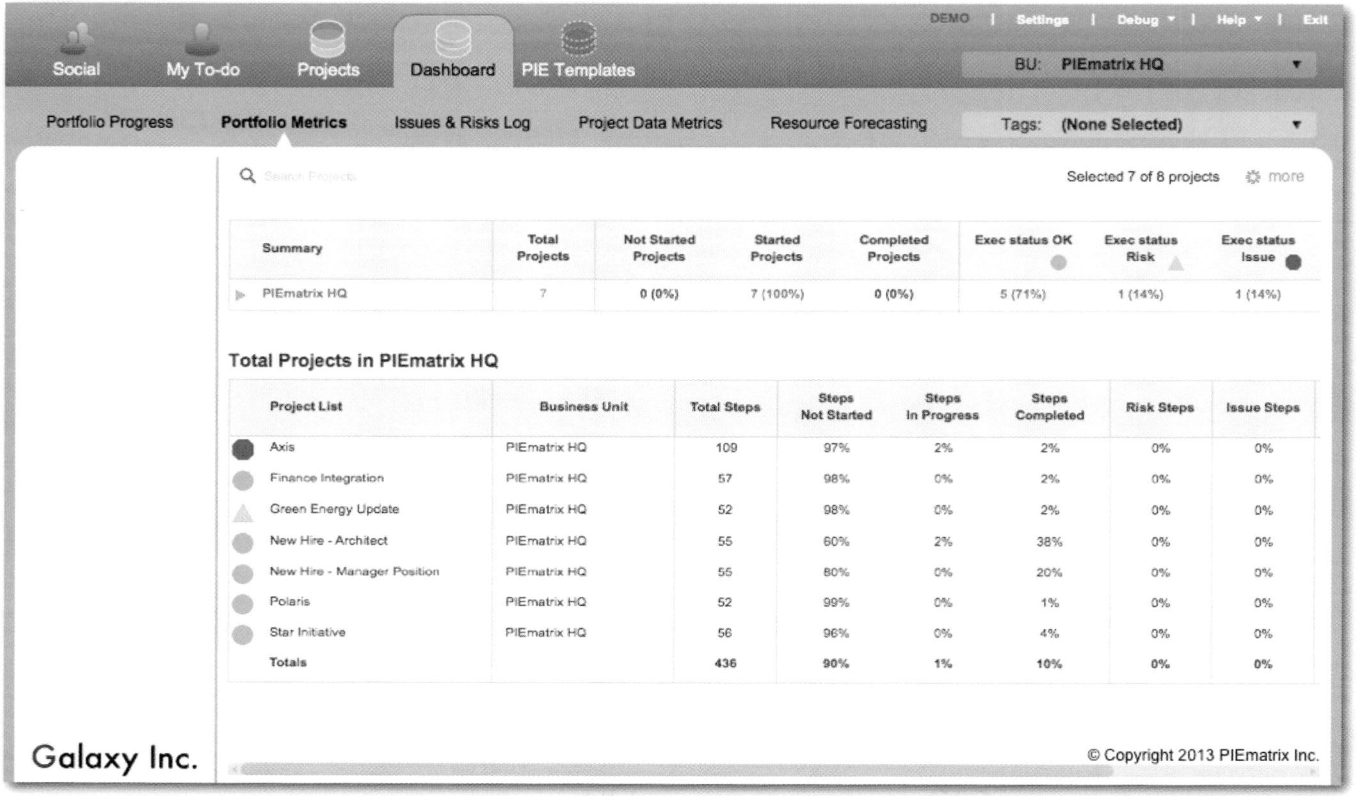

DEMO | Settings | Debug ▾ | Help ▾ | Exit

Social My To-do Projects Dashboard PIE Templates

BU: PIEmatrix HQ ▾

Portfolio Progress Portfolio Metrics Issues & Risks Log Project Data Metrics Resource Forecasting

Tags: (None Selected) ▾

Search Projects

Selected 7 of 8 projects more

Summary	Total Projects	Not Started Projects	Started Projects	Completed Projects	Exec status OK	Exec status Risk	Exec status Issue
PIEmatrix HQ	7	0 (0%)	7 (100%)	0 (0%)	5 (71%)	1 (14%)	1 (14%)

Total Projects in PIEmatrix HQ

Project List	Business Unit	Total Steps	Steps Not Started	Steps In Progress	Steps Completed	Risk Steps	Issue Steps
Axis	PIEmatrix HQ	109	97%	2%	2%	0%	0%
Finance Integration	PIEmatrix HQ	57	98%	0%	2%	0%	0%
Green Energy Update	PIEmatrix HQ	52	98%	0%	2%	0%	0%
New Hire - Architect	PIEmatrix HQ	55	60%	2%	38%	0%	0%
New Hire - Manager Position	PIEmatrix HQ	55	80%	0%	20%	0%	0%
Polaris	PIEmatrix HQ	52	99%	0%	1%	0%	0%
Star Initiative	PIEmatrix HQ	56	96%	0%	4%	0%	0%
Totals		**436**	**90%**	**1%**	**10%**	**0%**	**0%**

Galaxy Inc.

a role in getting the work done uses PIEmatrix. They follow their process, collaborate, and update their statuses in the platform. This activity is then automatically updated in the executive dashboard and metrics pages.

Show me what I need to do this week.

We'll first look at the team member's To-do List page. Figure 7–22 shows how simple it is for the team members to view their own work. Keep in mind that the team member can be anyone assigned to a project. This can be a business analyst, a part-time consultant, or an executive in a steering committee who's responsible for reviewing and signing off on certain deliverable documents. (PIEmatrix also has a built-in repository and workflow process of document files).

In the To-do List page, the user can group to-dos either by date grouping (as in Figure 7–22) or by project grouping. Once the user changes the step's check mark button to any state like In Progress, Completed, or Issue, the system updates any remote executive's dashboard within 10 seconds.

Show me how to best get my work done to minimize issues and risks.

Imagine all team members knowing not only what tasks to get done, but how to get the work done right with the help of guidance, tips, or required instructions. This is one of PIEmatrix's key strengths in the marketplace. Figure 7–23 shows what happens as we hover over an assigned step. A popup with text displays exactly what to do to ensure we get the step done the best way possible. The instructions can be setup as guidance or a concrete explanation to ensure compliance with regulations. In either case, it explains what to do and how to do it so nothing slips through the cracks. It's up to the organization's process experts to create the process content. The PIEmatrix platform comes with simple tools to help the organization establish repeatable processes. PIEmatrix also provides an easy way for the team members to send feedback to the process content author to help with continuous improvement. Another nice and important feature is the ability for the team members to create their own custom process steps and tasks on the fly.

The dashboard should not only reflect how things are being done on time and on budget, but also that things are indeed being done according to best standards and procedures. This is critical for any initiative or project type that is repeated. PIEmatrix's unique structure allows organizations to setup pre-established process templates that contain all the right steps, people roles, and document file templates. These pre-established templates can be from best practices, process standards, or critical procedures. Many organizations already have this kind of process content sitting on their servers as documents. However, PIEmatrix makes these standards come alive, ensuring they get integrated into daily executable steps. Organizations that do not already have process standards can use PIEmatrix story boarding features to build them from scratch. A project manager can select their best process template from a list in PIEmatrix. Kicking off a new project with repeatable standards is powerful, especially for novice managers.

Figure 7-22 Piematrix To-Do List—Real-Time Dashboard Data Is Automatically Derived When Team Members Execute Steps *(Reproduced by permission of PIEmatrix, Inc. All rights reserved.)*

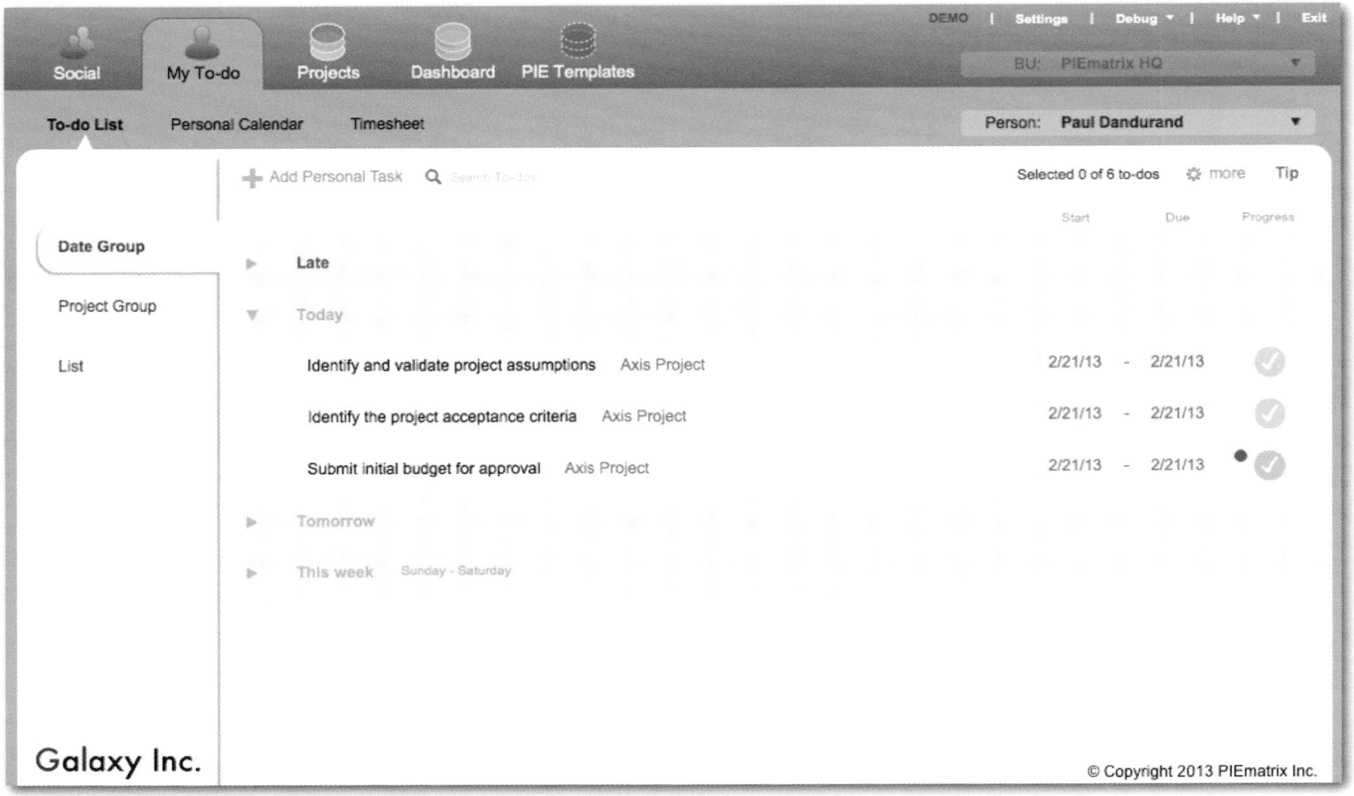

Figure 7-23 Piematrix to Do—Hover Over a Step to Show the Step's Instructions for Correct Execution *(Reproduced by permission of PIEmatrix, Inc. All rights reserved.)*

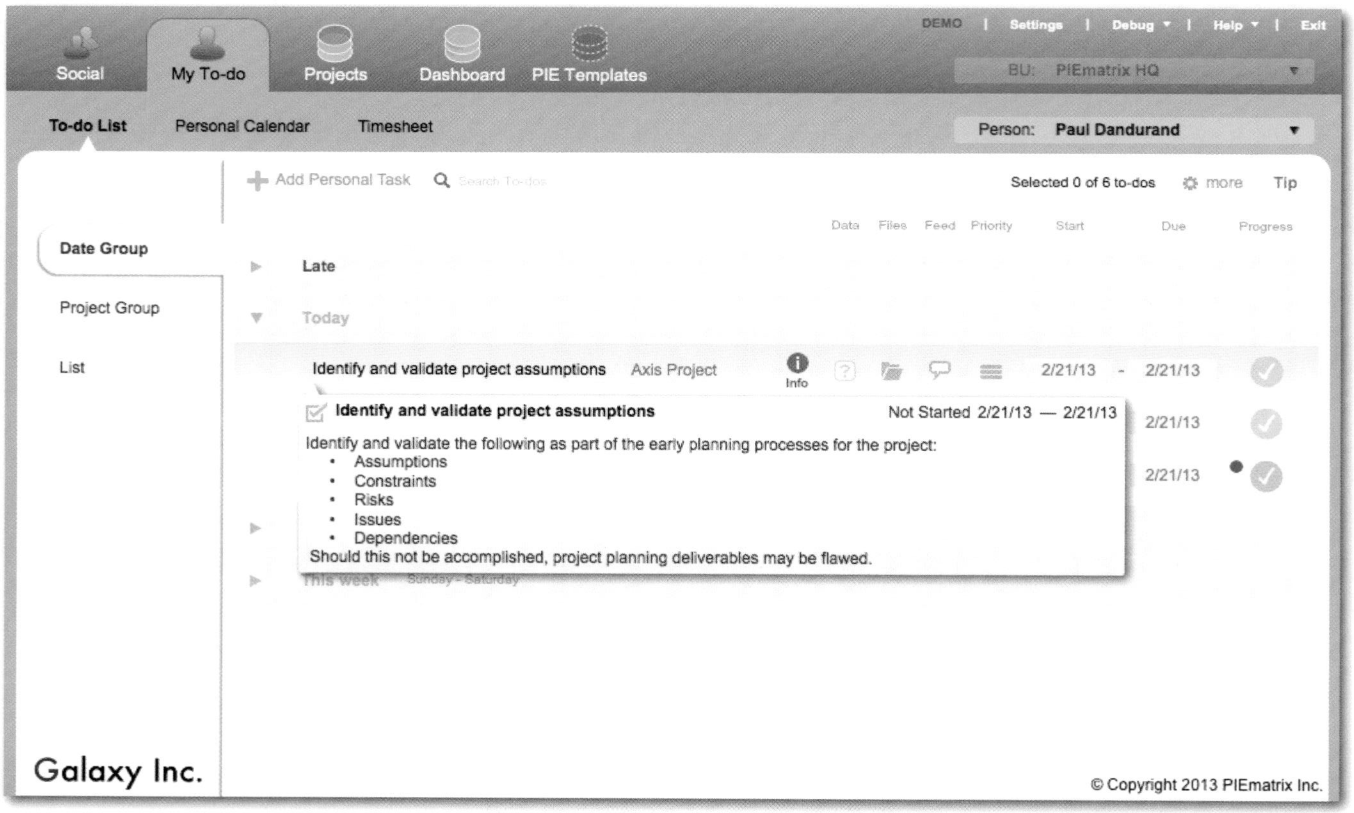

The bottom line is the executive can have more confidence that the information on the dashboard is based on executing the projects the best way possible.

Project—Governing and Executing Complex Projects in a Visual and Friendly Way

The PIEmatrix user interface makes it easier for us humans to follow complex project processes. This is done with their unique visual approach. This section will show how the team member or executive reviews and executes their part of the project process.

Launch my project.

A user can launch a project from multiple pages such as from the Dashboard, To Do, or the Project List page. Figure 7–24 shows a project being loaded from the project list page. In this example, the project has three process layers. We will select the Project Mgmt layer to load. This will load all the project's objects such as process steps, people assignments, dates, files, etc.

Show me the dates and progress for the project phase.

Once the layer loads, the user decides which phase ("pie slice") to view. In Figure 7–25, we selected the Plan phase to show its process boxes. Process boxes are the high-level steps of the phase. An executive can get to this view directly with one click when drilling down from the dashboard project page. The dates and the progress bars are turned on to show the progress bars and the schedule.

Show me the details of a project process.

Figure 7–26 shows what happens when we click on the process box Develop Project Charter. The box turns green and displays its action step list. To the right of each step are icons that can display information, such as dates and progress. The dark green checkmarks are set to Complete and the light green ones are set to In Progress. A project manager or executive can easily navigate between project pie slices and process boxes to view activity.

At this point, it's important to note that all the data (process boxes and steps) are derived from a predefined best-practice process template. As we can see, the content is generic for a number of projects of this type. The project team can easily customize the content with added custom process boxes, steps, and files. The project manager can even choose to make a predefined process step as "Not Needed." In doing so, PIEmatrix tracks the changes for accountability management.

Show me how to get my steps done right.

Previously we displayed the personal To-do page with a list of steps. This project view also shows the steps, but it shows them in a big picture view. We see all the steps for the team. This is helpful so we can see how our work fits into the larger scheme. As in the To-do page, a user can hover over a step to see its detailed instructions or tips on getting it done right.

Figure 7-24 Piematrix Project—Launch a Project and Select from Multiple Processes for Either Viewing or Execution *(Reproduced by permission of PIEmatrix, Inc. All rights reserved.)*

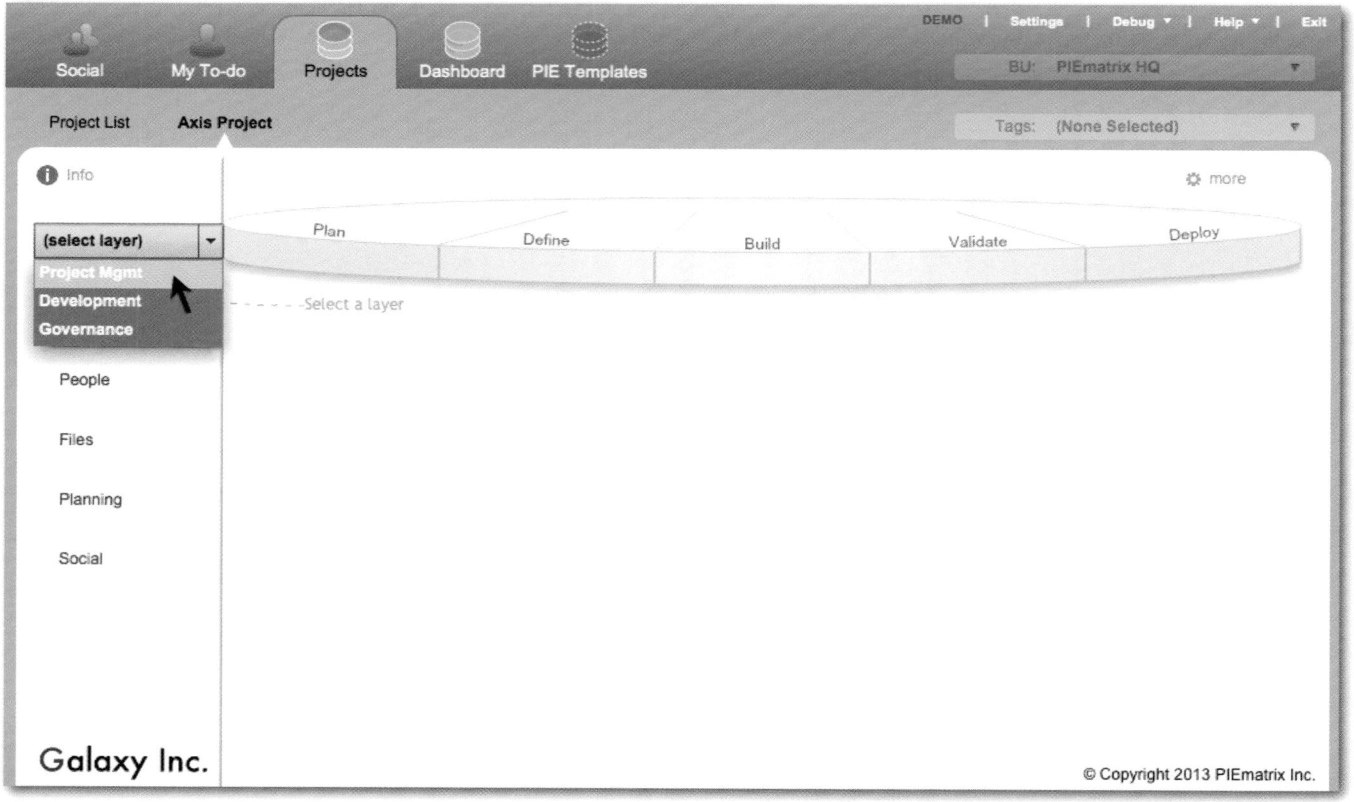

Figure 7-25 Piematrix Project—Selected Plan Phase and Turned on Progress Bars and Dates for Reporting

(Reproduced by permission of PIEmatrix, Inc. All rights reserved.)

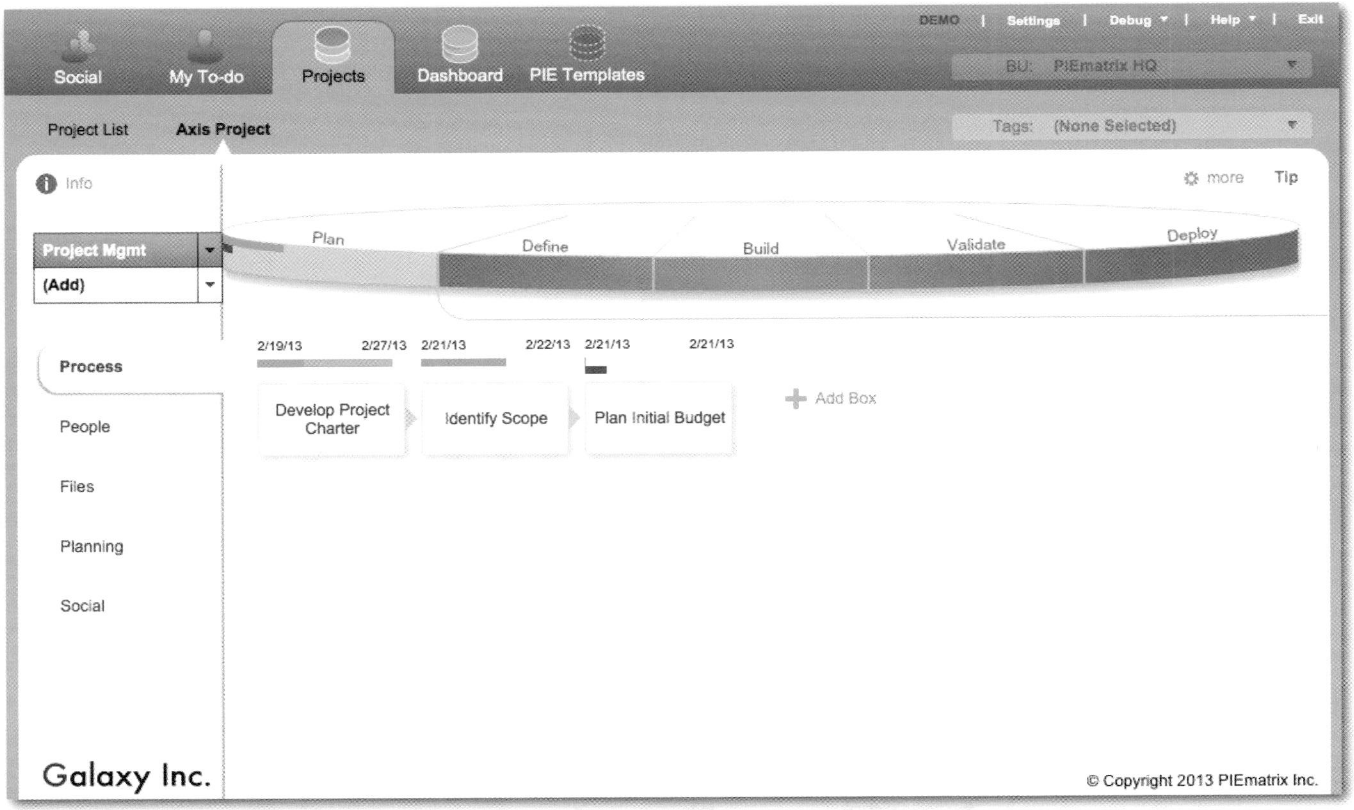

Figure 7-26 Piematrix Project—Selected Develop Project Charter Process Box to Show Its Steps *(Reproduced by permission of PIEmatrix, Inc. All rights reserved.)*

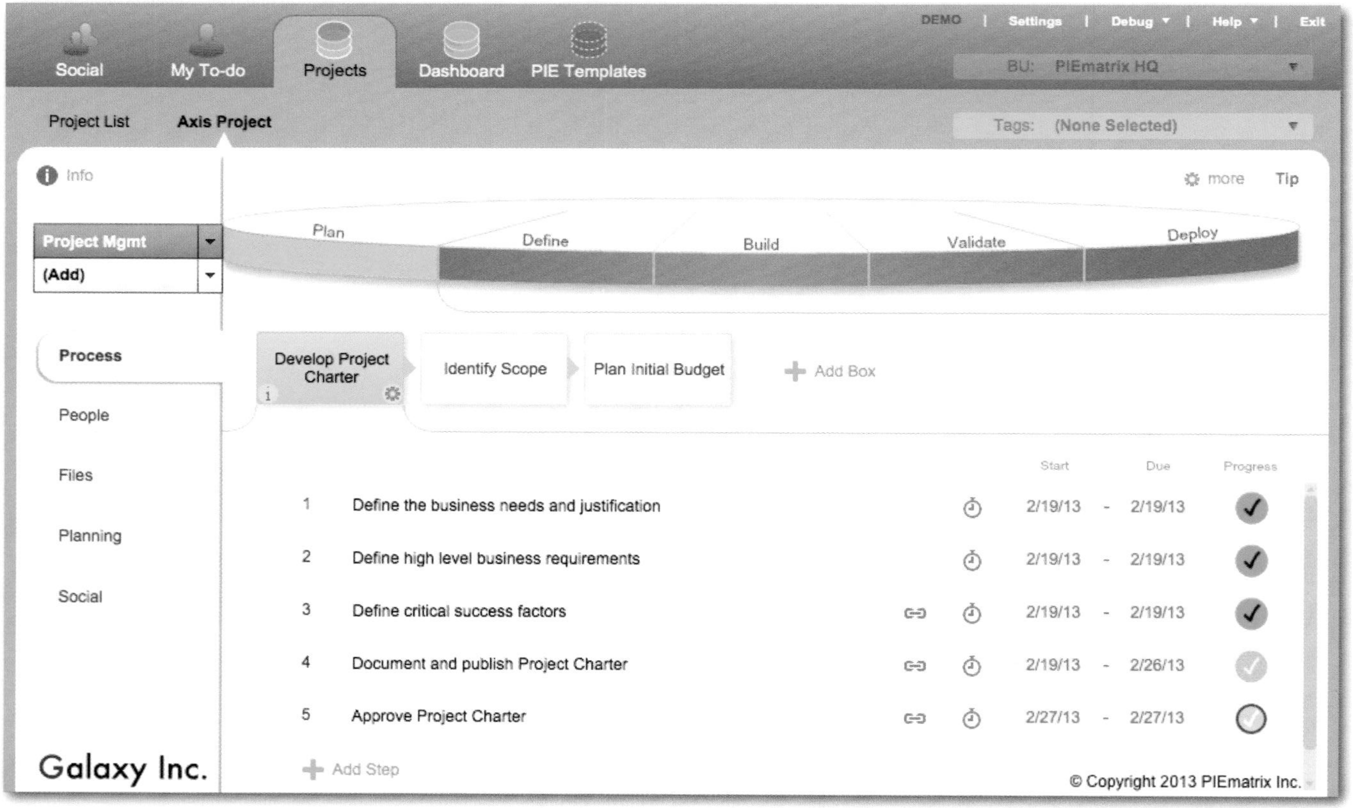

Figure 7–27 displays an example step description as we hover over the step "Define business needs and justification."

The project section of PIEmatrix has four other side tabs along with the Process tab. The information for a box changes as the user clicks between the side tabs. The People tab shows a people-centric view. The Files tab displays attached files and links for project assets. PIEmatrix has a built-in file repository system, which is important for allowing executives to access, review, and sign off on shared deliverable files. The Planning tab is described in the following section. The Social tab shows the news feed. The news feed includes messages and activity notices in a similar style to Facebook or LinkedIn.

Project—Planning the Project

PIEmatrix has a built-in planning tab for managing process steps, dates, and people assignments. This is done either in a Gantt view or a people list view. This is where the project manager sets up the schedule with dates. The executive comes to this view to review the actual schedule and also to compare it with the previously planned baseline. Getting here from the Dashboard view is just a couple of clicks away.

I want to view the detailed schedule for my process steps.

In our example, Figure 7–28 shows what happens when we click on the Planning tab while reviewing Develop Project Charter. Notice that the process box Develop Project Charter is still selected and the list of steps turn into a Gantt chart view. The steps now have progress bars that span their duration timeline (start to end date). Again, the dark green shows what's completed and the brighter greed shows what's in progress.

Show me the initial targeted baseline.

Projects don't always go on schedule. Some get behind while others get ahead. PIEmatrix allows the executive to view the initial baseline snapshot, which is a schedule set in a point in time. In the previous image, Figure 7–28, we display the current timeline for each step. The next image, Figure 7–29, shows the same steps, but this time with their initial baseline targets. The project manager had previously established this target to indicate the project's planned schedule.

Are we on target? Compare the current against the baseline snapshot.

Figure 7–30 displays a comparison between the current schedule (green and light blue colors) and the snapshot baseline schedule (dark blue color). In our project Axis, we can see the current schedule is already behind as compared with our initial plan.

Show me a higher roll-up view of the current and baseline.

We can view a higher level by unselecting the process box. In Figure 7–31 we deselected the process box Develop Project Charter. It now shows a Gantt view of the entire Plan phase. Notice the chart's bars represent the process boxes rather than the steps. Expanding a process box will then

Figure 7-27 Piematrix Project—Hover Over a Step to Show Its How-to Instructions (Process Standard for Execution) *(Reproduced by permission of PIEmatrix, Inc. All rights reserved.)*

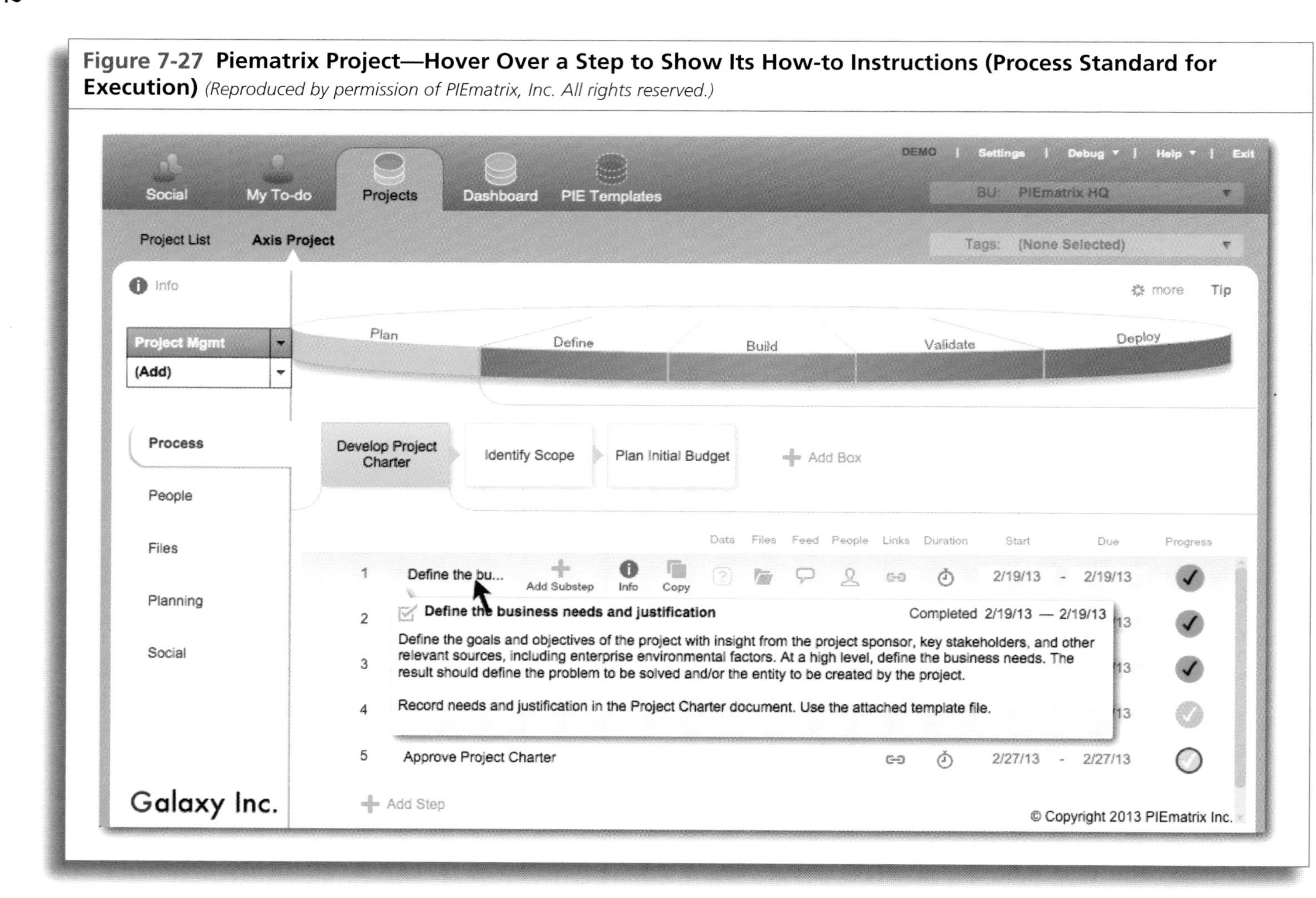

Figure 7-28 Piematrix Project Planning—Shows the Project's Current Schedule as a Gantt Chart *(Reproduced by permission of PIEmatrix, Inc. All rights reserved.)*

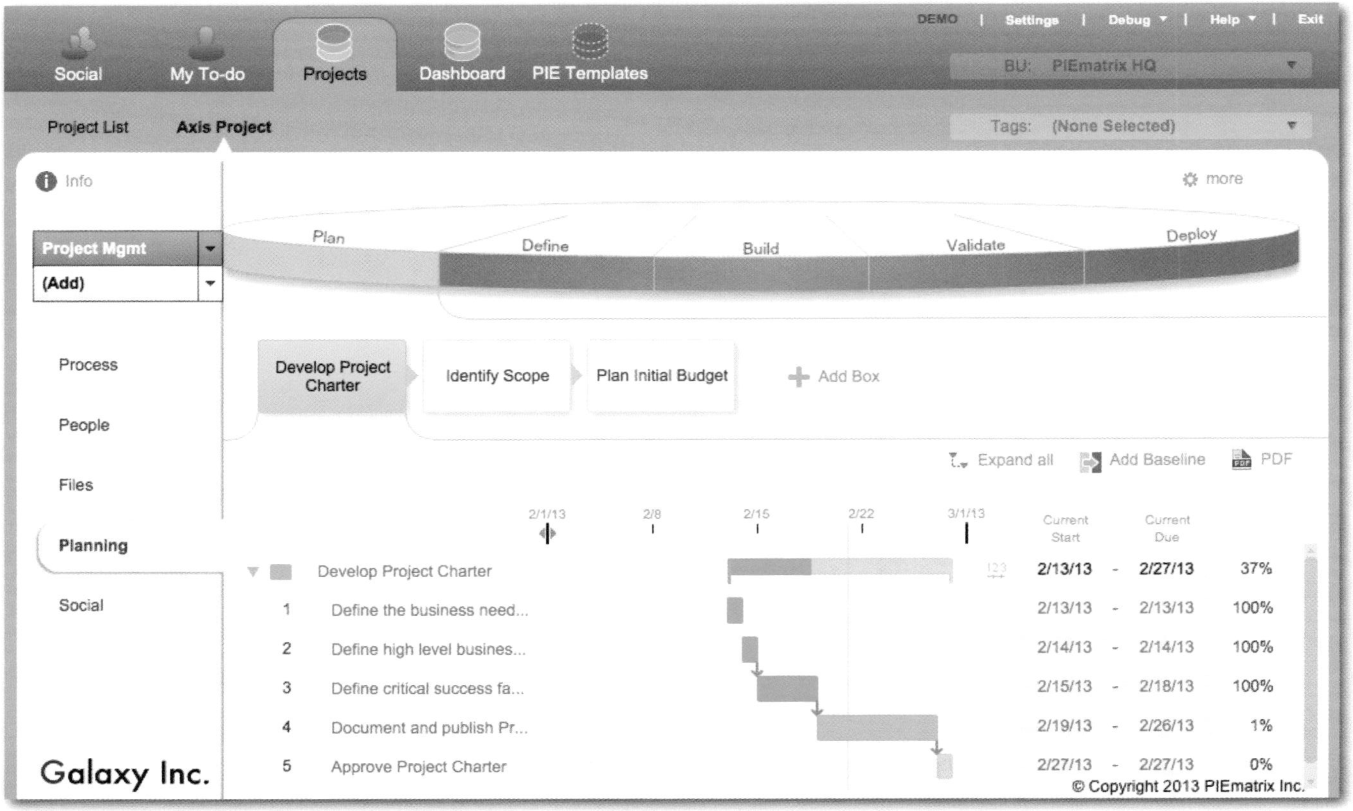

Figure 7-29 Piematrix Project Planning—Shows the Project's Initial Plan Schedule (Baseline Snapshot) as a Gantt Chart *(Reproduced by permission of PIEmatrix, Inc. All rights reserved.)*

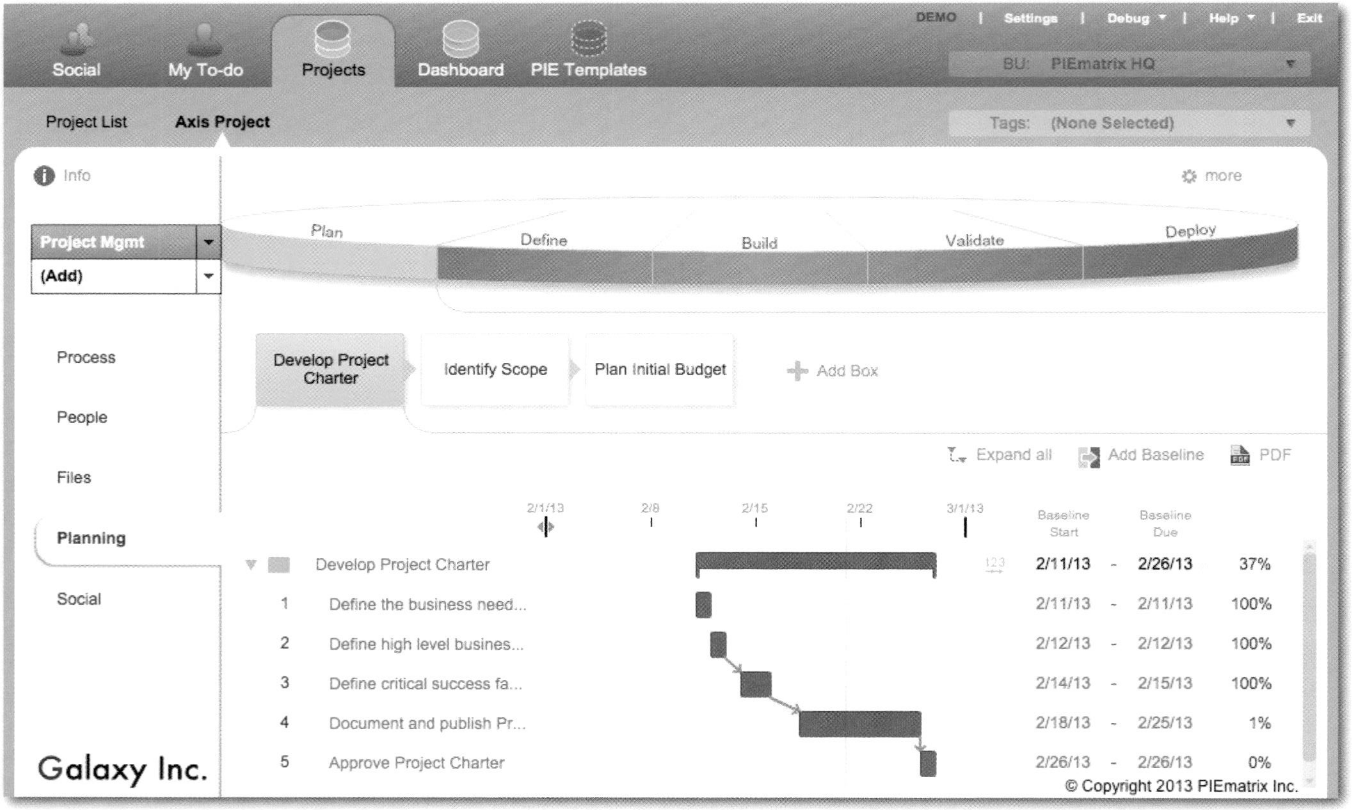
DEMO | Settings | Debug | Help | Exit
Social
My To-do
Projects
Dashboard
PIE Templates
BU: PIEmatrix HQ
Project List
Axis Project
Tags: (None Selected)
Info
more
Project Mgmt
(Add)
Plan
Define
Build
Validate
Deploy
Process
People
Files
Planning
Social
Develop Project Charter
Identify Scope
Plan Initial Budget
Add Box
Expand all
Add Baseline
PDF
2/1/13
2/8
2/15
2/22
3/1/13
Baseline Start
Baseline Due
Develop Project Charter 2/11/13 - 2/26/13 37%
1 Define the business need... 2/11/13 - 2/11/13 100%
2 Define high level busines... 2/12/13 - 2/12/13 100%
3 Define critical success fa... 2/14/13 - 2/15/13 100%
4 Document and publish Pr... 2/18/13 - 2/25/13 1%
5 Approve Project Charter 2/26/13 - 2/26/13 0%
© Copyright 2013 PIEmatrix Inc.
Galaxy Inc.

Figure 7-30 Piematrix Project Planning—Displays a Comparison between Current and Baseline (Dark Blue) Schedules *(Reproduced by permission of PIEmatrix, Inc. All rights reserved.)*

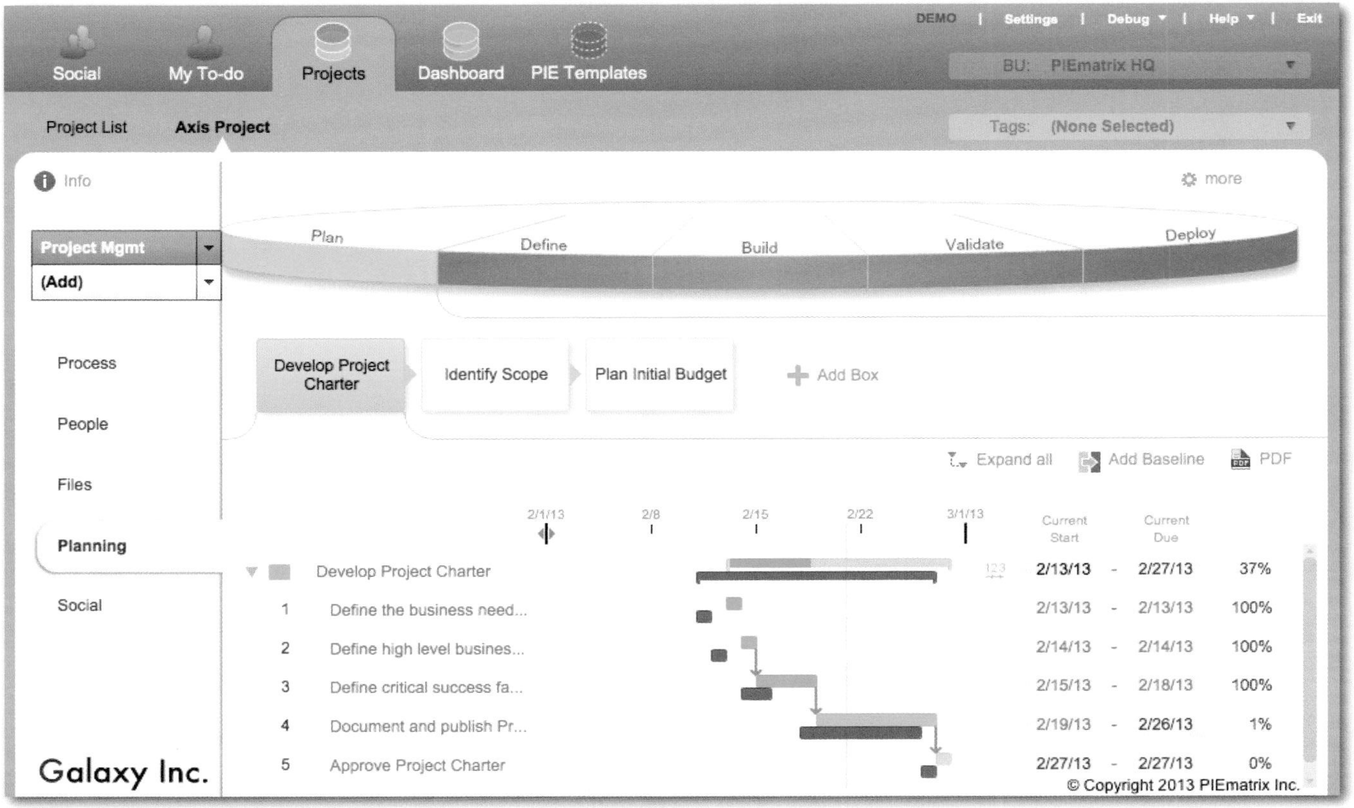

Figure 7-31 Piematrix Project Planning—Rollup Showing Gantt Chart for the Plan Phase

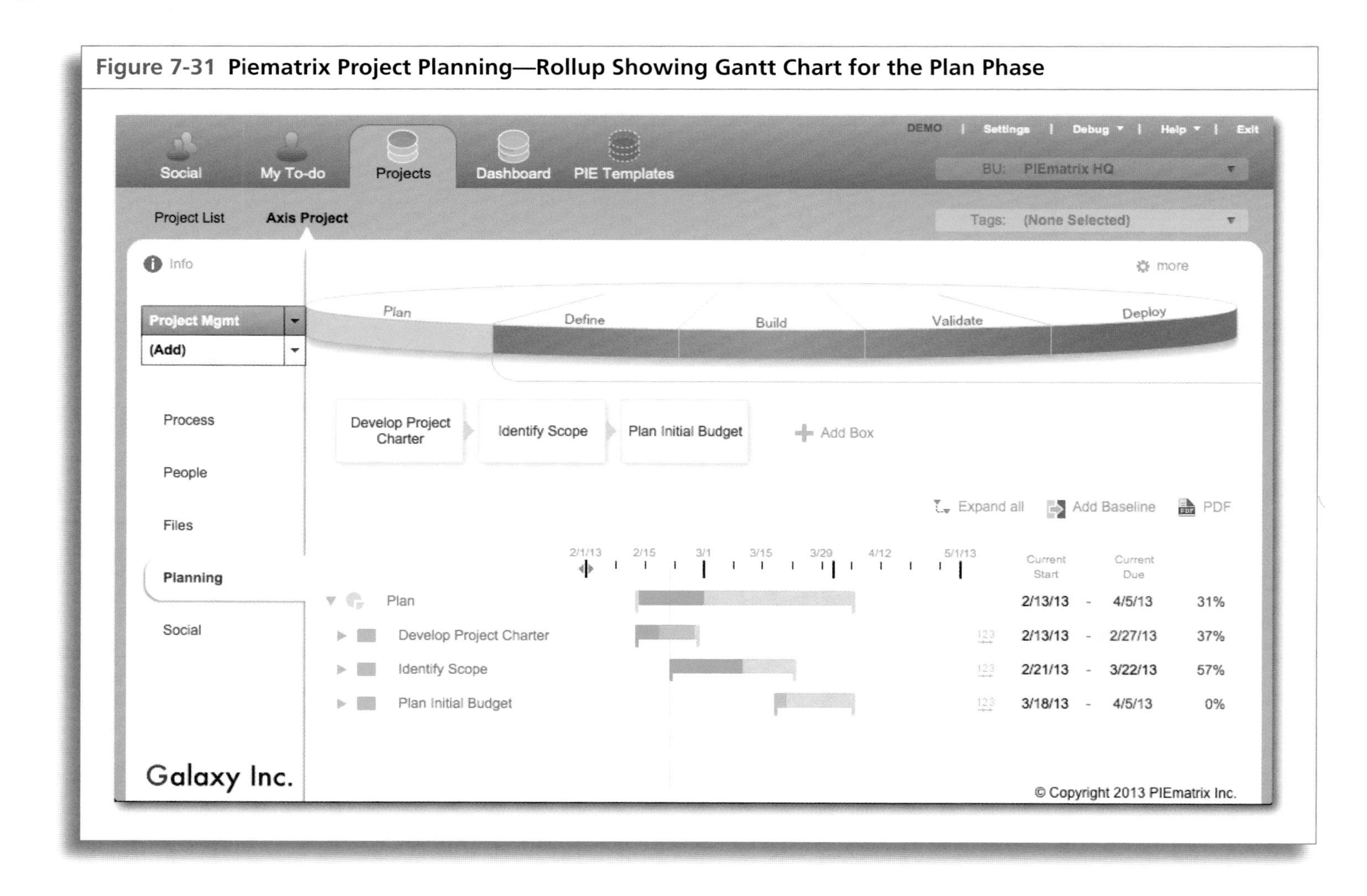

display its steps. PIEmatrix is consistent with showing the user only what is necessary to avoid noise overload. This contextual approach is very helpful for keeping the displays clean and focused.

As an executive or manager, we can increase the overall view of the project by deselecting the Plan slice button to see the progress bars across all major phases. Likewise, unloading the layer will show the entire project across all high-level process layers.

In summary, the executive can start at the dashboard level and work down from a high-level view to the detailed levels very quickly. And then he or she can work their way back up to high-level as needed.

Project—Breaking Down Silos

The PIEmatrix structure is perfect for process streams on large projects where different groups focus on different processes, yet they are still dependent on each other. We now introduce how a team and also an executive can see how different processes on a complex project can interconnect with each other.

Show me all project processes working in tandem on this project.

In Figure 7–32 we loaded three layers in the project view. As a project manager or executive we could load up any combination of process layers (work streams) and click on any cell (phase/layer intersection) to view details of that area.

Show me the progress of all the processes executing at the same time.

The Figure 7–33 shows the Plan slice cells selected down each process layer. We clicked the Progress button to display progress bars like you see in the Dashboard Portfolio Progress project detail view shown previously in Figure 7–19. Figure 7–33 is a dashboard-like view at the project level. From here we can quickly navigate to different phases and process streams with one click. We can also click once on the Planning tab to see the selection's Gantt view. This is all very simple and very fast.

Authoring—Where the Best Practice Content Comes From

Since PIEmatrix is process focused, the online application has an entire section built for creating and managing repeatable best practices, processes, or procedure content. An enterprise cannot only make these standards available for project teams; they can also keep them fresh with new updates in real-time. This PIEmatrix model is much better than the traditional documents on a server model where many struggle to keeping content up to date as well as getting everyone connected with the changes as they do their daily work.

Figure 7–34 displays a sampling of possible layers, or best practice content for different business areas. Each of these layer components contains phases, process steps, people roles, expected durations, dependency links, document templates, and more. The sky is the limit in terms of what

Figure 7-32 Piematrix Project—Displays Three Process Layers (or Work Streams) Ready for Viewing or Execution

(Reproduced by permission of PIEmatrix, Inc. All rights reserved.)

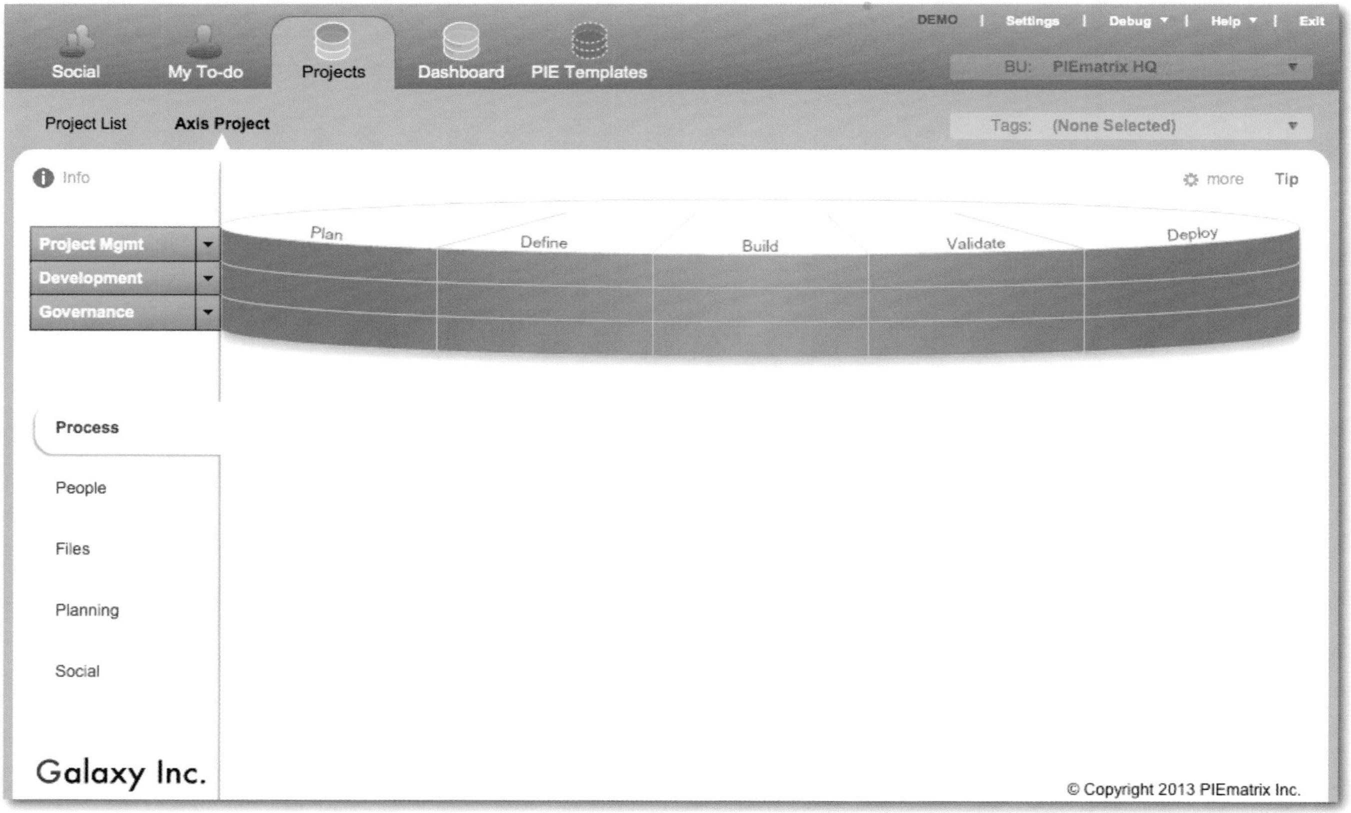

Figure 7-33 Piematrix Project—Displays Three Processes in Tandem under the Plan Phase

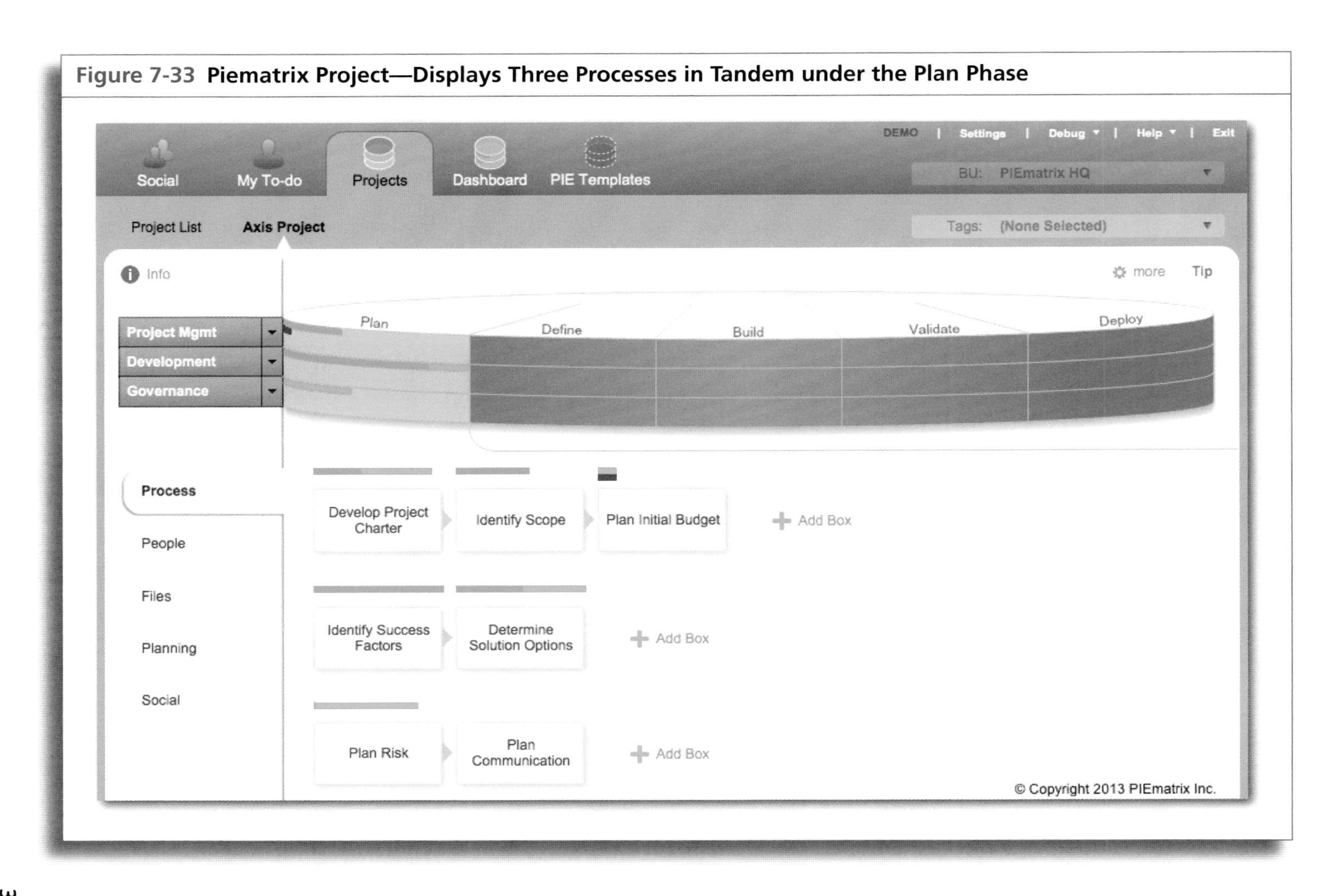

Figure 7-34 Piematrix Process Authoring—Sampling of Possible Process Areas Used as Pie Template *(Reproduced by permission of PIEmatrix, Inc. All rights reserved.)*

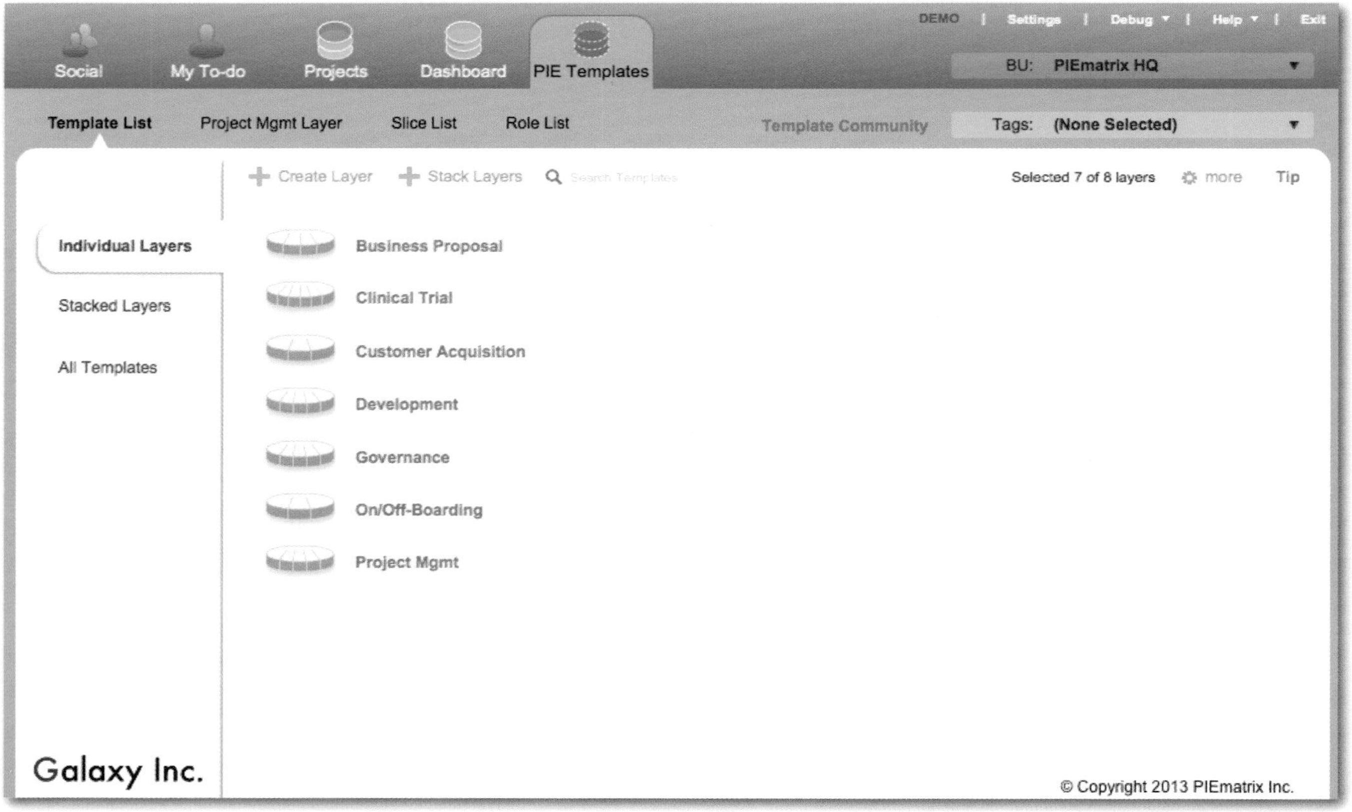

DEMO | Settings | Debug | Help | Exit
Social
My To-do
Projects
Dashboard
PIE Templates
BU: PIEmatrix HQ
Template List
Project Mgmt Layer
Slice List
Role List
Template Community
Tags: (None Selected)
Create Layer
Stack Layers
Search Templates
Selected 7 of 8 layers
more
Tip
Individual Layers
Stacked Layers
All Templates
Business Proposal
Clinical Trial
Customer Acquisition
Development
Governance
On/Off-Boarding
Project Mgmt
Galaxy Inc.
© Copyright 2013 PIEmatrix Inc.

processes an organization can build and deploy. The following is a sampling of process layer ideas:

- *Business proposal*. This can be a process for managing the intake of project proposals or ideas. The process could include how to submit a business case, review for approval, and then prioritize for execution.
- *Customer acquisition*. This could be a process for a complex sales life cycle that involves multiple stakeholders and would complement a CRM system.
- *On/off boarding*. This process layer could execute and manage the steps for hiring people. Another HR layer could be setup to execute yearly employee reviews.
- *Project management*. This layer can contain a set of practical steps that are based on the PMI PMBOK guidelines.

Other process examples could target sustainability, greening for energy reduction, pandemic response planning and execution, risk and safety controls, and so on.

In the process layers list page, clicking a layer launches it in edit mode. Figure 7–35 displays a sample layer ready for editing. Notice how this edit mode looks almost exactly as it does in project execution mode. PIEmatrix has an outline button that turns all of the visuals into a standard outline view for those who prefer building process content in an outline format.

From Authoring Back to the Executive Dashboard

Working our way backwards, our authored process layer becomes a ready-to-use process. This can be published for use across the enterprise or within a department. In any case, the repeatable best practice project content is ready for team execution and executive governance.

The teams execute and collaborate as they do their daily work. In parallel, executives govern their critical views as the activities are executed in real-time. More important, everyone sleeps better at night knowing the projects are being done with the right process steps that reduce cost, enhances efficiency, and keeps all out of trouble.

The next time an executive looks at a project portfolio dashboard, he or she should ask if the data is in real time, if it's process focused, and if the data is based on getting the job done right for the organization.

For information on PIEmatrix, go to www.piematrix.com.

Figure 7-35 Piematrix Process Authoring—Sampling of Possible Process Areas Used as Pie Template *(Reproduced by permission of PIEmatrix, Inc. All rights reserved.)*

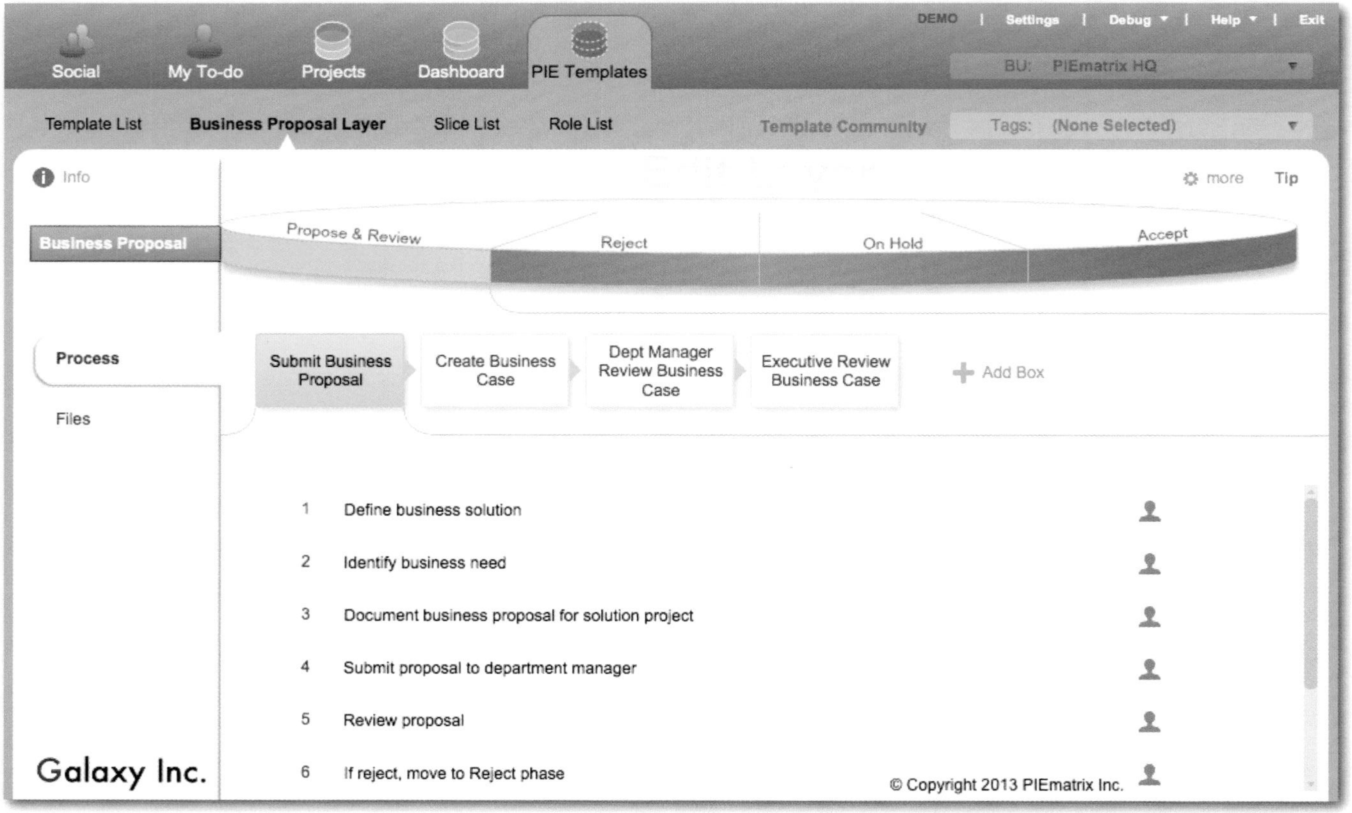

7.6 DASHBOARDS IN ACTION: INTERNATIONAL INSTITUTE FOR LEARNING

The International Institute for Learning (IIL) is one of the world's largest project management educational and consulting services providers. With a multitude of clients scattered across the globe and each at a possible different level of maturity in project management, IIL is often called upon to create different sets of dashboards for their clients.

Dashboards do not need to be highly complex. Even the simplest form of dashboard can provide effective information for decision makers. Figures 7–36 through 7–41 7–36, 7–37, 7–38, 7–39, 7–40 and 7–41 illustrate that simple dashboards can often be effective or even more effective than complex dashboards. Loading up a dashboard with unnecessary bells and whistles does not increase the quality of the information.

In the last line of Figure 7–41, $1′000,000 looks strange, but it is correct. In the international format, which is used by many countries; the apostrophe is used to separate millions and comma for thousands.

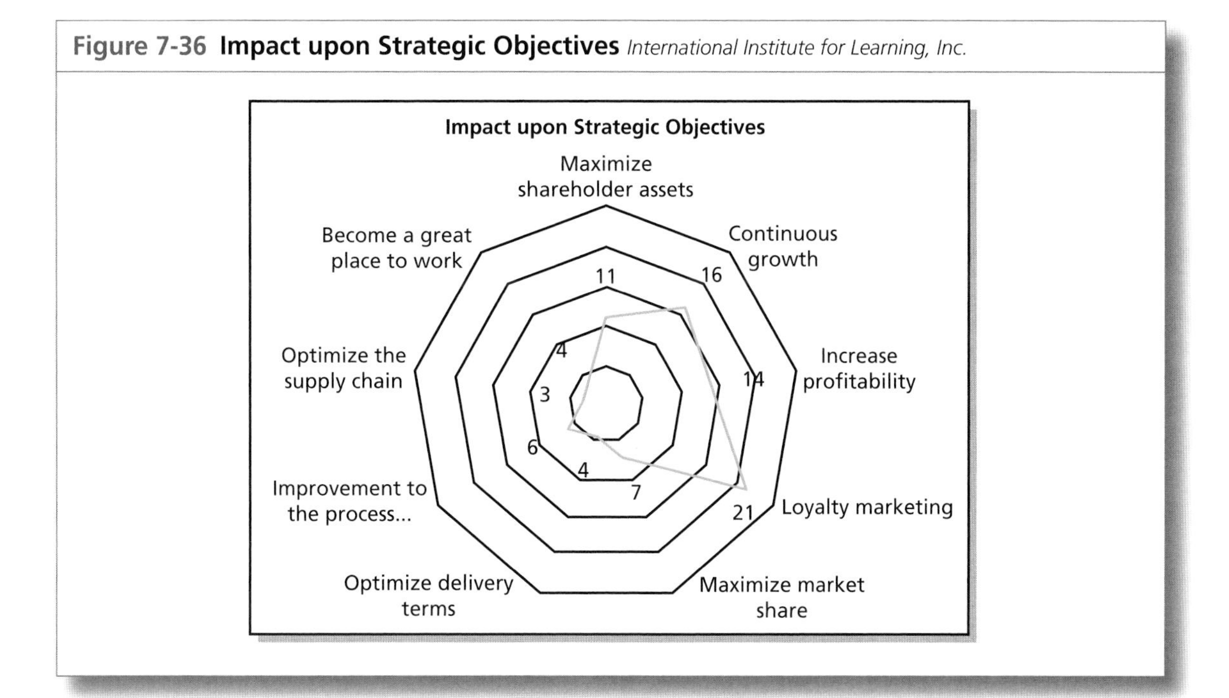

Figure 7-36 Impact upon Strategic Objectives *International Institute for Learning, Inc.*

Figure 7-37 Projects within the Business Area *International Institute for Learning, Inc.*

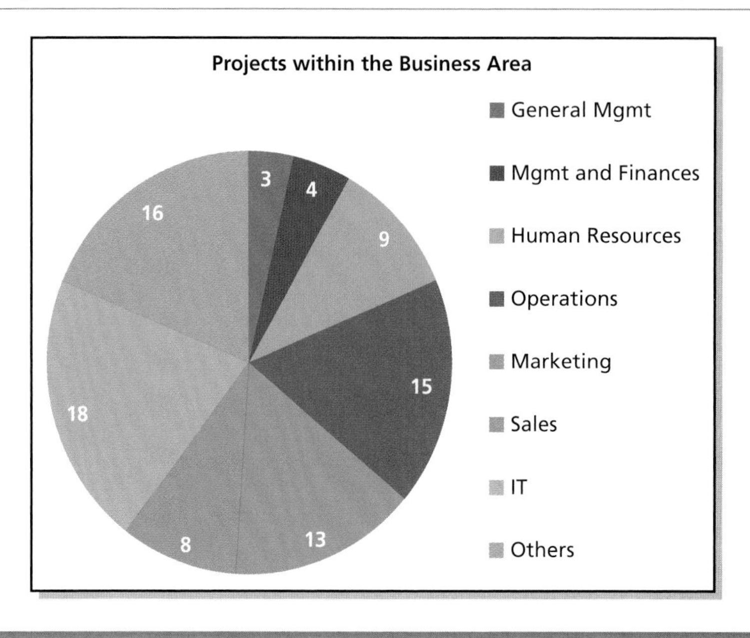

Figure 7-38 Project Origin *International Institute for Learning, Inc.*

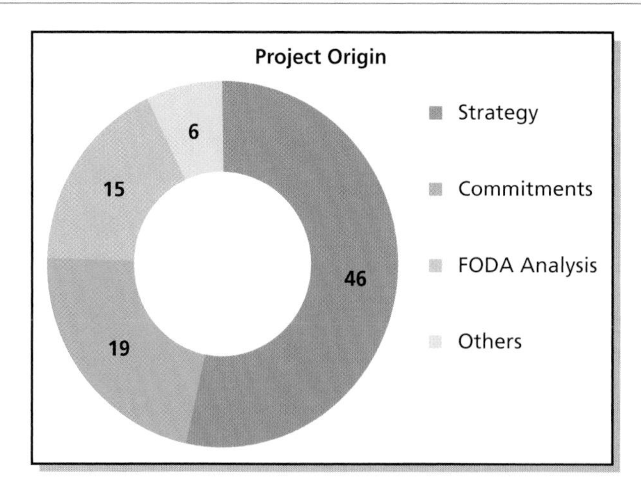

Figure 7-39 Project Status within the Business Unit *International Institute for Learning, Inc.*

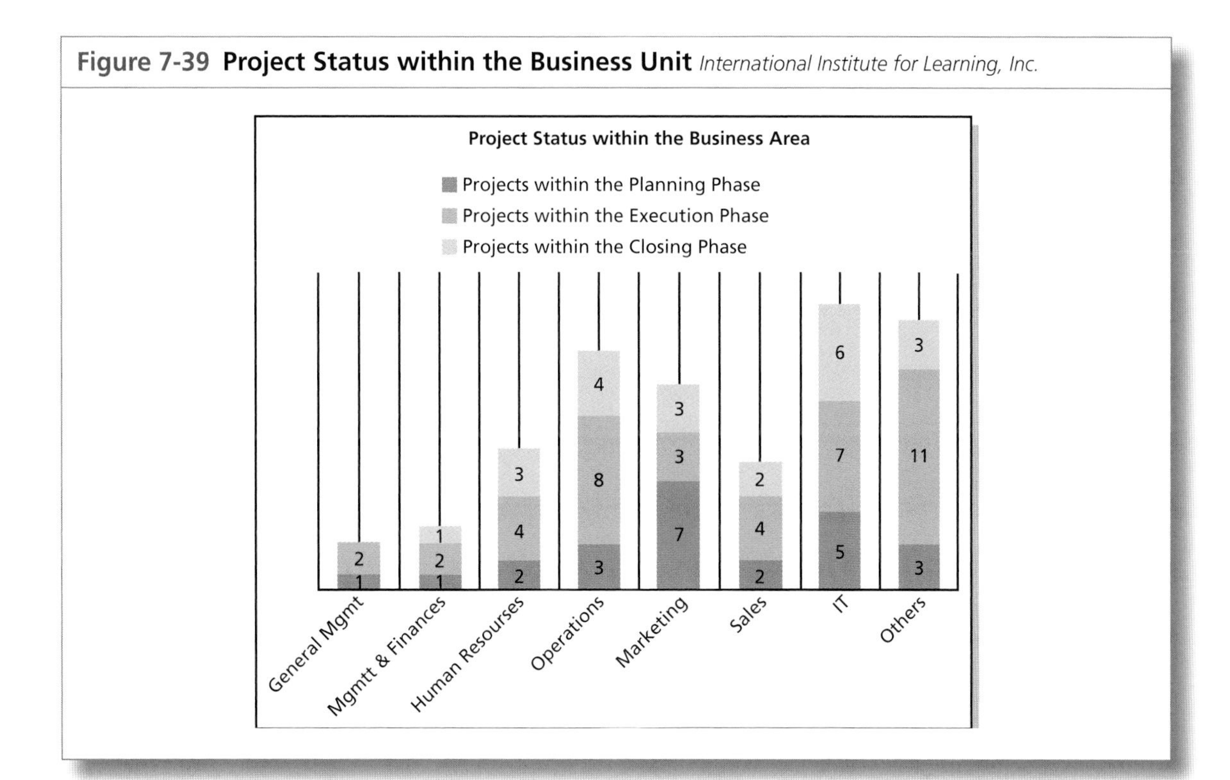

Figure 7-40 Projects by Year of Approval *International Institute for Learning, Inc.*

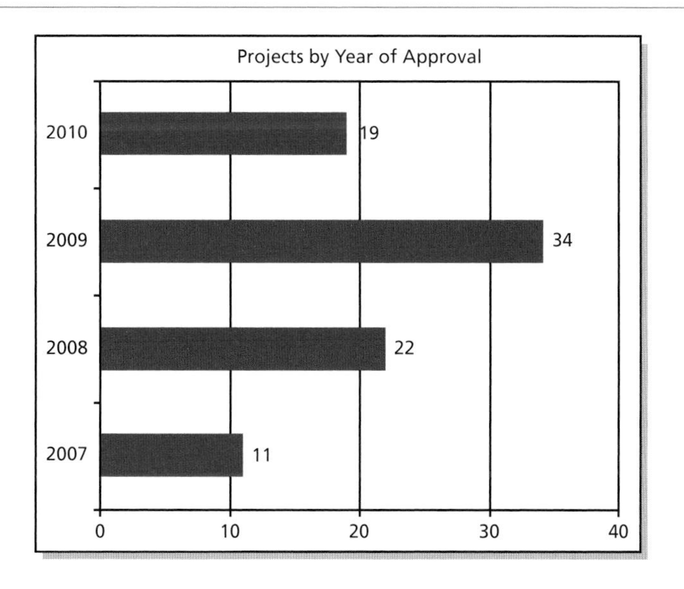

Figure 7-41 **Budget for the Projects** *International Institute for Learning, Inc.*

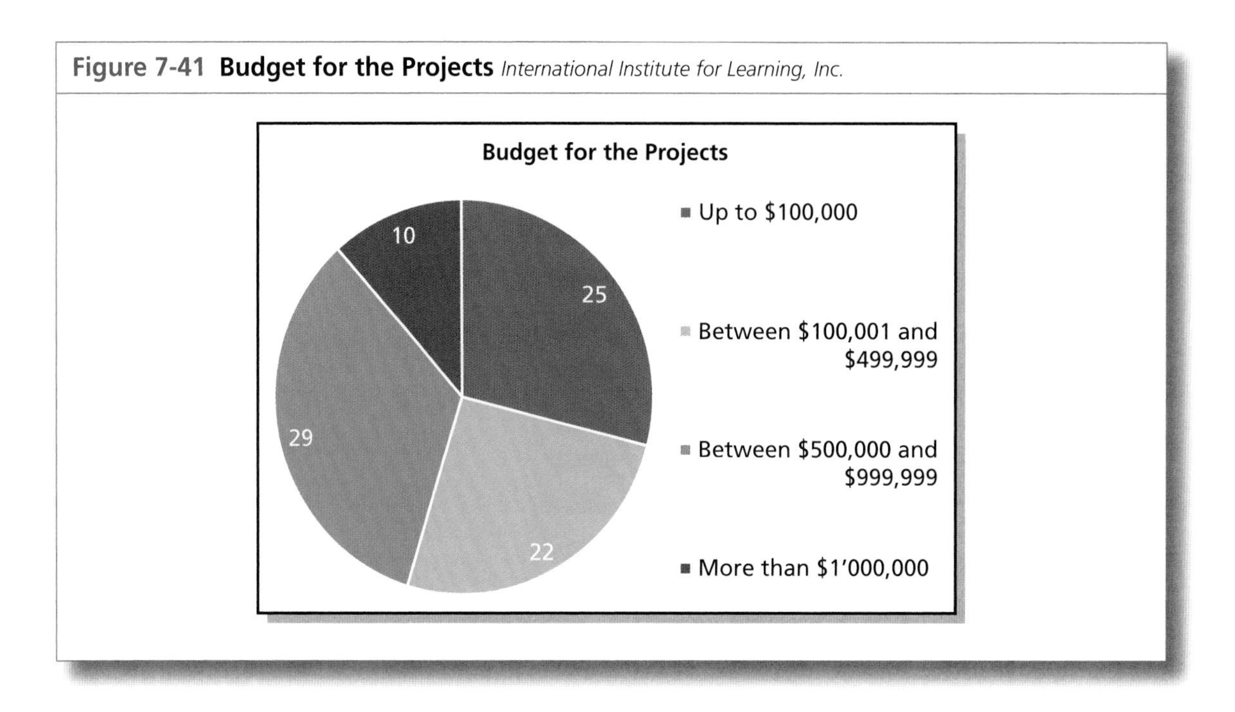

7.7 DASHBOARDS IN ACTION: WESTFIELD INSURANCE[4]

Westfield Insurance (westfieldinsurance.com), an insurance, banking, and related financial services group of businesses headquartered in Westfield Center, Ohio, has a delivery unit focused on project delivery via the use of "virtual teams" composed of technology and business team members. The status of enterprise projects are reported on monthly to the business and IT leaders. From that information and additional analysis of portfolio risk and project success, the project portfolio is summarized monthly in a dashboard format, as shown in Figure 7–42.

Figure 7–42 shows that effective dashboards are neither entirely artwork nor text material. A mixture of text and artwork often provides the best results.

4. Material provided by Janet Kungl, PMP, Program Manager, Westfield Insurance. © 2012 by Westfield Insurance.

Figure 7-42 June 2010 IT Services Metrics Dashboard ***©2012 by Westfield Insurance.***

Jun 2010-IT Service Metrics

IT Service: Program and Project Planning and Execution *Core Metrics:* Budget, Schedule, Quality, Customer Satisfacgtion, Scope

Scope of Service: Enables the delivery of new business capabilities through solution development

Portfolio Mix by Business Unit

Business Unit/Area	Current Projects	Est. % of Total Project Cost
Business Unit A	6	28.0%
Business Unit B	2	24.2%
Business Unit C	1	16.4%
Business Unit D	1	11.4%
Business Unit E	1	10.4%
Business Unit F	1	9.6%

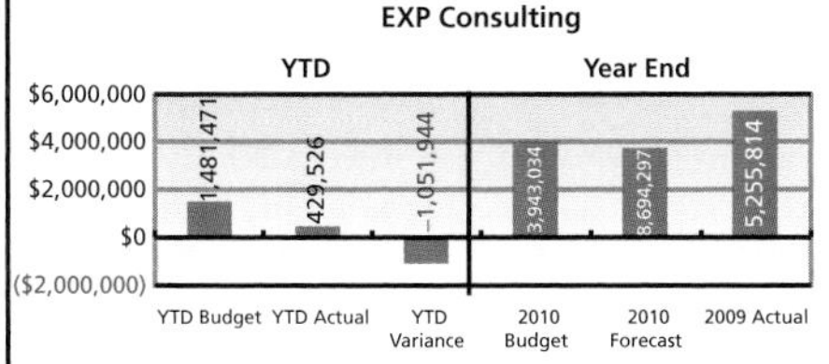

EXP Consulting

YTD | Year End

$6,000,000
$4,000,000
$2,000,000
$0
($2,000,000)

YTD Budget 1,481,471
YTD Actual 429,526
YTD Variance -1,051,944
2010 Budget 3,943,034
2010 Forecast 8,694,297
2009 Actual 5,255,814

Portfolio Execution Status

Status	June	May	April
Red	0	0	0
Yellow	0	2	1
Green	11	9	10
Not Started	0	0	0
Completed	1	1	1

Analysis

There are 12 current programs and projects in the portfolio. 11 are In Process and 1 is Completed.

The YTD underrum of $1,052M is attributable to an underspend in Project A ($625K), Project B ($156K), Project C ($167K) and Project D (S120K.)

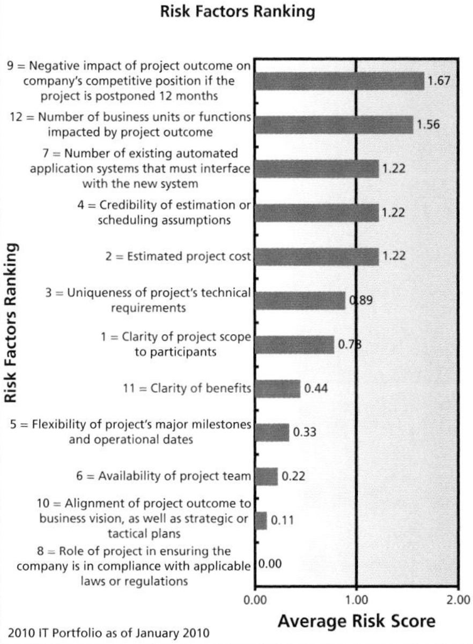

Risk Factors Ranking

Risk Factors Ranking

9 = Negative impact of project outcome on company's competitive position if the project is postponed 12 months 1.67
12 = Number of business units or functions impacted by project outcome 1.56
7 = Number of existing automated application systems that must interface with the new system 1.22
4 = Credibility of estimation or scheduling assumptions 1.22
2 = Estimated project cost 1.22
3 = Uniqueness of project's technical requirements 0.89
1 = Clarity of project scope to participants 0.78
11 = Clarity of benefits 0.44
5 = Flexibility of project's major milestones and operational dates 0.33
6 = Availability of project team 0.22
10 = Alignment of project outcome to business vision, as well as strategic or tactical plans 0.11
8 = Role of project in ensuring the company is in compliance with applicable laws or regulations 0.00

0.00 1.00 2.00

Average Risk Score

2010 IT Portfolio as of January 2010

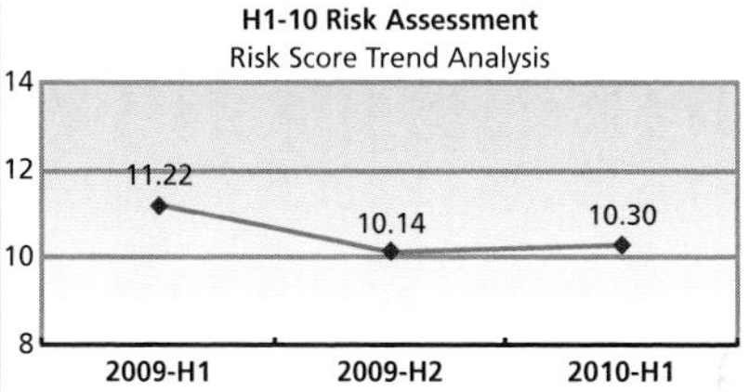

H1-10 Risk Assessment

Risk Score Trend Analysis

14
12
10
8

2009-H1 11.22
2009-H2 10.14
2010-H1 10.30

Analysis

The risk factors associated with the number of business units or functions impacted by project outcome, and the negative impact to the business if the project were delayed 12 months are the highest average risk factors.

As noted in previous assessments, this indicates that a high risk to the technology portfolio comes from the projects that are essential to our company's competitive position and that have broad impact to the business. We would expect these factors to remain high since they define the base reason for why we apply formal program and project management to initiatives in the technology portfolio.

This assessment's results were based on 10 projects; 4 at budget level and 6 at order of magnitude level. By comparison, the previous assessment was based on 7 projects; 6 at budget level and 1 at order of magnitude level.

As expected, we saw the overall risk rating trend upward as new order of magnitude projects were added to the technology portfolio. We also saw a moderate decrease in the risk associated with estimated project cost due to staffing portfolio projects with fewer consultants.

We expect the overall risk rating to decrease in a number of factors such as credibility of estimation assumptions and clarity of project scope as projects are further defined.

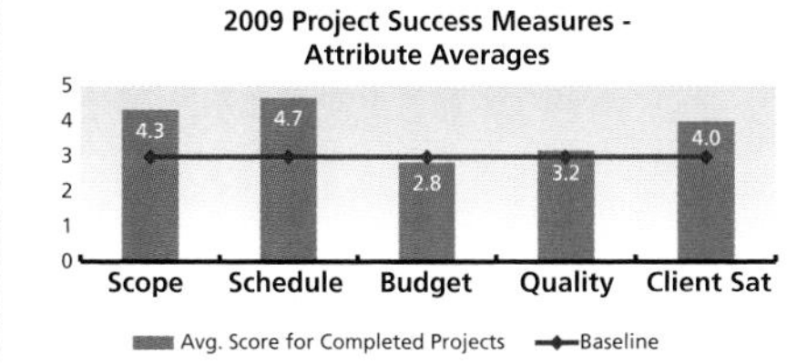

2009 Project Success Measures - Attribute Averages

5
4
3
2
1
0

Scope 4.3
Schedule 4.7
Budget 2.8
Quality 3.2
Client Sat 4.0

Avg. Score for Completed Projects —◆—Baseline

Success Measures

Attribute	2009 Count	2009 % of Completed Projects	2008 Count	2008 % of Completed Projects	2007 Count	2007 % of Completed Projects
Total Projects	6	-	17	-	8	-
Scope	6	100	17	100	8	100
Schedule	6	100	13	76	5	63
Budget	4	67	9	53	2	25
Qualit[1]	5	83	14	93	6	
Client Satisfaction[2]	6	100	11	73	6	100

1 - Quality ratings were only applicable for 15 of the 17 projects completed ub 2008, and for 6 of the 8 projects completed in 2007.

2 - Client satisfaction ratings were only available for 15 of the 17 projects completed in 2008, and for 6 of the 8 projects completed in 2007.

7.8 DASHBOARDS IN ACTION: MAHINDRA SATYAM[5]

Mahindra Satyam is a leading information, communications, and technology (ICT) company providing first-class business consulting, information technology and communication services. By leveraging deep industry and functional expertise, leading technology practices, and a global delivery model, Mahindra Satyam enables companies to achieve their business goals and transformation objectives.

Mahindra Satyam is powered by a pool of talented IT and consulting professionals across enterprise solutions, client relationship management, business intelligence, business process quality, operations management, engineering solutions, digital convergence, product life cycle management, and infrastructure management services, among other capabilities. They maintain development and delivery centers in the United States, Canada, Brazil, the UK, Hungary, Egypt, UAE, India, China, Malaysia, Singapore, and Australia as well as serving numerous clients, including several Fortune 500 companies.

Companies like Mahindra Satyam must possess the capability to develop multiple dashboard designs for a multitude of companies worldwide, and each at possibly at a different level of project management maturity. For companies at a more advanced level of maturity, highly detailed dashboards can be created. For companies that may be at the infancy stages of project management maturity, relatively simple dashboard designs may be usable. Figures 7–43, through 7–47 7–44, 7–45, 7–46, and 7–47 show typical dashboards that can be used for clients that are at various levels of project management maturity.

Figure 7-43 A Typical Dashboard for Projects per Customer *©2012 by Mahindra Satyam; reproduced with permission*

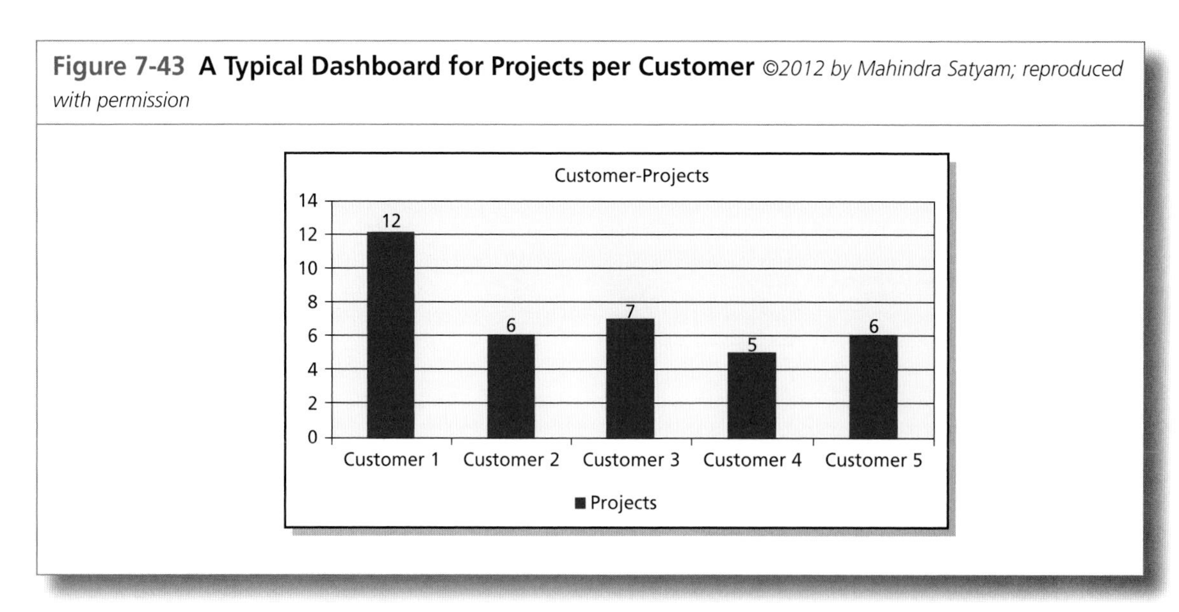

5. Material on Mahindra Satyam provided by Hirdesh Singhal and S. Mahadevan. of Mahindra Satyam's Process Management Group. © 2012 by Mahindra Satyam; reproduced with permission.

Figure 7-44 Dashboard for a Program Summary Report *©2012 by Mahindra Satyam; reproduced with permission*

Program Summary Report			
Program ID	PgM012010	Program Manager Name	Steve Martin
Program Name	UK Program 1	Program Manager ID	77747
Program Size - $(USD) Value*	$16,510,000,00	No. Customers	5
Program Size - No. Projects	48	No. Project manager's	36
Program Size - No. Associates	1800	Onsite Offshore Ratio	01:04.2
Source	Optima	Refresh Date	2-Aug-10

Customer Name	Size		Domain
	Dollar Value	Associate Count	
Customer 1	$5,200,000.00	718	Banking
Customer 2	$3,200,000.00	340	Banking
Customer 3	$4,200,000.00	390	Banking
Customer 4	$1,520,000.00	80	Insurance
Customer 5	$2,290,000.00	272	Insurance

Figure 7-45 Sample of a Typical Dashboard Report *©2012 by Mahindra Satyam; reproduced with permission*

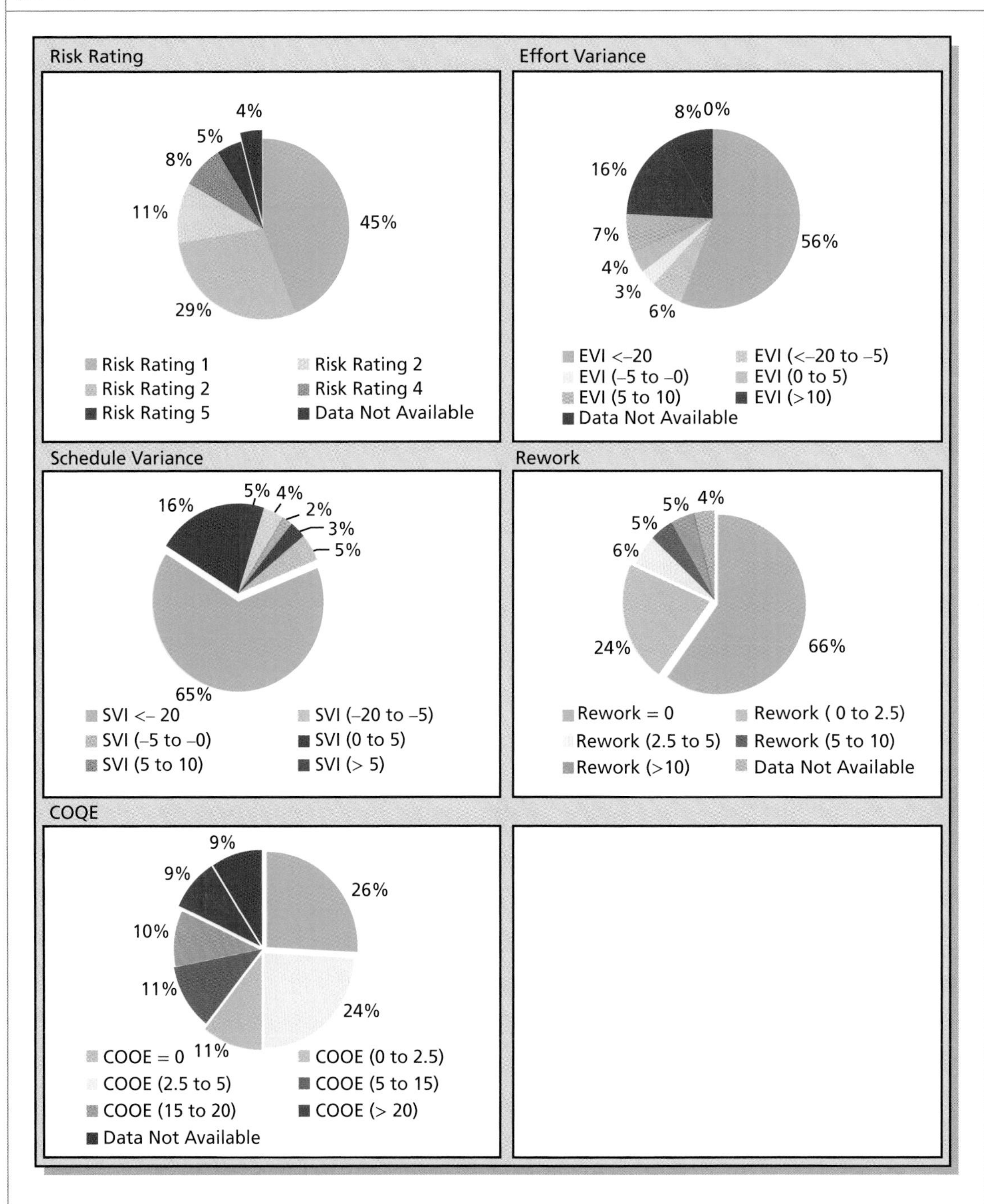

Figure 7-46 Sample of a Typical Dashboard Report *©2012 by Mahindra Satyam; reproduced with permission*

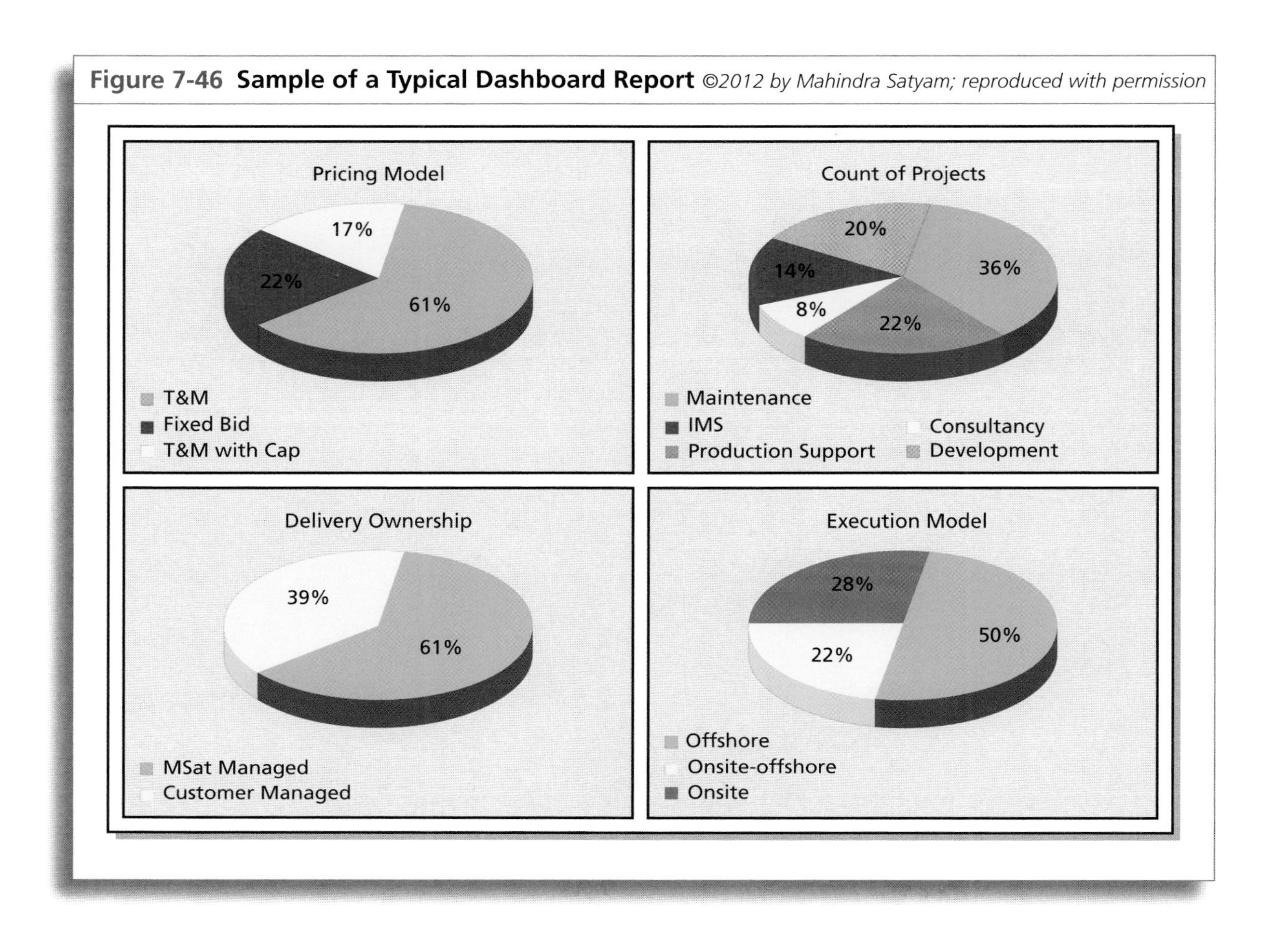

Figure 7-47 Sample of a Typical Dashboard Report *©2012 by Mahindra Satyam; reproduced with permission*

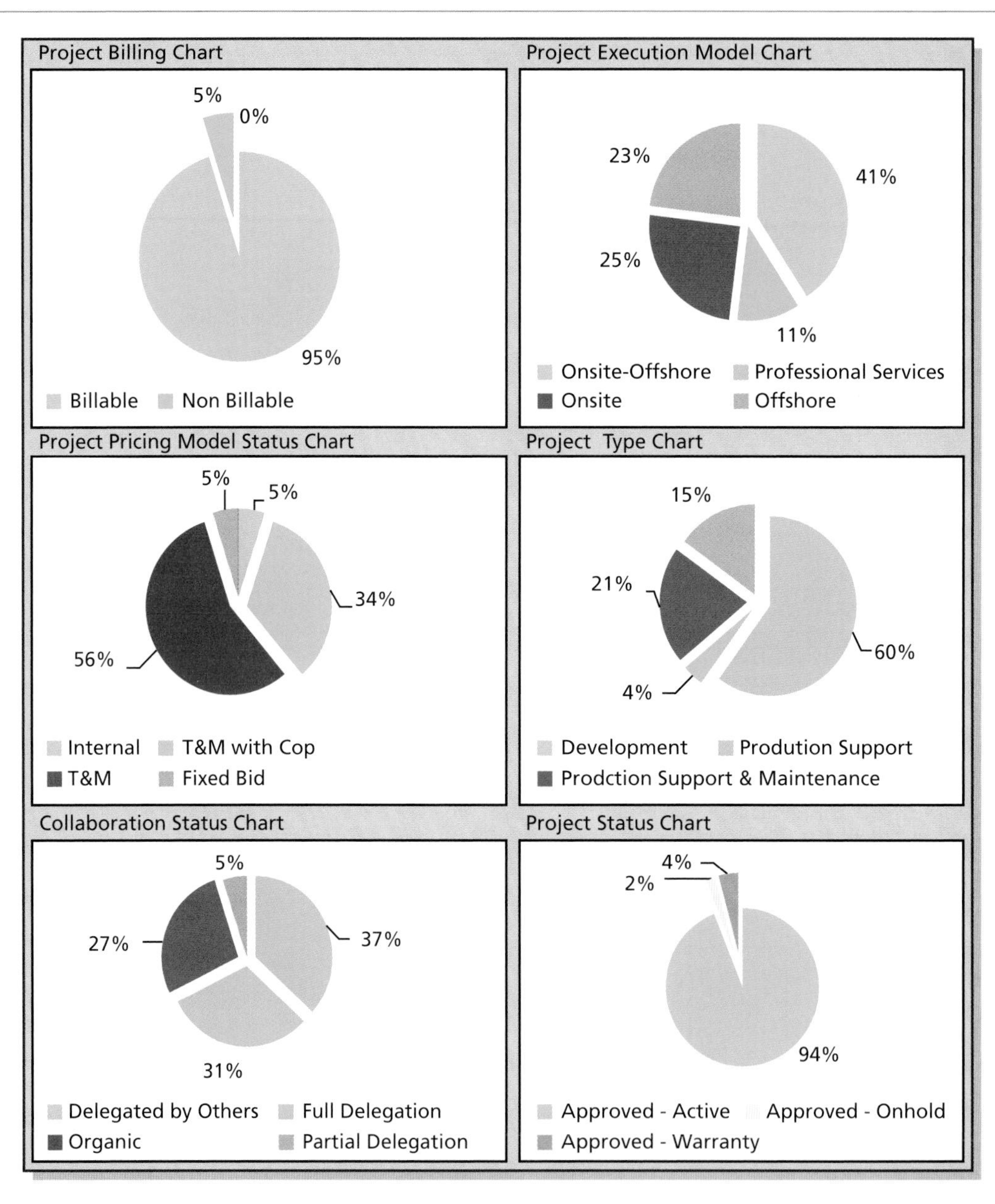

8 MEASUREMENT-DRIVEN PROJECT MANAGEMENT[1]

CHAPTER OVERVIEW

Measurement seems simple, and sometimes it is simple. Measurement concepts in a project or business environment are more complicated than most people think, however. To truly understand measurement, it is necessary to give up some commonly held beliefs and adopt new ways of thinking. This chapter will present measurement concepts, measurement definitions, and the measurement creation process, and demonstrate the process with a case study.

CHAPTER OBJECTIVES

- To understand the necessity for effective measurement
- To understand various measurement techniques
- To understand which techniques may be more appropriate than other for a given situation

KEY WORDS

- Measurement
- Information requirement
- Information solution

8.0 INTRODUCTION

The chapter is divided into four parts. The first part presents an explanation of measurement concepts and addresses some common misconceptions. The second part defines some of the words of measurement that are not the definitions used by most people. The third part presents a process for the creation of a measure in a project environment. In a project context, measurement gives people the information they need to make decisions and manage a project to a successful conclusion. The process transforms abstract needs into an information requirement and provides a useful

1. Chapter 8 was prepared by John Sponholtz; ©2010 by John Sponholtz; Reproduced by permission. Material has been adapted from John Sponholtz's work in progress book *Measurement-Driven Project Management*.

information solution to those information requirements. The process has four steps:

- Identify Information Requirements
- Analyze Information Requirements
- Create Indicators
- Integrate Measurement into the Project Processes

The fourth part of this chapter introduces some additional information about measurement categories.

8.1 MEASUREMENT CONCEPTS[2]

There are many misconceptions about measurement. This section presents some concepts that form the basis for the rest of the chapter and presents statements designed to help people think properly about measurement.

If It Matters, It Is Detectable

If no one knows of its existence, how can it be of concern? The very fact that something matters means that it is detectable. If it is detectable, there will be evidence of its existence and some means of detection.

If It Is Detectable, It Can Be Measured

If it is detectable, it can be measured directly or indirectly. Many people mistakenly believe that something must be a familiar, everyday item measured by familiar, simple techniques. It is natural to view the world this way because this way of thinking is sufficient for most of our measurement needs. There are, however, more complex measurement problems that require more complex measurement techniques. The solution is to become aware of the more complex situations and learn the more complex techniques.

If It Can Be Measured, It Can Be Managed

By definition, a measurement provides information that allows people to make better decisions because it reduces uncertainty. Some information will be more valuable than other information. Measurement analysis will indicate the value of the measurement.

---

2. For a more detailed description of some of these management concepts, see Douglas W. Hubbard, *How to Measure Anything: Finding the Value of Intangibles in Business*, 2nd edition, Hoboken, NJ: John Wiley & Sons, 2010, pp. 32–35.

It Has Probably Been Done Before

Very few things are totally new. Most "new" things have been done somewhere else, have been done in some other field of study, are a different combination of familiar components, or are a new application of something already in existence.

There Is More Available Data Than You Think

Many times people fail to do research because they believe that more information is not available or because of time, cost, or difficulty. They do not consider the consequences that will result if they do not do the research. It is simply easier to look at the immediate problem than it is to consider the future consequences. Many people do not understand the benefit of initial research. They do not see the benefit of preventing the problem because they did do the research. Therefore, they believe that they must let the problems occur. This can result in avoidable ad hoc project management and heroic effect.

Initial research has many benefits. The Internet provides a great deal of useful information. Some of the information may not be readily available and other times the amount of information may seem to be overwhelming and redundant. While these factors may be true, ultimately the benefit of researching an issue is greater than the negative consequences of not researching the issue.

You Don't Need As Much Data As You Think

There is a widely held belief that people need all possible information in order to make a decision or at least as much information as possible. The natural assumption is that more information will eliminate uncertainty or make the decision easier, but this is a myth. Many times additional information is redundant and does not provide any new insight. Even if the additional insight becomes available, it may not useful to the decision. In the end, it is impossible to eliminate all uncertainty. Fortunately, it is not necessary to totally eliminate uncertainty or even most of the uncertainty. You only need to reduce uncertainty enough to make a decision.

What Gets Measured, Gets Done

This is critical to achieving business and project goals. If we do not have a way to accurately measure our beginning point, our end point, and our progress, we tend to lose our way and chaos is introduced. Chaos reduces consistency and efficiency. It reduces consistency because without a clear goal and plan, we perform activities that are not aligned with the goal or the other activities in our project. It reduces efficiency because we do extraneous work that takes more time and effort than an efficient process. In the end, just measuring things provides an accurate knowledge of reality and focuses our progress (or lack of progress) and expenditures in relationship to our goal.

You Have to Think Differently Than Most People

This may be the most difficult part of measurement. One reason is the element of uncertainty, which exists in all measurement. It is natural to avoid uncertainty, but this leads to problems because uncertainty is always present, and avoiding uncertainty is denying an essential part of reality. One way that we have to think differently is to acknowledge reality and make it an essential part of our thinking about every aspect of the project. Another way that we have to think differently is to use reverse logic to identify reality. Instead of trying to increase certainty, it is sometimes easier to reduce uncertainty.

8.2 DEFINITIONS

The identification, creation, and modification of many measures require an ability to transform abstract concepts into usable measures. Words become very important and useful definitions may vary from normal usage. The definition of the following words will make measurement work easier.

Information Requirement

An information requirement is information that will be needed throughout the project to make decisions. It has an impact on the business objectives and project objectives. It is what the user needs to know.

Entity

An entity is an object of concern. In project management an entity may be a process, product, project, resource, or concept. Examples of project entities are effort and risk.

Attribute

An attribute is a property or characteristic of an entity that can be measured. In a project, effort can be an entity that has an attribute of hours. Risk is another entity that has the attributes of probability and impact.

Process

A process is a set of activities that transforms inputs into outputs.

Measurement

A measurement is an observation that identifies or describes reality and is expressed as a quantity. Measurement is a relationship or mapping between the thing being measured and numbers.

Some numbers may surprise you. Consider the following two numeric scales:

- Nominal numbers (sometimes called categorical numbers) are numbers that provide no information other than identification. In general, the only meaningful operation with nominal numbers is to compare two nominal numbers to see whether they are equal, whether they refer to the same object, or whether they refer to the same category. An example is a defect identified by category. A category one defect might be a defect in materials, a category two defect might be a design defect.
- Ordinal numbers place things in order. They tell us that one value is greater than another value but not by how much. An example is the assignment of a defect to security level one, which would indicate a security problem greater than level two but would not indicate how much greater.

Other more familiar scales include:

- Interval numbers have equal increments for equal quantities of the attribute. The numbers begin at one (zero is not included). An example is the number of paths through a process in which the minimum number of paths is one.
- Ratio numbers have equal increments for equal quantities of the attribute. The numbers begin at zero (zero is included). An example is the number of counted defects. It is possible to have zero defects.

Measurements have the characteristics of uncertainty, accuracy, and precision.

Uncertainty

It is not possible to truly understand measurement without understanding uncertainty. In principle, all measurements have an element of uncertainty because it is not possible to know anything with absolute certainty. As a matter of common practice, we do know some things precisely such as the number of days since the project charter was signed. Many project-related decisions involve obvious uncertainty that is critical to management, however, so it is a mistake to manage uncertainty by trying to measure uncertain things in a precise way. This leads to thinking that causes many problems in projects.

One common problem is goal-based thinking. People like precise goals. They are easy to understand, they are motivational, and inherently there is nothing wrong with goal-based thinking. However, many project goals have an element of uncertainty and cannot be measured with precision. Consider project initiation. At the beginning of the project, we have very little actual information and we are compelled to make some of the most important decisions about the project. It is natural to attempt to bring clarity to the project by thinking in terms of very precise goals. For example, some people estimate task duration as a precise length of time because this "eliminates" uncertainty, but human nature tends to transform the single point estimate into a precise goal that may not be realistic. Matters are made worse when no one questions the uncertainty of the statement after it is stated.

This leads to heroic effort with good intentions, but this is not the same as a realistic estimate. For example, "I don't really know how long the task will take, but I said the duration would be three days; therefore, I will work hard and complete the task in three days." Three days has unconsciously become more of a goal than an estimate. Realistic measurement has become a secondary consideration to heroic effort. Worse yet, it is likely that the stated goal is not realistic because duration estimates do have uncertainty. For example, there are variables beyond the person's control, even if the estimate is padded. Furthermore, this approach isolates every task from the context of the overall project. Project success now directly depends on the success of every task, which places the project at high risk. The thinking has become, "If all of the tasks are on time, the project will be done on time." What is the probability that every project task will be done on time?

The principle of uncertainty is good because it identifies the uncertainty that is inherent in every project, opens our thinking, and allows us to manage projects realistically. Simply recognizing the uncertainty in measurement is important because it means that it is not necessary to eliminate uncertainty. It is only necessary to reduce uncertainty enough to manage the project. Good measurement is stated as a range that includes uncertainty. In other words, "I don't know the exact duration of the task, but it's somewhere between two and five days." This estimate accurately expresses reality because it acknowledges the presence of uncertainty. Good measurement also has a confidence interval, which is a statistical expression of uncertainty that expresses how much we currently know. A 90 percent confidence interval is a range that has a 90 percent chance of containing the true value. For example, "I have a 90 percent confidence interval of two days to five days." In other words, "I don't know the exact duration of the task, but I am almost certain that it's somewhere between two and five days." A measurement with a range and a confidence interval provides a great deal of information because it states what is known, what is not known, and the level of confidence in the measurement. With this information, a project manager can apply other techniques such as risk management to successfully manage the project. Techniques that determine ranges and confidence levels are beyond the scope of this text. The critical point is the value of understanding the concept of measurement uncertainty.

Accuracy

Accuracy is a characteristic of a measurement where measurement results are not consistently above or below the true value. Results may vary within a wide range, but the range will probably contain the true value.

Precision

Precision is a characteristic of a measurement where results are consistent even if they are far from the true value. Results will vary within a narrow range, but the range may not anywhere near the true value. Ideally, a measurement should be both accurate and precise, but this is not always possible. If a measurement is not accurate and precise, accuracy is usually preferred because it has a high probability of containing the true value. The concept of accuracy is the basis of good measurements that are expressed in a range with a confidence interval.

Measure

A measure is a variable to which a quantity is assigned to represent one or more attributes. A measure is the result of measurement. Typical measures include but are not limited to metrics, indicators, or key performance indicators.

Indicator

An indicator is a measure that provides insight to an information requirement and supports decision making. There are simple indicators and complex indicators. Simple indicators consist of a single measure. Complex indicators consist of multiple simple measures used within a formula or algorithm.

Information Solution

An information solution is the combination of an indicator and the related decision criteria that address the information requirement.

8.3 MEASUREMENT PROCESS

As stated earlier, an indicator is a measure that provides insight into an information requirement. When an indicator is combined with decision criteria, the result is an information solution that meets information requirements and supports project decisions. For any measurement to be useful, it must meet the information requirements of the project. The process of transforming information requirements into information solutions includes the following steps:

- Conduct Preliminary Research
- Identify Information Requirements
- Analyze Information Requirements
- Create Indicator
- Integrate Measurement into the Project Processes

This chapter will use a case study to demonstrate the measurement process.

Preliminary Research

There are two goals for the preliminary research. The first goal is to understand the environment, business objectives, and project objectives. This is important because it provides the context of the project and the measurement activities. It helps to identify many uncertainties related to the project and provides information that reduces uncertainty. The second goal is to select the subject matter experts. These are the people who provide direct information to define concepts and measures. The group should also include key decision makers. It is important to keep the number of people in this group as small as possible. It is usually necessary to educate this group about the concepts of measurement.

Case Study: Customer Loyalty Project, Part 1

Antique Trucks is a magazine for people who are interested in the history of trucks, the history of the trucking industry, and restoring antique trucks. The people in this hobby are passionate about trucking history and like as much accurate information as possible. Subscribers come from all parts of society and income levels. They value integrity and personal relationships, and freely share information. The magazine staff knows the subscribers very well. Many of the readers and authors frequently contact the office by email, letters, and phone calls sharing suggestions, and comments. They enjoy helping the magazine become more successful.

The magazine has won national quality awards from professional organizations. Subscribers have stated that the quality of the magazine is a key element of customer satisfaction and will actively recommend the magazine to other people. They will even call the office if the magazine is a "few days late." Many new subscribers will order all of the back issues, making the magazine a collector's item.

Based on preliminary research, the project manager and sponsor have identified the following decision makers and subject matter experts, illustrated in Table 8–1. Jane Editor is the magazine's editor and has decision

TABLE 8-1 Decision Makers and Subject Matter Experts

ROLE	PERSON
Sponsor	Jane Editor
Project Management	Tom Project
Accounting	Mary Business
Printing	Jack Production
Customer Representative	David Customer

authority over the production of the magazine, including content and page layout. Tom Project is the project manager and has authority over the day-to-day activities of the project. Mary Business is the accountant and will provide information regarding money and customer renewals. Jack Production is the production manager at the printing company that prints and distributes the magazine issues. David Customer is the chairman of the advisory board and will represent the subscribers.

Identify Information Requirements

An information requirement is information that will be needed throughout the project to make decisions and has an important impact on the project's objectives. Typical sources of information requirements are uncertainty, risks, and issues. Uncertainty is the absence of complete certainty and is characterized by a lack of information. Risk is an uncertain event or condition that, if it occurs, will have an impact on one or more project objectives. Risks include individual risks and overall project risk. An issue is an event that has occurred or a condition that exists. Often, but not always, an issue is a risk that has actually occurred. There is a strong link between risk and measurement because of the concept of uncertainty. Risk management is a good beginning point for identifying information requirements, but it is not always adequate. Risk management alone will not be sufficient to establish the measurement practice for a project.

Many of the specific information requirements will be identified at the beginning of the project. It is possible that the identified information requirements will change during the project's life cycle, and new information requirements will be identified. The probability of change may be very high on a value-driven project. If the information requirements change, the measures and indicators will also change. A proper change control and configuration management system will be required to control the changes and store the versions of information requirements and measures.

The primary tool used to identify information requirements is the facilitated workshop. Usually more than one workshop session will be required.

Its purpose is to define decision criteria and the information needs of the project so that measures can be defined to meet those needs. The workshop must include the subject matter experts and decision makers related to the project. The needs of the project manager are critical in the workshop. The project manager is a primary user of measures, will integrate the measures into the project processes, and will modify them during the project's life cycle. The project manager will gather data, analyze data, provide forecasts, and produce project reports.

The workshop members will identify the information requirements of the project using various techniques. Possible sources of information include organizational values, business objectives, project objectives, risk assessments, project constraints and assumptions, quality assurance, quality control, acceptance criteria, external requirements, project deliverables, and historical data. The information requirements should be stated in the terminology of project management because the measures will need to be integrated into the project management processes. Information requirements must provide enough information to create the measure, so the measure must be well formed. In general, the criteria for well-formed measurements are that the measurement must clearly communicate the measure, and the measurement process, must be repeatable and must be traceable.

The workshop members should begin by defining or evaluating the business strategy, business objectives, and project objectives. Business objectives and project objectives provide critical data required to align the project with the strategic goals of the organization. Business and project objectives provide fundamental information that is used to identify, categorize, and prioritize information requirements and, ultimately, to create the measures that will be integrated into the project processes. Business objectives provide the reason, business case, or justification for the project. Project objectives provide the goal, product, service, result, or benefit of the project.

Another important consideration is the project success criteria. This is an important measurement tool because the project team can measure the current condition of the project in relationship to project success criteria. It is important to clearly define the success criteria because they will become the basis for important project measures. The traditional definition of project success is the triple constraints, which include scope, time, and cost.

The time criterion means that the project will end on schedule. The cost criterion means that the project will end on budget. The scope criterion means that all the work of the project will be completed. However, the traditional interpretation of the triple constraints tends to focus on very precise goals. Good measurement techniques would state these goals as ranges to accommodate the uncertainty inherent in the project. Another potential problem with the traditional triple constraints is that there is no direct consideration of the delivery of measured business value. Recently the value-driven project has been introduced. A value-driven project is a project with a goal of achieving an organizational value such as a designated amount of market share or level of brand recognition. The success criteria for a value-driven project are the achievement of a value plus the triple constraints.[3]

The workshop members should also define project characteristics including project type, criticality, capacity, and stability. Project type includes general concepts such as industry. It also includes concepts unique to the organization. Some examples of project type are product development, IT, or administrative.

Project criticality aligns the project to the corporate strategy. One method includes the following categories: mission critical, business objectives, process improvement, and administrative improvement.

Organizational capacity describes the ability of the project team to accomplish the project objectives based upon the number of resources and their measured level of skill related to the project requirements.

Requirements stability is an estimate of how often and how much the project requirements will change throughout the project.

Case Study: Customer Loyalty Project, Part 2

Business objectives: Management is focused on long-term goals and values. A core value of the magazine is customer loyalty because the long-term success of the magazine depends on retaining subscribers while increasing the number of new subscribers. For example, the cost of advertising in other magazines for new subscribers can be higher than the direct profit from the new subscriptions. As a result, profit from new subscriptions may not occur until the first renewal. The business objective of this project is to improve customer loyalty.

Project objectives: Feedback from customers indicates that they would like to have some changes made to the magazine. The requested changes have been prioritized by an advisory board of subscribers from the most valuable to the least valuable from the viewpoint of the subscribers. Management believes that these changes will improve the business objective of customer loyalty. Management has decided to implement the "Customer Loyalty" project. An assessment has been conducted based upon data from similar projects and a cost/benefit analysis. The assessment has found that the project needs to increase customer loyalty by 2 percent to justify the cost of the project. It also found that an increase greater than 5 percent is not justified at this time because other projects in the portfolio are limiting the resource capacity of the magazine. (There will be other projects in the future to increase customer loyalty.) The project objective is to make some or all of the changes to the magazine requested by the subscribers in order to increase customer loyalty by 2 to 5 percent. Management is not willing to lower the quality of the magazine to achieve the project objectives.

---

3. Harold Kerzner and Frank P. Saladis, *Value-Driven Project Management*, Hoboken, NJ: John Wiley & Sons/IIL, 2009, pp. 50–63.

TABLE 8-2 Project Success Criteria

SUCCESS CRITERIA	DEFINITION
Value	Increase customer loyalty by 2 to 5 percent.
Cost	To be determined by the project plan.
Time	To be determined by project plan.
Scope	Scope is variable. The project will deliver sufficient changes to increase customer loyalty by 2 percent beginning with the highest-priority requested change and proceeding to the lowest priority. Delivered changes will stop when measured customer loyalty reaches 5 percent. Each issue of the magazine will be published with one of the requested changes and will be a deliverable. Each issue deliverable will have two milestones. The first milestone will be called the production milestone and will be achieved after the magazine has been produced, has been checked for defects, and is ready to go to the printer. The second milestone will be achieved when the printer has printed the issue, It has been checked for defects, and it is ready to be shipped to the subscribers.

TABLE 8-3 Project Characteristics

CHARACTERISTIC	DEFINITION
Project Type	Publishing
Project Size	Moderate
Project Criticality	Business Improvement
Organizational Capacity	Management is very confident that it has sufficient capacity, including personnel, to complete the project. The magazine staff has many years of experience. The process to make the required changes has been documented, assessed, and verified. Everyone agrees that the changes are realistic. The staff has made three similar changes to the magazine in the last seven years.
Requirements Stability	Management believes that there is virtually no probability that the requirements will change during the project.

Management has determined that the project is a value-driven project defined by corporate value plus the triple constraints (scope, time, and cost). They have defined the success criteria as shown in Table 8–2. Table 8–2 also represents the designated goal for each criterion. Management has identified the project characteristics shown in Table 8–3.

A facilitated discussion was conducted by the project manager to identify the information requirements. The discussion used a question-based format to identify the most critical questions being asked by management about the project. As a result of this discussion, the information requirements in Table 8–4 were identified.

TABLE 8-4 Information Requirements

INFORMATION REQUIREMENTS
Will the project meet quality requirements?
What is the status of customer loyalty?
Are the deliverables being completed?
Is project spending meeting budget objectives?
Is project spending meeting schedule objectives?

On a real project, these information requirements would likely be more complicated, and there would likely be other information requirements such as individual risks and overall project risk. The case study has been limited for purposes of simplicity and clarity.

Analyze Information Requirements

The analysis of the information requirements has two parts: categorize the information requirements and prioritize the information requirements. Categorizing the information requirement provides additional information and ensures that all of the information requirements are considered. Prioritizing the information requirements ensures the appropriate selection of the measures that will have the greatest impact on the defined project objectives.

Categorize Information Requirements

It is important to measure all aspects of the project, so the workshop members should identify information requirements from multiple categories. Categories should be created based upon a number of considerations such as market conditions, organizational characteristics, business objectives, project objectives, project success criteria, and project characteristics.

Special consideration must be given to the information categories required by the project manager because these are the areas of key concern required to manage the project on a daily basis.

There are several advantages to using information categories. First, information categories address the management needs of the project manager. Second, many organizations have established organizational categories. Using these established categories ensures that all organizational concerns will be represented and avoids the selection of too many measurements from one category. Third, using information categories makes it easier to examine the relationship between the information requirements and creates the best combination of information requirements. Fourth, information about the category provides a natural source of additional data that

TABLE 8-5 Information Categories

INFORMATION REQUIREMENT	CATEGORY
Will the project meet quality requirements?	Quality
What is the status of customer loyalty?	Corporate Value
Are the deliverables being completed?	Scope
Is the project on budget?	Cost
Is the project on schedule?	Schedule

Note: The case study may not be entirely realistic. It has been simplified for this text.

may otherwise be overlooked. Fifth, the information in a category is usually structured and, therefore, provides a useful framework to guide discussions that help define the final measurement. For example, information about a category may provide insight that enables the measurement committee to merge several information requirements into one information requirement that meets several needs.

Case Study: Customer Loyalty Project, Part 3

The workshop members placed the information requirements into the categories illustrated in Table 8-5.

Prioritize Information Requirements

The workshop members must prioritize the information requirements. The project stakeholders will usually identify a large number of information requirements. Many times it is not practical or useful to monitor all of the information requirements. Therefore, it is important to prioritize the information requirements to ensure that the most appropriate information requirements are selected. There are a number of qualitative and quantitative methods that can be used to prioritize the information requirements. One method is a scoring system that is used to prioritize information requirements on a scale, such as one to five. One is the highest priority and five is the lowest priority. Management can use techniques such as voting system to select the priorities.

Case Study: Customer Loyalty Project, Part 4

The workshop members prioritized the information requirements illustrated in Table 8-6 using a scoring system. Customer loyalty was rated one because it is the project objective. Quality was rated two because it has high value to the customers. Scope was rated number three because management felt comfortable that the deliverables were very achievable. Cost was

TABLE 8-6 Prioritized Information Requirements

INFORMATION REQUIREMENT	CATEGORY	PRIORITY
What is the status of customer loyalty?	Corporate Value	1
What is the quality of the magazine?	Quality	2
Are the deliverables being completed?	Scope	3
Is the project on budget?	Cost	4
Is the project on schedule?	Schedule	5

rates four because of other projects in the portfolio. Schedule was rated five because it is flexible.

Create Indicator

The appropriate measurements are created in this step. The goal is to define the data that must be used and create the measurements that will become indicators. Simple measurements will be identified and combined into complex measurements, which will become indicators. Ultimately, the indicators provide insight into an information requirement and will be used in combination with the decision criteria to manage the project. One well-known example of this is the Schedule Performance Index (SPI), which is created from a combination of other measurements. Sometimes it is possible to use predefined measures, but they should be used with caution. One problem is that predefined measures are generic, but project information requirements are specific. Another problem is the lack of a common interpretation of predefined measures. For example, it is common to measure actual progress against the project schedule to determine if the project is on schedule, but evidence suggests that there very little agreement on what it means to be "on schedule." The best measures will be created measures or predefined measures that have been modified to meet the information requirements of the project. At a minimum, review predefined measures to make sure that there is a consensus and that they meet your actual information requirements.

Measures are the basic building block of the final indicator that is critical to project success. Good measurement must meet the criteria for well-formed measures defined by Deming, Juran, and other experts.

First, well-formed measures must communicate clearly. It will allow trained personnel to know precisely what is being measured, why it is being measured, how it is being measured, when it is being measured, and what is included and excluded. Trained personnel should be able interpret the results and reach similar conclusions.

Second, well-formed measurements must be repeatable. Different trained personnel should be able to repeat the measurement and get essentially the same results.

Third, well-formed measures must be traceable. The data must be identified in terms of measurement tools, time, source, measurement process, environment, and authorized personnel. The traceability data will be especially relevant in assessing and improving process performance, making changes to the measurement, and the creation of new measurements throughout the project's life cycle.

These criteria can be used to create an indicator for the projects. The specifications for the indicators are presented in the case study and meet the three criteria stated previously. The description aligns the solution to the knowledge requirement and provides clear definitions of the primary terms. A simple measure and its method is specific measurement method of a single attribute. The method is repeatable and traceable, as indicated by a clear description of the unit of measure, frequency of collection, responsible party, and tools. Other elements can be added to the specification as needed.

Case Study: Customer Loyalty Project, Part 5

The workshop members began by selecting the measures for the Customer Loyalty Project. One indicator was selected for each knowledge requirement as indicated in Table 8–7.[4]

The first indicator is the Customer Loyalty indicator presented in Table 8–8. Management measures subscription renewals as a product of customer loyalty. Happy subscribers become loyal subscriber when they renew their subscription.

TABLE 8-7 Selected Indicators

KNOWLEDGE REQUIREMENT	INDICATOR
What is the status of customer loyalty?	Customer Loyalty
What is the quality of the magazine?	Defect Density
Are the deliverables being completed?	Deliverable Progress
Is the project on budget?	Cost Performance Index
Is the project on schedule?	Schedule Performance Index

4. For additional information on indicator specifications, see John McGarry et al., *Practical Software Measurement: Objective Information for Decision Makers*, Reading, MA: Addison-Wesley, 2002, p. 160.

TABLE 8-8 Customer Loyalty Indicator

CUSTOMER LOYALTY	
Description	
Information Requirement	Is customer loyalty increasing?
Measurement Category	Corporate Value
Definitions	
Customer	A customer is a person or organization that has received one or more magazines as a result of subscription payment. Customers are the people and organizations that are included in the Customer Log. The editor has the sole authority to decide any issues relating to the interpretation of this definition.
Subscription	A subscription is a consecutive series of magazine issues continuously delivered to a specific customer for one year. The editor has the sole authority to decide any issues relating to the interpretation of this definition.
Subscription Renewal	A subscription renewal is a subscription to a customer who has a current subscription, has paid for a new subscription, and will continue to receive the magazine.
Loyal Customer	A loyal customer is defined as a person or organization that has a current subscription to the magazine and is renewing their subscription. A loyal customer must meet the definition of customer stated in the above paragraph. A loyal customer must have a current subscription as defined in the preceding paragraph. A loyal customer must have a subscription renewal as defined in the above paragraph. The editor has the sole authority to decide any issues relating to the interpretation of this definition
Customer Loyalty	Customer loyalty is measured on the basis of all subscriptions for the magazine. The subscription renewal rate compares the cumulative actual subscription renewals to the cumulative possible subscription renewals. This rate is then compared to a reference subscription rate prior to the initiation of the project.
Simple Measure	
Measurement Name	Actual Subscription Renewals
Method	Count the cumulative number actual subscription renewals since the project charter was signed. The count can be done by a person or a computer.
Unit of Measure	Actual Renewals
Frequency of Collection	The major deliverables of the project are each issue of the magazine. For each deliverable there is one milestone at the point the magazine is ready to be sent to the printer (production milestone) and a second at the point the magazine is sent to the distributer (print milestone). The defect density is assessed at the point the magazine is sent to the printer.
Responsible	Editor
Tools	Customer Log Microsoft Excel

(Continued)

TABLE 8-8 Customer Loyalty Indicator (*Continued*)

CUSTOMER LOYALTY	
Simple Measure	
Measurement Name	Possible Subscription Renewals
Method	Count the cumulative number of possible subscription renewals since the project charter was signed. The count can be done by a person or a computer.
Unit of Measure	Possible Renewals
Frequency of Collection	The major deliverables of the project are each issue of the magazine. For each deliverable there is one milestone at the point the magazine is ready to be sent to the printer (production milestone) and a second at the point the magazine is sent to the distributer (print milestone). The defect density is assessed at the point the magazine is sent to the printer.
Responsible	Editor
Tools	Customer Log Microsoft Excel
Complex Measure	
Indicator Name	Customer Loyalty Indicator
Algorithm	1. Calculate: Project Subscription Renewal Rate = (Cumulative Actual Renewals/ Cumulative Possible Renewals) from the beginning of the project. 2. Calculate: Reference Subscription Renewal Rate (Actual Renewals/Possible Renewals) for the subscription period immediately prior to the project charter. 3. Calculate: Percent Improvement = Current Subscription Renewal Rate—Reference Subscription Renewal Rate.
Decision Criteria	Any number equal to or greater than +2% is required to justify the cost of the project. Any number less than +2% in a subscription period will initiate an assessment.
Sample	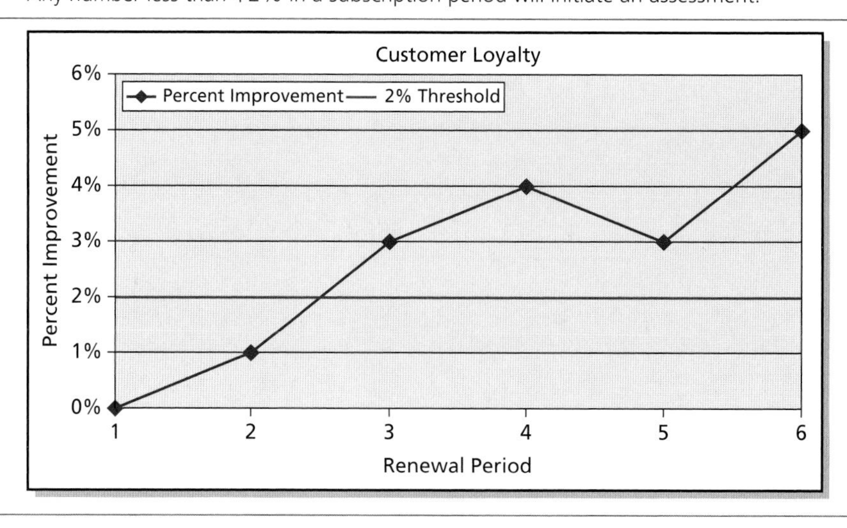

TABLE 8-8 Customer Loyalty Indicator (*Continued*)

CUSTOMER LOYALTY	
Analysis Process	
Analysis Frequency	Major deliverables of the project are each issue of the magazine. For each deliverable there is one production milestone at the point the magazine is ready to be sent to the printer and a second print milestone at the point the magazine is sent to the distributer. The defect density is assessed at the print milestone.
Project Phase	Execution
Interpretation	Increasing customer loyalty is defined as an increase in the Percent Improvement. Decreasing customer loyalty is defined as a decrease in the Percent Improvement.
Data Source	Customer Log
Tool	Microsoft Excel
Analyst	Editor, Project Manager

The defect density indicator is presented in Table 8–9 is an example of a simple measure that is an indicator. The editor evaluates the quality of the magazine by counting the defects found in each issue of the magazine.

In this measure, the project manager needs to know if the project is on schedule and on budget. The well-formed measure is defined in Table 8–10.

Table 8–11 presents the Deliverable Progress indicator, which compares the planned deliverables with the actual deliverables. This can be used as a schedule indicator. It can also indicate that the project has insufficient capacity.

Integrate Measurement into Project Processes

The next step is to integrate the measures into the project management process. This process determines how the data is collected and analyzed within the context of the project plan.

Assign Measurements to Proper Project Processes

Ultimately the indicators, metrics, and key performance indicators (KPIs) must be integrated into the project processes. In order for measurement to work, specific procedures must be created and responsibility assignments must be in place. There are two categories of measurement procedures that must be integrated into the project processes. The first category is the collection and storage of data which is further divided into two categories. The subdivisions are manual and electronic data collection and storage. The second category is data analysis and reporting. The measurement solutions must provide sufficient detail so that multiple trained people can perform the measurement work and get essentially the same results.

TABLE 8-9 Defect Density	
Defect Density	
Description	
Information Requirement	What is the quality of the magazine?
Measurement Category	Quality
Definitions	
Defect	A magazine defect is a variation from the Magazine Specification Standard.
Defect Density	The number of defects present in each issue of the magazine.
Simple Measure/Indicator	
Measurement Name	Defect
Method	Count the actual number of defects found in a magazine issue
Unit of Measure	Defect
Frequency of Collection	The major deliverables of the project are each issue of the magazine. For each deliverable, there is one milestone at the point the magazine is ready to be sent to the printer (production milestone) and a second at the point the magazine is sent to the distributer (print milestone). The defect density is assessed at the point the magazine is sent to the printer.
Responsible	Editor
Tools	Specification Standard Project Management Information System
Sample	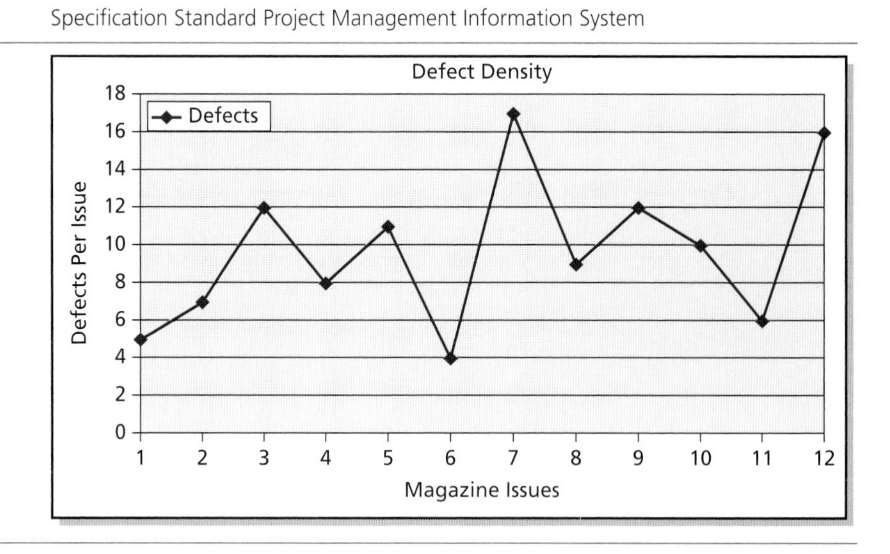
Analysis Process	
Analysis Frequency	The indicator is analyzed at the print milestone for each issue.

TABLE 8-9 Defect Density (*Continued*)

Project Phase	Execution
Interpretation	A defect density greater that 18 will trigger an assessment.
Data Source	Magazine issue
Tool	Project Management Information System
Analyst	Editor

TABLE 8-10 SPI and CPI

SCHEDULE PERFORMANCE INDEX & COST PERFORMANCE INDEX	
Description	
Information Requirement	Is project spending meeting budget and schedule objectives?
Measurement Category	Cost
Definitions	
Planned Value	Budgeted Cost of Work Scheduled
Actual Cost	Actual Cost of Work Performed
Earned Value	Budgeted Cost of Work Performed
Simple Measure	
Measurement Name	Planned Value
Method	Calculate the budgeted cost of work scheduled
Unit of Measure	Dollars
Frequency of Collection	Week
Responsible	Project manager
Tools	Microsoft Project Microsoft Excel
Simple Measure	
Measurement Name	Actual Cost
Method	Calculate the actual cost of work performed
Unit of Measure	Dollars
Frequency of Collection	Week
Responsible	Project manager
Tools	Microsoft Project Microsoft Excel

(*Continued*)

TABLE 8-10 SPI and CPI (*Continued*)

SCHEDULE PERFORMANCE INDEX & COST PERFORMANCE INDEX	
Simple Measure	
Measurement Name	Earned Value
Method	Calculate the budgeted cost of work performed
Unit of Measure	Dollars
Frequency of Collection	Week
Responsible	Project manager
Tools	Microsoft Project Microsoft Excel
Complex Measure	
Indicator Name	SPI & CPI Indicator
Algorithm	CPI = Earned Value/Actual Cost SPI = Earned Value/Planned Value
Decision Criteria	CPI—Values greater than 1.1 and less than.9 require an assessment. SPI – Values greater than 1.1 and less than.9 require an assessment.
Sample	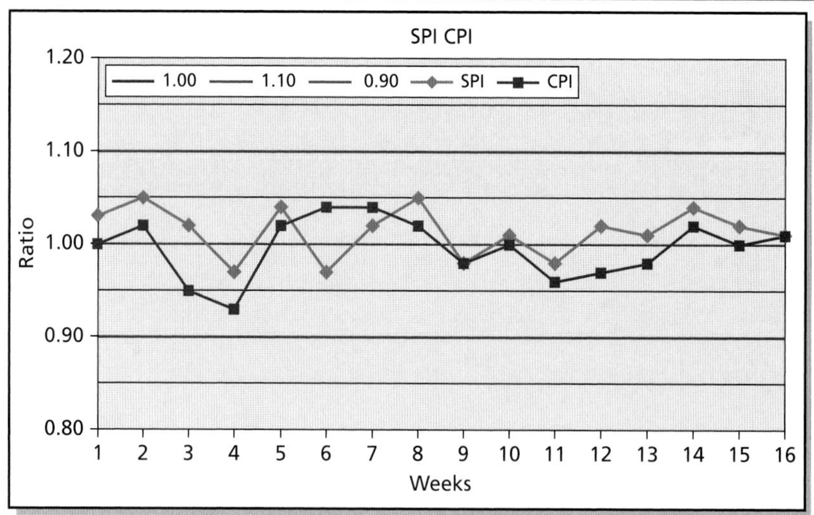
Analysis Process	
Analysis Frequency	Week
Project Phase	Execution
Interpretation	CPI – Values less than one indicate a negative trend. CPI – Value greater than one indicate positive trend. SPI – Values less than one indicate a negative trend. SPI – Value greater than one indicate positive trend

TABLE 8-10 SPI and CPI (*Continued*)

SCHEDULE PERFORMANCE INDEX & COST PERFORMANCE INDEX	
Data Source	The data source for planned value is the project plan. The data source for actual cost is the work performance information. The data source for earned value is planned value and actual cost.
Tool	Project management software
Analyst	Project manager

TABLE 8-11 Deliverable Progress

DELIVERABLE PROGRESS	
Description	
Information Requirement	Are the deliverables being completed?
Measurement Category	Scope
Definitions	
Deliverable	A verifiable product, service or result produced to complete a process, phase or project.
Simple Measure	
Measurement Name	Planned Deliverables
Method	Count the cumulative planned deliverables for each issue
Unit of Measure	Deliverable
Frequency of Collection	Print milestone
Responsible	Project manager
Tools	Microsoft Project Microsoft Excel
Simple Measure	
Measurement Name	Actual Deliverables
Method	Count the cumulative actual deliverables for each issue
Unit of Measure	Deliverable
Frequency of Collection	Print milestone
Responsible	Project manager
Tools	Microsoft Project Microsoft Excel

(*Continued*)

TABLE 8-11 Deliverable Progress (*Continued*)

DELIVERABLE PROGRESS	
Complex Measure/Indicator	
Indicator Name	Deliverables Indicator
Method	Count the cumulative number planned deliverables Count the cumulative number of actual deliverables
Unit of Measure	Dollars
Frequency of Collection	Week
Responsible	Project manager
Tools	Microsoft Project Microsoft Excel
Sample 	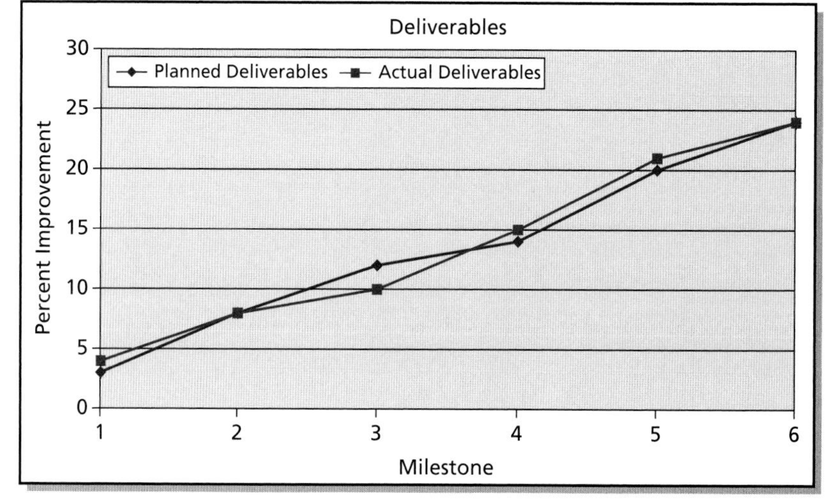
Analysis Process	
Analysis Frequency	Analysis occurs at the point of the print milestone.
Project Phase	Execution
Interpretation	Actual deliverables should equal the number of planned deliverables. If actual deliverables are less than planned deliverables, an assessment is conducted.
Data Source	Project Plan Work Performance Information
Tool	Project management software Progress Reports
Analyst	Project manager

Most projects require a combination of manual and electronic data collection and measurement. Because of time and expense considerations, it is very important to choose the appropriate combination of measures that covers all areas of the project as well as manual and electronic measurements. Manual data collection tends to take more time and money than electronic data collection so some projects take few or no manual measurements. This can be a serious mistake because there will usually be some data that can only be measured manually. Good measurement involves two concepts: compliance and consistency. The measurement solution must have sufficient methodology detail to allow a trained person to know what is being measured, why it is being measured, and how it is being measured. This allows the collection agent to comply with the measurement solution. Another concept is consistency. The measurement must be done the same way each time because no measure is perfect. Accuracy, precision, and bias can be minimized, or at least considered, if the measurement is consistent.

Electronic data collection has become more common as project activities have been computerized, and most project management software has included some common measures plus the ability to create custom measures and reports. This is usually less expensive and time-consuming than manual measurement. It is helpful to exercise some caution with common measures provided by the software because the software may not use standard formulas. Another consideration is that the common formulas may have to be modified to meet the information requirements of the project. Electronic data collection is ideally suited to the collection of work performance data, so it is tempting accept the information from team members. Once again, this can be a problem. Simply extracting data from reports provided by team members without verification leads to low-quality data. At least some of the reports should be verified by independent inspection.

Data storage is an important consideration because information, including measurement data, is the basis for project management and business decisions. Data must be stored in a protected repository and access must be managed. Unprotected data tends to get lost or modified. Configuration management is helpful because it protects data and tracks versions of the data. Only people trained in measurement should input data and measures into the repository and interpret the raw output of the indicators because many people do not fully understand the concept of measurement or project management. This will help to ensure high-quality data and the proper interpretation of the information.

Data analysis and reporting is the second category of measurement procedures that must be integrated into the project processes. These activities analyze the collected data, report the results, and make recommendations to the decision makers. A great deal of the project measurement time will be spent in data analysis because this is the activity that transforms the simple measures and raw data into the indicators. The transformation usually has four elements, which are planed values, actual values, variance, and decision criteria. The planned values come from the stakeholders and project management plan and include things such as schedule, budget, quality specifications, and project value. The work of the project is the source of actual values and includes completed deliverables and work performance information. The element of variance is a formula or algorithm that compares planned values to actual values. The last element is the decision criteria and is usually provided by management considerations. The variance value is compared to the decision criteria, and recommendations are made based upon the output of the comparison. The recommendation may be an automated response that requires a predetermined activity, a non-automated response that requires a thoughtful decision or information that does not require any response or immediate decision. With these considerations, the deliverables and project performance measured will be selected. It will be necessary to schedule these activities and allocate sufficient funds in the budget.

Define Roles and Responsibilities

As mentioned earlier, data collection and measurement will be manual or electronic. Manual data collection and measurement should include a requirement that the collection personnel to be trained but note that the required training may only take a few minutes. The important point is that the person consistently follows the measurement method to ensure results that are compliant and consistent with the information solution. For example, if the task is to measure a steel rod, the person should measure the rod the same way each time, from left to right (or right to left) with the same measuring device using the same line of sight. Quality management may require that the person then take multiple sample measurements, which would be averaged to minimize uncertainty. Some further action may be required to address bias. Electronic data collection and measurement is usually done by computer and is less expensive and time-consuming, but there is a problem. Computers can't think. A common problem is that people automatically believe whatever is on the computer screen without any further interpretation or thought. A better practice is to have a person trained in measurement monitor some or all electronic measurement activities to ensure proper interpretation.

8.4 ADDITIONAL INFORMATION ON MEASUREMENT CATEGORIES

There are many ways to create measures and measurement categories. One obvious source for measurement categories is the Knowledge Areas in *A Guide to the Project Management Body of Knowledge*.[5] The nine Knowledge

5. Project Management Institute, *A Guide to the Project Management Body of Knowledge*, 4th edition. Newtown Square, PA: Project Management Institute, 2008, pp. 403–407.

TABLE 8-12 Measures Related to Knowledge Areas

KNOWLEDGE AREA	TYPICAL MEASURES
Project Integration Management	Development Maturity Productivity Efficiency
Project Scope Management	Requirement Stability Deliverable Progress
Project Time Management	Critical Path Performance Milestone Progress
Project Cost Management	Budget Compliance Estimate Adequacy
Project Quality Management	Defect Reduction Progress Process Compliance
Project Human Resource Management	Personnel Adequacy Resource Utilization
Project Communications Management	Milestone Completion
Project Risk Management	Risk Status
Project Procurement Management	Baseline Changes Contract Compliance

Areas provide a great deal of useful information and a convenient means to use common measurement categories. There are hundreds of potential measures that can be developed and mapped to the nine Knowledge Areas. It is important to understand that it is not practical or desirable to use all of the potential measures. Table 8–12 presents some example measures that that can be mapped to the nine Knowledge Areas.

8.5 FINAL COMMENTS

It is impossible to present all of the concepts, knowledge, skills, tools, and techniques required for good project measurement in the space of one chapter. Concepts were summarized and some things were excluded such as estimating, process analysis, other measurement techniques, and tools. A more comprehensive treatment will be presented in the upcoming book *Measurement-Driven Project Management.*

INDEX

Note: Page references in *italics* refer to figures and tables.